Natural Attenuation of Trace Element Availability in Soils

Other Titles from the Society of Environmental Toxicology and Chemistry (SETAC):

Use of Sediment Quality Guidelines and Related Tools for the Assessment of Contaminated Sediments
Wenning, Batley, Ingersoll, Moore, editors
2005

Ecological Assessment of Aquatic Resources: Linking Science to Decision-Making
Barbour, Norton, Preston, Thornton, editors
2004

Amphibian Decline: An Integrated Analysis of Multiple Stressor Effects
Linder, Krest, Sparling, editors
2003

Metals in Aquatic Systems:
A Review of Exposure, Bioaccumulation, and Toxicity Models
Paquin, Farley, Santore, Kavvadas, Mooney, Winfield, Wu, Di Toro
2003

Silver: Environmental Transport, Fate, Effects, and Models:
Papers from Environmental Toxicology and Chemistry, 1983 to 2002
Gorusch, Kramer, La Point, editors
2003

Contaminated Soils: From Soil–Chemical Interactions to Ecosystem Management
Lanno, editor
2003

Environmental Impacts of Pulp and Paper Waste Streams
Stuthridge, van den Heuvel, Marvin, Slade, Gifford, editors
2003

Porewater Toxicity Testing: Biological, Chemical, and Ecological Considerations
Carr and Nipper, editors
2003

Reevaluation of the State of the Science for Water-Quality Criteria Development
Reiley, Stubblefield, Adams, Di Toro, Erickson, Hodson, Keating Jr, editors
2003

For information about SETAC publications, including SETAC's international journals, *Environmental Toxicology and Chemistry a* *Integrated Environmental Assessment and Management*, contact the SETAC Administrative Office nearest you:

SETAC Office
1010 North 12th Avenue
Pensacola, FL 32501-3367 USA
T 850 469 1500 F 850 469 9778
E setac@setac.org

SETAC Office
Avenue de la Toison d'Or 67
B-1060 Brussels, Belgium
T 32 2 772 72 81 F 32 2 770 53 86
E setac@setaceu.org

www.setac.org

Environmental Quality Through Science®

Natural Attenuation of Trace Element Availability in Soils

Edited by
Rebecca Hamon
Mike McLaughlin
Enzo Lombi

Coordinating Editor of SETAC Books
Joseph W. Gorsuch
Gorsuch Environmental Management Services, Inc.
Webster, New York, USA

SETAC®

Taylor & Francis
Taylor & Francis Group
Boca Raton London New York

CRC is an imprint of the Taylor & Francis Group,
an informa business

Published in collaboration with the Society of Environmental Toxicology and Chemistry (SETAC)
1010 North 12th Avenue, Pensacola, Florida 32501
Telephone: (850) 469-1500 ; Fax: (850) 469-9778; Email: setac@setac.org
Web site: www.setac.org
ISBN: 1-880611-60-0(SETAC Press)

SETAC Press is an imprint of the Society of Environmental Toxicology and Chemistry.

Printed in the United States of America on acid-free paper
10 9 8 7 6 5 4 3 2 1

International Standard Book Number-10: 1-4200-4282-3 (Hardcover)
International Standard Book Number-13: 978-1-4200-4282-5 (Hardcover)

Library of Congress Cataloging-in-Publication Data

Special Symposium on the Natural Attenuation of Trace Elements in Soils (2001 : Guelph, Canada)
Natural attenuation of trace element availability in soils / editors, Rebecca Hamon, Mike McLaughlin, Enzo Lombi.
p. cm.
Papers from a Special Symposium on the Natural Attenuation of Trace Elements in Soils held during 6th International Conference on the Biogeochemistry of Trace Elements in Guelph, Canada in August 2001.
Includes bibliographical references and index.
ISBN 1-4200-4282-3 (alk. paper)
1. Soil remediation--Congresses. 2. Groundwater--Purification--Congresses. 3. Soils--Trace element content--Congresses. 4. Hazardous wastes--Natural attenuation. I. Hamon, Rebecca. II. McLaughlin, Mike. III. Lombi, Enzo, 1968- IV. International Conference on the Biogeochemistry of Trace Elements (6th : 2001: Guelph, Canada) V. Title.

TD878.S678 2006
628.5'5--dc22 2006049036

Visit the Taylor & Francis Web site at
http://www.taylorandfrancis.com

and the CRC Press Web site at
http://www.crcpress.com

and the SETAC Web site at
www.setac.org

SETAC Publications

Books published by the Society of Environmental Toxicology and Chemistry (SETAC) provide in-depth reviews and critical appraisals on scientific subjects relevant to understanding the impacts of chemicals and technology on the environment. The books explore topics reviewed and recommended by the Publications Advisory Council and approved by the SETAC North America, Latin America, or Asia/Pacific Board of Directors; the SETAC Europe Council; or the SETAC World Council for their importance, timeliness, and contribution to multidisciplinary approaches to solving environmental problems. The diversity and breadth of subjects covered in the series reflect the wide range of disciplines encompassed by environmental toxicology, environmental chemistry, and hazard and risk assessment, and life-cycle assessment. SETAC books attempt to present the reader with authoritative coverage of the literature, as well as paradigms, methodologies, and controversies; research needs; and new developments specific to the featured topics. The books are generally peer reviewed for SETAC by acknowledged experts.

SETAC publications, which include Technical Issue Papers (TIPs), workshop summaries, newsletter (*SETAC Globe*), and journals (*Environmental Toxicology and Chemistry* and *Integrated Environmental Assessment and Management*), are useful to environmental scientists in research, research management, chemical manufacturing and regulation, risk assessment, and education, as well as to students considering or preparing for careers in these areas. The publications provide information for keeping abreast of recent developments in familiar subject areas and for rapid introduction to principles and approaches in new subject areas.

SETAC recognizes and thanks the past coordinating editors of SETAC books:

C.G. Ingersoll, Columbia Environmental Research Center
US Geological Survey, Columbia, Missouri, USA

T.W. La Point, Institute of Applied Sciences
University of North Texas, Denton, Texas, USA

B.T. Walton, US Environmental Protection Agency
Research Triangle Park, North Carolina, USA

C.H. Ward, Department of Environmental Sciences and Engineering
Rice University, Houston, Texas, USA

Table of Contents

Foreword

Natural Attenuation of Trace Element Availability in Soils brings together the expertise of a diverse, international group of experts who participated in a Special Symposium on the Natural Attenuation of Trace Elements in Soils organised for the Sixth International Conference on the Biogeochemistry of Trace Elements held in Guelph, Canada, in August 2001.

"Natural attenuation" was first used to describe the decrease in bioavailability and toxicity of organic contaminants over time in the environment. Natural attenuation is caused by dilution, dispersion, biodegradation, volatilisation, and "irreversible" sorption of a contaminant in the environment. These processes, with the exception of biodegradation, can also decrease the bioavailability of inorganic contaminants and nutrients. Other terms such as "aging" and "fixation" have been used to describe the same processes but are more ambiguous because a decrease in toxicity or bioavailability may occur rapidly (such as in the case of formation of metal precipitates) or be reversible. However, in the context of this book, natural attenuation and the other terms are used interchangeably.

The mechanisms of natural attenuation are still not completely understood. Even if there is a partial agreement in regard to the main processes involved, their relative importance is still the subject of a lively debate. Similarly, the reversibility of natural attenuation processes is still very unclear. These debates are not only interesting for scientific reasons: natural attenuation of contaminants, or lack thereof, has significant repercussions both at legislative level and in terms of soil/environment health and fertility. Changes in bioavailability are increasingly important in view of a more pronounced emphasis on bioavailability as the basis for environmental regulations. Similarly, enhanced natural attenuation processes have been proposed as a possible strategy for soil remediation. In both cases, the reversibility of these processes is of fundamental importance. Also, natural attenuation plays an important role in the decreased availability of plant nutrients over time and is largely responsible for the low efficiency of fertilisers.

This book covers 3 main areas of research. Chapter 1 through Chapter 3 focus on methods that have been applied to assess natural attenuation. In particular, chemical extractions, isotopic dilution techniques, and biological assessment of natural attenuation are described and discussed. Chapter 4 through Chapter 8 analyse processes that could lead to natural attenuation. Diffusion and partitioning of contaminants, structural dynamics of mineral surfaces, and biological and redox processes are elucidated. These chapters provide a mechanistic understanding of the fundamental processes responsible for the natural attenuation of contaminants and nutrients. The remaining 3 chapters discuss the implications of this process in terms of risk assessment and remediation of inorganic contaminants and bioavailability of nutrients.

It is our hope that this publication will increase the awareness of the importance of natural attenuation in governing inorganic contaminant and nutrient bioavailability in soils, assist understanding of the potential processes controlling natural attenuation, and clarify the consequences in terms of contaminant risk assessment and remediation and soil fertility.

1 Natural Attenuation of Trace Element Availability Assessed by Chemical Extraction

Bal Ram Singh

1.1 INTRODUCTION

The potential mobility and availability of trace elements depend on their total concentration in the soil, in soil solution, and in exchangeable forms. Retention and release reactions of solute with different components of the soil matrix govern the chemical behavior of trace elements. For elements freshly added to soils, the partitioning between the soil solution and the solid phase gradually changes with time until it reaches a state of pseudoequilibrium. The rate at which this equilibrium is attained is not only a time-dependent process, but it is also governed by the nature of the trace element and soil properties. These issues are discussed in more detail in Chapter 4 to Chapter 6. Predictions of persistence, the potential mobility, and the bioavailability of trace elements in contaminated soils currently require both chemical and biological methods (Kennedy et al. 1997), especially to quantify the fraction of trace elements available for biological uptake. However, chemical surrogates for assessing bioavailability have long been sought due to their simplicity and rapidity, in comparison to biological methods. Chemical methods include single and sequential extraction methods, which have both been used to quantify the concentration of trace elements available for plant uptake (He and Singh 1993a; Singh et al. 1995; Narwal and Singh 1998; Almås et al. 1999; Kukier and Chaney 2000). Sequential extraction procedures can provide more detailed information on the association of a trace element with soil components in comparison to a single extraction (Pickering 1986). The initial stages of a sequential extraction procedure have similarity to a single extraction and have been used to assess the potential short-term biological uptake of trace elements. The objective of this chapter is to discuss the use of chemical extraction procedures to assess natural attenuation of trace element availability in soils. Examples from different experiments, conducted under different soil and environmental conditions, will be cited to illustrate the effectiveness of different chemical extraction procedures in assessing the attenuation of trace element mobility or availability in soils.

1.2 CHEMICAL EXTRACTION TECHNIQUES

Chemical methods have been widely used to predict the fraction of trace elements that is mobile or accessible to plant roots. It is this fraction that poses an environmental concern when risk of contaminant metals in soils is assessed, especially where the soil–plant pathway is the main contributor to potentially harmful effects on animals and human beings. An ideal chemical extraction procedure should remove both the active, immediately available form and the fraction of the reserve form that replenishes the former over time (Levesque and Mathur 1988). However, in reality, there is as yet no single extraction procedure that can simulate the action of trace element uptake by plants, and hence different investigators have used many different extraction procedures (He and Singh 1993b; Kennedy et al. 1997; Christensen and Huang 1999). In contrast to single extraction techniques, sequential extraction procedures can provide more detailed information. In the sequential extraction technique, the initial extractants attempt to access immediately available forms of elements, that is, they extract elements from the solution phase and displace elements from the readily reversible exchange sites (physical and electrostatic sorption). The later stage extractants attempt to displace elements from more strongly sorbed phases (chemisorption) and from oxidizable phases (breakage of chemical bonds by redox agents) (Salbu et al. 1998). The goal of sequential fractionation procedures is to provide information useful to predict both short-term biological uptake and to reveal the trace element fraction that may become available over a longer period of time.

1.2.1 Single Extractants for Assessment of the Bioavailable Fraction

Many investigators have attempted to find a single extractant that can isolate the labile fraction of trace elements that is related to plant availability in a range of different soil types, but no one so far has succeeded in finding an extractant applicable to most soils and under all soil conditions. A large number of different extracting solutions have been trialed in these attempts to assess plant-available trace elements in soil. Some selected examples of these extractants are presented in Table 1.1. Single extractants may broadly be divided into 3 main classes: 1) weak neutral salt solutions ($MgCl_2$, $CaCl_2$, NH_4NO_3), 2) dilute solutions of either weak acids (e.g., acetic acid) or strong acids (e.g., HCl, HNO_3), and 3) chelating agents (e.g., DTPA, EDTA). The first type of extractants is able to release into solution metals that are associated with the exchange sites on the soil solid-phase and, hence, can be considered as bioavailable, though they may not extract the entire bioavailable fraction (McLaughlin et al. 2000). One of the arguments for using neutral salts over acid extractants has been that they do not change the soil pH and they simulate the ionic strength of the soil. Thus, they may correlate better with the solution bathing the plant roots. However, plant roots can change the chemistry of the rhizosphere, including rhizosphere pH, and may be able to access less readily exchangeable forms of elements than are released during neutral salt extraction. The chelating agents, such as diethylenetriaminepentaacetic acid (DTPA) and ethylenediamine-N,N,N′,N′-tetraacetic acid (EDTA), form complexes with free metal ions in solution and thus reduce their activity in solution. In response, metal ions desorb from soil surfaces or

TABLE 1.1
Example of some single extractants used to assess labile fractions of metals in soils

Extractant	Predominant fraction proposed to be extracted	References (some examples)
H_2O	Aqueous	Chang et al. (1984), Oughten et al. (1992)
1 M NH_4OAc (pH 4.8–7.0)	Aqueous and exchangeable	Andersson (1975), He and Singh (1993b), Oughten and Slabu (1994)
1 M NH_4NO_3	Aqueous and exchangeable	Symeonides and McRae (1977), Gupta and Aten (1993), He and Singh (1993b)
0.005 M DTPA	Aqueous and exchangeable	Lindsay and Norwell (1978), Arnesen and Singh (1998), Kukier and Chaney (2000)
0.1 M $NaNO_3$	Aqueous and organic bound	Gupta and Aten (1993)
1 M NaOAc in 25% v/v HOAc	Aqueous and fraction bound to Fe and Mn oxides	Tessier et al. (1979), Oughten et al. (1992)

from labile solid phases to replenish the free metal ions in solution. The amount of complexed metals that accumulate in solution during the extraction is a function of both the activity of metal ions in the soil and the ability of the soil to replenish those ions (Lindsey and Norwell 1978). However, chelating agents can also solubilize various proportions of soil solid phase minerals, which may result in extraction of associated elements from nonplant-available pools, leading to overestimates of their plant availability.

The ability to predict the plant-available fraction of trace elements using each of these methods has been found to vary considerably. Gupta and Aten (1993) assessed the bioavailability of Cd, Cu, and Zn for lettuce and ryegrass grown on sandy soils by extracting them with 2 M HNO_3, 0.52 M NH_4OAc + EDTA, 0.05 M $CaCl_2$, 0.1 M NH_4NO_3, 0.1 M $NaNO_3$, 0.1 M KNO_3, and 0.1 M $CaCl_2$. They found better correlations with 0.1 M $NaNO_3$ compared to other extractants and, hence, recommended this extractant for bioavailability studies of trace elements in contaminated soils. He and Singh (1993b) analyzed extractable Cd in a large number of samples (n = 133) from long-term cultivated soils with widely varying properties using the following extractants: 1 M NH_4OAc, 0.005 M DTPA, 1 M NH_4NO_3, 0.1 M $CaCl_2$, 0.2 M HCl, and 0.5 M NH_4OAc + 0.02 M EDTA. Although they found significant relationships between concentrations of extractable Cd in soil and plant Cd concentrations, none of these extractants gave a good assessment of plant-available Cd for all samples used in this study. However, of the extractants used, NH_4NO_3 was found to be the most reliable. The same extractant (NH_4NO_3) was found to give good correlations with the concentrations of Cd, Cu, Zn, Ni and Mn in samples collected from naturally metal-rich field soils (Mellum et al. 1998). In assessing the phytoavailability of Ni in soils, Echevarria et al. (1998) found DTPA to be useful for assessing labile Ni and suggested this could be considered a possible method for routine analysis. These authors showed that DTPA extracted isotopically

TABLE 1.2
Example of extractants commonly used to assess plant available fraction of metals

Extractants	Metal	Bioassay	References (examples)
0.25 M $MgCl_2$	Zn	Navy beans	Neilsen et al. (1987)
0.1 M $CaCl_2$	Cd and Zn	Oats and grasses	Sauerbeck and Stypereck (1985), He and Singh (1993b)
1 M NH_4NO_3	Cd	Radish, wheat, carrot	Symeonides and McRae (1977), Singh et al. (1995)
0.005 M DTPA	Cu, Zn, Fe, Mn, Cd	Sorghum, wheat, carrot, lettuce	Lindsay and Norwell (1978), Singh et al. (1995)
0.1 M $Ca(NO_3)_2$	Cd, Ni, Zn, Fe	Ryegrass, tobacco	Boisson et al. (1998)
0.01 M $Sr(NO_3)_2$	Ni	Oats, wheat, red beet	Kukier and Chaney (2000)

exchangeable Ni from silt loam, reinforcing the validity of this chemical to assess the phytoavailability of Ni. Arnesen and Singh (1998) also found positive and significant correlations between Ni and Zn concentrations in plants grown in a naturally metal-rich soil (alum shale) and their respective amounts extracted with DTPA. Kennedy et al. (1997) have provided an excellent review of extraction methods used to predict "available" metal fractions in soils. Examples of extractable concentrations of specific elements in soils and their uptake by different plant species are given in Table 1.2.

1.2.2 Sequential Extraction Schemes

The chemical forms of a trace element determine its behavior in the environment and its remobilization potential (Ma and Rao 1997). The use of solutions with a gradual increase from least aggressive to most aggressive extraction power may provide additional information about trace element fractions that can be released from solid-phase associations and become available for organism uptake because of changes in the soil environment. Many of the schemes used in more recent years are based on that originally developed by Tessier et al. (1979). A large number of different sequential extraction procedures have been used to extract trace elements in soils, sediments, sludges, and dissolved solids in waters (Kennedy et al. 1997). Some selected schemes for different trace elements are presented in Table 1.3. The schemes vary in the number of fractions extracted, as well as the order and type of reagents used.

Reagent selectivity and the extent of trace element redistribution during the extraction process have been examined. Some studies demonstrated poor selectivity (Kheboian and Bauer 1987) or extensive redistribution (Rendell et al. 1980; Li and Shuman 1996; Bunzl et al. 1999), whereas others showed high selectivity and limited redistribution (Kim and Fergusson 1991). Bunzl et al. (1999) used an

TABLE 1.3
Some sequential extraction techniques used to fractionate metals in soils

Metal	Soluble	Exchangeable (nonspecific)	Adsorbed (specific)	Carbonate bound	Fe and Mn oxides	Organic bound	Residual	References
Cd		$^{1}MgCl_2$		$^{2}NaOAc$	$^{3}NH_2OH$, HCl+HOAc	$^{4}H_2O_2$-HNO_3	^{4}HF-$HClO_4$	Tessier et al. (1979)
Cu, Fe, Mn	$^{1}H_2O$	$^{2}Ca(NO_3)_2$	$^{2}Pb(NO_3)_2$		^{6}Fractions[a]	$^{5}K_4P_2O_7$	^{8}HF-HNO_3	Miller et al. (1986)
Cd, Cu, Ni, Pb, Zn	$^{1}H_2O$	^{2}1 M NH_4OAc pH 7	^{3}1 M NH_4OAc pH 5		$^{4}NH_2OH$, HCl+ 25%HOAc	$^{5}H_2O_2$-HNO_3	$^{6}HNO_3$	Salbu et al. (1998)
Cd, Cr, Cu, Ni, Pb, Zn		$^{1}KNO_3$	$^{2}H_2O$	$^{4}Na_2$ EDTA		$^{3}NaOH$	$^{5}HNO_3$	Chang et al. (1984)
Cd	^{1}0.1 M $Sr(NO_3)_2$		^{2}1 M NaOAC		^{4}0.2 M oxal. acid[b]	^{3}5% NaOCl, pH 5.0	^{5} Microwave digest	Ahnstrom and Parker (1999)
Cd, Cu, Pb Zn	^{1}M NH_4OAc pH 7					$^{2}H_2O_2$ pH 2	$^{3}HNO_3$	Edwards et al. (1999)

Note: Numbers on the left side of an extractant refer to the sequence the extractant was used in the respective extraction scheme.

[a] Three fractions of metal oxides extracted separately: (i) Mn oxides: (NH_2OH-HCl); (ii) amorphous Fe oxides: $(NH_4)_2C_2O_4 + H_2C_2O_4$ (oxalate reagent); (iii) crystalline Fe oxides: oxalate reagent + UV radiation.

[b] 0.2 M $(NH_4)_2C_2O_4$ + 0.1 M ascorbic acid (pH 3.0).

extended Tessier scheme (Tessier et al. 1979), where subsequent to fraction IV (bound to organic matter) the residue was extracted with concentrated HNO_3, and they referred to this fraction as "persistently bound." They used this scheme to study the chemical partitioning of trace elements in a slag-contaminated soil, as well as in the pure slag and an uncontaminated soil. They showed that during the various extraction steps, substantial redistribution of trace elements between the slag and soil particles can occur. In many cases, elements released during the extraction with acid hydroxylamine or acid hydrogen peroxide are partially readsorbed by solid constituents of the mixture and will therefore be found in subsequent fractions. Under these conditions, partitioning of an element in a slag-contaminated soil may not necessarily give any relevant information on the form of this element in the slag or slag and soil mixture because of the redistribution processes. It is, however, possible that redistribution and poor selectivity might also be the result of poor extraction efficiency of a given reagent (Han and Banin 1997; Ahnstrom and Parker 1999). In a study of Cd reactivity in soils, Ahnstrom and Parker (1999) optimized a sequential extraction procedure for Cd fractionation. Cadmium was fractionated into 5 operationally defined fractions: 0.1 M $Sr(NO_3)_2$ (F1, soluble-exchangeable); 1 M NaOAc adjusted to pH 5.0 (F2, sorbed-carbonate); 5% NaOCl, pH 8.3 (F3, oxidizable); 0.4 M oxalate + 0.1 M ascorbate (F4, reducible); and 3 HNO_3:1 HCl (F5, residual). They found that dissolution products of major components of the targeted phases (e.g., high Ca in F2) generally reflected the desired reagent specificity and selectivity. The redistribution was generally minimal (<3%), though it did reach 12% for F3 of the sludge-amended soil. Salbu et al. (1998) modified the scheme by Tessier et al. (1979) and proposed that the reagents be used in the following order to differentiate between binding mechanisms: reversible physical sorption by extracting with H_2O (F1) followed by 1 M NH_4OAc at soil pH (F2); reversible electrostatic sorption using a pH effect by extracting with 1 M NH_4OAc at pH 5.0 (lower than soil pH, F3); irreversible chemisorption by using redox reagents, which was provided sequentially by using 0.04 M $NH_2OH.HCl$ in 25% v/v HOAc (F4), H_2O_2 in 1 M HNO_3 (F5), and 7 M HNO_3 (F6). As in other schemes, overlap between phases (e.g., F2 and F3 or F4 and F5 fractions) is also a problem in this scheme (Salbu et al. 1998).

1.3 ATTENUATION OF ELEMENT SOLUBILITY AND AVAILABILITY IN CONTAMINATED SOILS

In spite of the limitations in reagent specificity and selectivity, and the difficulties of overlap between fractions, sequential extraction schemes can be useful when applied on a comparative basis to study temporal and conditional changes in trace element solid-phase associations. Trace elements retained by the solid-phase can be mobilized into the solution phase by changes in soil pH, temperature, organic matter decomposition, redox potential, leaching, ion-exchange processes, and microbial activity. On the other hand, these changes can also result in immobilization of elements from the solution phase into the solid phase. Some examples are cited here to show how the attenuation of trace element availability is affected by these changes, and how it can be assessed by single or sequential extraction techniques.

1.3.1 Assessing Attenuation by Single Extractants

The dominant attenuation process for trace elements is immobilization in the solid phase, such that their ability to be transferred into the soil solution is reduced. It is the mobile fraction of trace elements that is accessible to plants and microbes. Different soil additives (lime, organic matter, aluminosilicates, phosphates, etc.) have been used to immobilize trace elements in soils. Chemical extraction procedures are able to predict the changes in trace element mobility or bioavailability in soils, as shown by the examples cited below.

Kukier and Chaney (2000) examined the effects of various soil amendments on the extractability of Ni, and its uptake by redbeet, wheat, and oat plants, in Ni-contaminated muck soils. Nickel extracted by either the DTPA or 0.01 M $Sr(NO_3)_2$ methods was significantly affected by soil amendments (lime and hydrous ferric oxide (HFO)). The DTPA-extractable Ni was up to 300 times higher than that extracted by $Sr(NO_3)_2$ (Table 1.4), but the latter extraction was more sensitive to the amendment-induced changes in the soil-binding capacity for Ni. Lime applied alone was only slightly more effective in reducing Ni extractability than HFO, but both amendments applied together had a much larger mitigating effect. This phenomenon results from the increase in trace element sorption by HFO when pH becomes more alkaline. The results of both extraction procedures demonstrated that applied amendments caused a redistribution of the Ni within the soil matrix, leading to a decrease in the most labile forms of soil Ni. Nickel concentrations in the shoots of oat were significantly decreased by lime application, but the effect of HFO on Ni concentrations in oat was not significant (Table 1.4). Nevertheless, both amendments at least partially ameliorated Ni toxicity in oat. The closest relationship between DTPA-extractable Ni and concentrations of Ni in plant shoots was obtained for redbeet ($R^2 = 0.91$), followed by wheat ($R^2 = 0.52$) and oat ($R^2 = 0.42$). The corresponding R^2 values for $Sr(NO_3)_2$ were slightly lower for all species. The predictive value of both extractants was similar for the same species but varied among species.

TABLE 1.4
The effect of soil amendments on extractable soil Ni in a high-Ni muck soil

Soil amendment	DTPA extractable Ni ($mg \cdot kg^{-1}$)	0.01 M $Sr(NO_3)_2$ extractable Ni ($mg \cdot kg^{-1}$)	Oat yield (g/pot)	Ni in shoots ($mg \cdot kg^{-1}$)
Control	1451a	16.3a	2.00a	78.0a
Limestone	1059b	5.57b	2.26a	50.3b
HFO[a]	1125c	6.57c	0.65c	59.1ab
Lime + HFO[a]	755d	2.54d	1.32b	61.1ab

Note: Means followed by the same letter in the same column are not significantly different at $P<0.05$.

[a] Hydrous ferric oxide.

Source: Extracted from Kukier U, Chaney RL. 2000. *Can J Soil Sci* 80:581–593.

The effectiveness of single extractants in predicting the lime-induced decrease in Cd availability to wheat, carrot, and lettuce grown in 2 naturally metal-rich soils was investigated by Singh et al. (1995). The soils used were developed on alum shale (sulfide-bearing rocks formed in anoxic environments) and were naturally rich in several elements, of which Cd was of greatest concern. They found that the concentrations of both DTPA and NH_4NO_3-extractable Cd decreased with increased pH, caused by liming in both soils. The same effect of lime was also seen in Cd concentrations in crops grown on the soils, with significant decreases in Cd concentrations found in wheat and carrot. The regression equations between soil pH and extractable Cd showed that soil pH was an important factor affecting extractable Cd. Both DTPA and NH_4NO_3 were found equally effective in relation to predicting the effect of lime on the Cd concentrations in plants from both soils.

The mobility, bioavailability, and phytotoxicity of Cd and Zn were investigated using single soil extractions and vegetation experiments in silty soils near a smelter at Evin, France, by Mench et al. (1994). Thomas basic slag (TBS), hydrous manganese oxide (HMO), and steelshot (SS) were used as chemical agents to immobilize elements in soils and thus decrease element availability to plants. Although only the TBS treatment increased soil pH, all the agents decreased the extractability of Cd and Zn by water and $Ca(NO_3)_2$ (Table 1.5). No clear effect of these agents was seen on EDTA-extractable Cd and Zn. Hydrous manganese oxide and ST decreased the water-soluble fraction of Cd by >80% as compared to untreated soil. The corresponding decrease in $Ca(NO_3)_2$-extractable Cd was 95% and 66% by HMO and ST,

TABLE 1.5
Extractable fractions of the metals and correlation between the metals extracted and metal uptake by ryegrass

	Water (mg·kg^{-1})		M Ca $(NO_3)_2$ (mg·kg^{-1})		EDTA (mg·kg^{-1})		
Treatment	**Cd**	**Zn**	**Cd**	**Zn**	**Cd**	**Zn**	**Soil pH**
Untreated	0.15a	1.8a	1.2a	14.7a	11.4a	267a	7.8
TBS (1%)	0.06b	0.7b	1.0b	7.4b	11.0a	212b	8.5
HMO (1%)	0.03c	0.7b	0.05d	2.2d	10.6a	268a	7.8
ST (1%)	0.02c	0.4c	0.04c	3.4c	7.6b	185c	7.8
	Correlation coefficient (r^2)						
Ryegrass	0.64^A	0.58	0.94^B	0.64^A	0.09	0.29	

Note: TBS = Thomas basic slag, HMO = hydrous manganese oxide, ST = steelshot; Means followed by the same letter in the same column are not significantly different at $P < 0.05$.

^{A} r^2 at 0.05.
^{B} r^2 at 0.01.

Source: Extracted from Mench M, Vangronsveld J, Didier V, Clijsters H. 1994. *Environ Pollut* 86:279–286.

respectively. In all treatments, EDTA extracted about 63% and $Ca(NO_3)_2$ about 7% of the total Cd. Concentrations of Cd and Zn in the latter extractant showed a strong correlation with Cd and Zn uptake by ryegrass (Table 1.5). The results suggest the chemical agents used, especially HMO and ST, were effective in mitigating Cd uptake by plants and that the $Ca(NO_3)_2$-extractable Cd predicted plant availability of soil Cd. Similar positive effects of beringite and ST on Cd and Ni extractability, in a long-term sludge-treated field at Bordeaux, France, were observed by Boisson et al. (1998). Single extractants can provide an empirical method for evaluating changes in the potential availability of soil contaminants for plant uptake for a wide range of soils and can be used for a number of elements, including radionuclides (Kennedy et al. 1997).

1.3.2 Assessing Attenuation by Sequential Extraction

The solubility and bioavailability of elements in contaminated soils are influenced by their speciation and distribution between various fractions in the soil. Sequential extraction techniques have been used to determine the changes caused by the application of various additives (Edwards et al. 1999) or alteration in the soil environment (Kashem and Singh 2004). The use of extract solutions with a gradual increase in displacement or dissolution strength in a sequential scheme gives information about trace element fractions that may be released from soil phase association and become available to plants, or fractions that may be trapped within the crystal lattice of soil mineral particles and become immobilized due to chemical changes in the soil environment. Thus, sequential extractions can give information on the potential for both mobilization and immobilization of elements over time due to changes in the soil environment (Kennedy et al. 1997). Examples cited below show how attenuation of metal mobility and availability to plants, caused by changes in soil conditions (pH and redox levels) and soil amendments (organic and inorganic), can be assessed by sequential extraction techniques.

Results from a growth chamber study showed that the flooding (submergence) of metal-contaminated soils resulted in highly significant decreases in NH_4OAc-extractable amounts of Cd, Zn, and Ni in soils as compared to nonflooded soils (Kashem and Singh 2004). The data for Zn are shown in Figure 1.1. The percentage decreases in concentrations of Cd, Zn, and Ni in this fraction (F1) were 97%, 87%, and 47%, respectively. In contrast, the concentrations of Cd and Zn in the Fe- and Mn-oxide (F3) fraction increased by 29% and 14%, respectively, in the flooded soils. Contrary to Cd and Zn, Ni concentration in the carbonate (F2) fraction increased by 29% but decreased by 13% in the Fe and Mn-oxide (F3) fraction in flooded soils. The organically bound (F4) fraction of Cd and Zn in flooded soils was increased by >11%. Changes in the concentrations of these metals in the residual (F5) fraction were only minor. It seems Cd and Zn transformed from F1 to F3 and F4 fractions, and Ni transformation took place both from F1 and F3 to F2 fraction. This transformation resulted in higher immobilization of Cd and Zn compared to Ni. The observation that metals transform from the mobile to the immobile fraction of soils after submergence was also reflected in decreased metal uptake by rice plants. Kashem and Singh (2001a) found that elemental concentrations in polished rice at

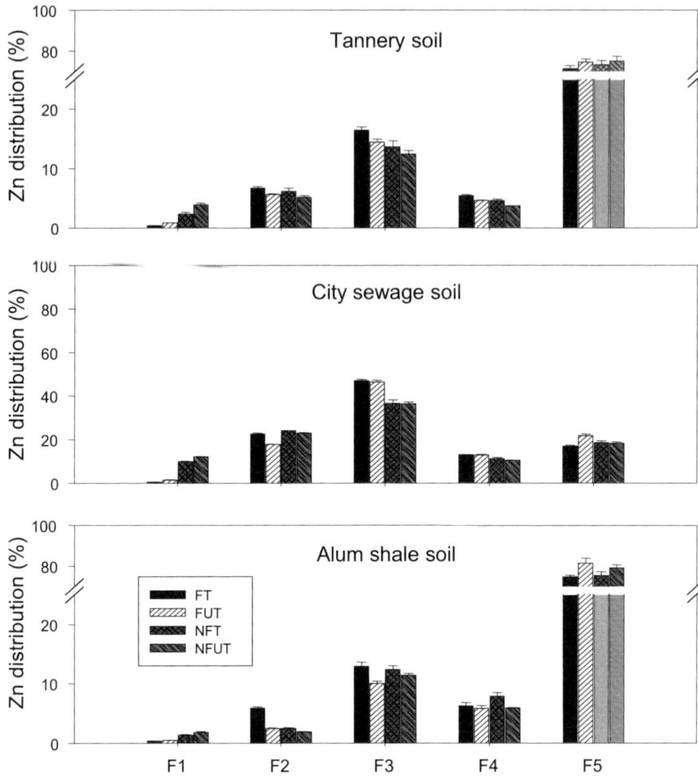

FIGURE 1.1 Effect of flooding and organic matter on percentage distribution of Zn among 5 fractions in contaminated soils. FT = flooding with organic matter (OM); FUT = flooding without OM; NFT = no flooding with OM; NFUT = no flooding without OM. The vertical bars represent the standard deviation of 3 replicates. The fractions F1, F2, F3, F4, and F5 were extracted by I M NH_4OAc (pH 7), 1 M NH_4OAc (pH 5), 0.04 M NH_2 OH HCl in 25% HAc, 30% H_2O_2, and 7 M HNO_3, respectively. (Redrawn from Kashem and Singh 2004. With permission.)

maturity, under flooding conditions, decreased by 79% for Cd, 65% for Ni, and 16% for Zn as compared to nonflooded conditions. The decrease in mobility of elements in soils (e.g., NH_4OAc-extractable), and in their uptake by rice plants, were associated with decreased Eh and increased pH in flooded soils, consistent with other findings that these chemical changes generally decrease element solubility (Kashem and Singh 2001b).

Edwards et al. (1999) used a 3-stage sequential extraction scheme to determine metal bioavailability and species distribution in metal-contaminated soils amended with synthetic zeolites. Ammonium acetate (1 M, pH 7) was used to liberate water soluble and exchangeable elements, which are held through electrostatic interaction on negatively charged sites in the soil structure. Incorporation of each type of zeolite in the soils led to a decrease in metal concentrations in the NH_4OAc-extractable fraction. The magnitude of decrease in extractable Cd concentration by 2 types of

TABLE 1.6
Fractions of Cd and Zn in Trelogan-contaminated soil

Treatment	NH_4OAc	H_2O_2	HNO_3	Total
	Cd ($mg \cdot kg^{-1}$)			
Unamended	37	16	20	73
Zeolite P 1%	15	13	45	73
Zeolite A 1%	14	13	44	71
	Zn ($mg \cdot kg^{-1}$)			
Unamended	8500	4850	3500	16850
Zeolite P 1%	3840	800	13000	17640
Zeolite A 1%	4200	2500	10000	16700

Note: Collected from the site of metal refinery, an old lead–zinc mine spoil, and a field which has been treated with sludge and different rates of zeolite. The synthetic zeolites were obtained from Crossfield Chemicals, Warington, England.

Source: Extracted from Edwards R, Rebedea I, Lepp NW, Lovell AJ. 1999. *Environ Geochem Health* 21:157–173.

zeolite was >60% in Trelogan soil (lead–zinc mine spoil), and the corresponding decrease in Zn concentration in this soil was about 50% (Table 1.6). Hydrogen peroxide was used to determine the concentration of elements associated with organic matter in the soil. Although the decrease in Cd concentration in the H_2O_2-extractable fraction was rather small in zeolite-amended Trelogan soil, the concentration of Zn in this fraction decreased by 50% to 84%, depending on zeolite type. The residual soil element fraction, extracted by HNO_3 digestion, increased in all cases where zeolite caused a decrease in the element concentration of the NH_4OAc-extractable or H_2O_2-extractable fractions (Table 1.6). This suggests that zeolites form stable complexes with elements, and their complexes are not broken down under conditions found in the soil environment. Zeolite addition increased soil pH due to exchange of Na^+ ions from the zeolite for H^+ ions in the soil solution. This caused reductions in the concentrations of Cd, Cu, and Zn in soil solution by about 66%, 50%, and 60%, respectively, in zeolite-treated Prescot soil (Table 1.7). However, the decrease in soil solution metal concentration was far more significant than could be accounted for by pH increase alone. Sequential extraction methods are widely used to look at changes in soil phase associations, and the information obtained from early stages of these sequences can be used to predict potential short-term bioavailability by relating them to plant concentrations. The information from later stages in the sequences can be useful for predicting potential longer-term release of trace elements from soil.

TABLE 1.7
Metal concentration in soil solution of treated and untreated prescot soil

Treatment	pH	Cd	Cu	Zn
Unamended	4.3 ± 0.3	0.9 ± 0.1	52 ± 3	30 ± 3
Zeolite P 1%	7.0 ± 0.1	0.3 ± 0.1	24 ± 2	12 ± 2
Zeolite A 1%	7.0 ± 0.4	0.4 + 0.1	26 ± 2	12 ± 2

Note: Contaminated by atmospheric deposition after 90 d treatment with synthetic zeolites

Source: Extracted from Edwards R, Rebedea I, Lepp NW, Lovell AJ. 1999. An investigation into the mechanism by which synthetic zeolite reduce labile metal concentration in soils. *Environ Geochem Health* 21:157–173.

Partitioning of ^{109}Cd and ^{65}Zn in a naturally metal-rich alum shale soil (developed on sulfide-bearing rocks) into mobile and nonmobile fractions was done by Almås et al. (1999) by using the sequential extraction scheme of Salbu et al. (1998). The fractions of this scheme are as described earlier in Subsection 1.3.2. Elements recovered by H_2O or by exchange reactions are referred to as "mobile fractions" (F1, F2, and F3). The nonmobile or inert fractions (F4, F5, F6, and F7) refer to fractions where redox agents release chemi-sorbed elements. An index of the ratio of mobile and nonmobile fractions was described as the mobility factor (MF), shown as follows (Salbu et al. 1998):

$$\text{Mobility factor (MF)} = \frac{\Sigma \text{ Mobile fractions } (F1+F2+F3)}{\Sigma \text{ Immobile fractions } (F4+F5+F6+F7)}$$

The relative distribution of ^{109}Cd and ^{65}Zn showed that their concentration decreased with time in the 3 mobile fractions, and increased in the nonmobile fractions (Figure 1.2). As both elements showed generally the same pattern, results for ^{65}Zn only are presented in Figure 1.2. About 30% and 20% of ^{109}Cd and ^{65}Zn, respectively, were associated with the mobile fraction after 0.5 h, but these values decreased to 20% and 10% after 1 year (8750 h). Among the nonmobile fractions, the F4 fraction accounted for 40% to 55% of the total, and hence this fraction retained most of the solid phase ^{109}Cd and ^{65}Zn. The exchange of ^{109}Cd and ^{65}Zn between mobile and nonmobile fractions was estimated by calculating the mobility factor (Table 1.8). The diffusion of ^{109}Cd and ^{65}Zn toward the nonmobile fraction is indicated by the declining value of MF with increasing time. The low value of MF indicates low mobility of the element. The MF was also affected by temperature as the elevated temperature significantly decreased the MF value (Table 1.8) (Almås et al. 1999). However, the uptake of ^{109}Cd and ^{65}Zn and their stable isotopes was

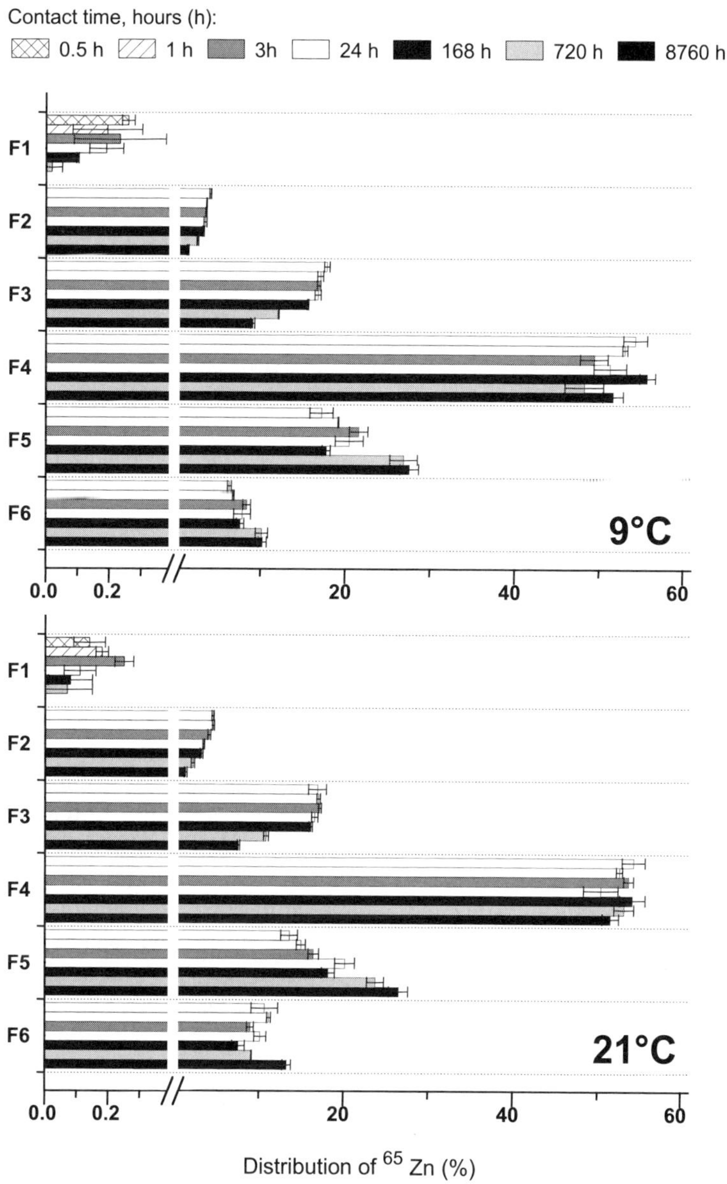

FIGURE 1.2 Effect of equilibration time (aging) and temperature on relative distribution (mobilization or immobilization) of ^{65}Zn in a naturally metal-rich alum shale soil. The bars represent the standard deviation of 3 replicates. The fractions F1, F2, F3, F4, F5, and F6 were extracted by H_2O, I M NH_4OAc (pH 7), 1 M NH_4OAc (pH 5), and 0.04 M NH_2 OH HCl in 25% HAc, 30% H_2O_2, and 7 M HNO_3, respectively. Elements recovered by fractions F1, F2, and F3 are referred as "mobile," whereas those recovered by F4, F5, and F6 as "immobile." (Redrawn from Almås Å, Singh BR, Salbu B. 1999. *J Environ Qual* 28:1742–1750. With permission.)

TABLE 1.8
Mean values of mobility factor (MF) for ^{109}Cd and ^{65}Zn at different times after contact with alum shale soil at 9 and 21 °C

	^{109}Cd		^{65}Zn	
Contact time (h)	**9 °C**	**21 °C**	**9 °C**	**21 °C**
0.5	0.493	0.438	0.283	0.274
720	0.296	0.271	0.171	0.150
8760	0.269	0.250	0.116	0.096

Source: Extracted from Almås Å, Singh BR, Salbu B. 1999. *J Environ Qual* 28:1742–1750.

higher in ryegrass grown at 21 °C than that grown at 9 °C. Fractionation of Cd and Zn in soil by sequential extraction assisted the prediction of plant uptake of these elements (Almås and Singh 2001).

1.4 OTHER EXPERIMENTAL PARAMETERS OF IMPORTANCE

When interpreting soil test results, it is not only the choice of chemical extractions and analytical measurements which is important, but sampling procedure, sample preparation and handling, soil-to-solution ratio and contact time, reproducibility of the method used, and the interpretation of data also play very important roles. Because extraction methods are useful tools to assess attenuation of trace elements, and because they are affected by the parameters described earlier, it was thought pertinent to discuss these parameters briefly.

Sampling techniques: The soil sample used must be representative of the area and hence topography, uniformity of soil type, and soil sampling techniques, including the number of replicates from the sampled area, should be taken into consideration. Some recommendations for soil sampling were made by Petersen and Calvin (1986) and Rubio and Ure (1993).

Preparation of sample: Sample preparation and handling can have significant effects on chemical and physical properties of soils (Rubio and Ure 1993; Brown 1999). For instance, if soils are dried at over 50 °C and ground to fine powder, extractable concentrations of some elements may not relate well to plant concentrations in the field.

Soil–solution ratio and contact time: If the soil-to-extractant ratio is too low (<1:5), element readsorption may occur during extraction. Kukier and Chaney (2000) reported that the amount of Ni, Fe, Cd, Zn, and Mn increased when the soil–solution ratio increased from 1:5 to 1:10, and then leveled off at ratios of 1:20 and 1:30. The contact time between soil and extractant should be such that it allows a steady state to be established, and it can range from 1 to 24 h depending upon the vigor of the shaking.

Precision and reproducibility of method: Reproducibility of extraction methods for trace elements is a big problem in many laboratories. To avoid bias in extraction data, it is advisable to use an internal laboratory reference soil for each set of soils. However, there are no certified reference samples for trace elements which can be used. Efforts are being made to establish an internationally acceptable protocol for extraction procedures for trace elements.

Interpretation of data: Ion exchange processes may be altered in the field by changes in soil moisture, leaching, plant uptake, and microbial activity. Hence, single extractants, which aim to assess the exchangeable fraction but use soil samples taken to the laboratory, may not provide a reliable estimate of bioavailability in the field. Nevertheless, extraction methods can provide good approximation of element availability in soils. For sequential extractions, nonspecificity in extraction of soil solid phases, readsorption during extraction, and cumulative errors in adding sequential concentrations also complicate the interpretation of data obtained by these procedures.

1.5 CONCLUSIONS

Trace element contaminated soils are a threat to biota and groundwater, hence the attenuation of trace element availability or mobility in soils is important to protect humans, animals, and the environment. The dominant attenuation process for trace elements in soils is immobilization on the solid phase. Chemical extraction techniques are effective tools to predict natural attenuation of trace element availability in soils. The cited examples of attenuation of trace element availability show that single extractants of neutral salts (e.g., $MgCl_2$, $CaCl_2$, NH_4NO_3) involving ion exchange processes are able to predict the amount of trace element associated with the exchange sites on the soil solid phase, considered to be bioavailable. Additional advantages for using neutral salts are that soil pH is not changed and, if a weak salt solution is used, the ionic strength of the soil is simulated, so that results obtained with these methods may correlate better with biological uptake of trace elements. The chelating agents (DTPA, EDTA), which form complexes with free metal ions in solution but are also able to replenish the free metal ions in solution, have also shown their potential for predicting available fraction of trace elements in soils. Although none of these methods is able to simulate the action of trace element uptake by plants, they can provide empirical information on the potential availability of trace elements for root uptake. Thus, refinement or development of extraction methods for individual elements in order to assess their bioavailability is encouraged.

In spite of the limitations in specificity and selectivity of reagents, and the difficulties of overlap between phases, sequential extraction procedures provide information useful for predicting both short-term biological uptake and longer-term bioavailability of trace elements (by predicting release of elements into labile forms).

Both single and sequential extraction methods cited here show wide variability in predicting the bioavailable fractions of trace elements as affected by various attenuation processes in soils. Results obtained showed good correlations between trace element amounts extracted with single extractants or with initial stages of sequential extraction schemes and plant uptake of these elements. However, the extractants were condition- and site-specific. The use of certified reference soils and interlaboratory comparisons may help in validating extraction methods for wider use. In addition, combining these methods with biological methods and modeling may improve their prediction efficiency.

REFERENCES

Almås Å, Singh BR, Salbu B. 1999. Mobility of ^{109}Cd and ^{65}Zn in soil influenced by equilibration time, temperature and organic matter. J Environ Qual 28:1742–1750.

Almås Å, Singh BR. 2001. Plant uptake of cadmium ^{109}Cd and ^{65}Zn at different temperature and organic matter level. J Environ Qual 30:869–877.

Andersson A. 1975 Relative efficiency of nine different soil extractants. Swed J Agric Res 5:125–135.

Ahnstrom ZAS, Parker DR. 1999. Development and assessment of a sequential extraction procedure for the fractionation of soil cadmium. Soil Sci Soc Am J 63:1650–1658.

Arnesen AKM, Singh BR. 1998. Plant uptake and DTPA-extractability of Cd, Cu, Ni and Zn in a Norwegian alum shale soil as affected by previous addition of dairy and pig manures and peat. Can J Soil Sci 78:531–539.

Boisson J, Mench M, Sappin-Didier V, Solda P, Vangronsveld J. 1998. Short-term in situ immobilization of Cd and Ni by beringite and steelshots application to long-term sludged plots. Agronomie 18:347–359.

Boissan J, Mench M, Vaugronsveld, J, Rutleus, A, Kopponcu, P, Delcoe, T. 1999. Immobilization of trace metals and arsenic by different soil additives: Evaluation by means of chemical extractions. Commun Soil Sci Plant Anal 30:365–387.

Brown AJ. 1999. Soil sampling and soil handling for chemical analysis. In: Peverill KL, Sparrow LA, Reuter DJ, editors. Soil analysis: An interpretation manual. Melbourne, Australia: CSIRO Publishing. p 35–53.

Bunzl K, Trautmannsheimer M, Schramel P. 1999. Partitioning of heavy metals in a soil contaminated by slag: a redistribution study. J Environ Qual 28:1168–1173.

Chang AC, Page AL, Warnecke JEGE. 1984. Sequential extraction of soil heavy metals following sludge application. J Environ Qual 13:33–38.

Christenson TH, Huang PM. 1999. Solid phase cadmium and the reactions of aqueous cadmium with soil surfaces. In: McLaughlin, MJ, Singh, BR. editors. Cadmium in soils and plants. Dordrecht, The Netherlands: Kluwer Academic Publishers. p 65–96.

Echevarria G, Morel JL, Fardeau JC, Levlere-Cessac E. 1998. Assessment of phytotoxicity of nickel in soils. J Environ Qual 27:1064–1070.

Edwards R, Rebedea I, Lepp NW, Lovell AJ. 1999. An investigation into the mechanism by which synthetic zeolite reduce labile metal concentration in soils. Environ Geochem Health 21:157–173.

Gupta SK, Aten C. 1993. Comparison and evaluation of extraction media and their suitability in a simple model to predict the biological relevance of heavy metal contamination in contaminated soils. Int J Environ Anal Chem 51:25–46.

Han F, Banin A. 1997. Long-term transformations and redistribution of potentially toxic heavy metals in arid-zone soils incubated: I. Under saturated conditions. Water Air Soil Pollut 95:399–423.

He QB, Singh BR. 1993a. Cadmium distribution and extractability in soils and its uptake by plants as affected by organic matter and soil type. J Soil Sci 44:641–650.

He QB, Singh BR. 1993b. Plant availability of cadmium in soils. I. Extractable cadmium in newly and long-term cultivated soils. Acta Agric Scand Sect B Soil Plant Sci 43:134–141.

Kashem MA, Singh BR. 2001a. Metal availability in contaminated soils: II. Uptake of Cd, Ni and Zn in rice plants as affected by moisture level and organic matter. Nutr Cycl Agroecosyst 61:257–266.

Kashem MA, Singh BR. 2001b. Metal availability in contaminated soils: I. Effects of flooding and organic matter on the changes in pH, Eh and solubility of Cd, Ni and Zn. Nutr Cycl Agroecosyst 61:247–255.

Kashem MA, Singh BR. 2004. Transformation in solid phase species of metals as affected by flooding and organic matter. Commun Soil Sci Plant Anal 35:1435–1456.

Kennedy VH, Sanchez AL, Oughton DH, Rowland AP. 1997. Use of single and sequential chemical extractants to assess radionuclide and heavy metal availability from soils for root uptake. Analyst 122: R89–R100.

Kheboian C, Bauer CF. 1987. Accuracy of selective extraction procedure for metal speciation in model aquatic sediments. Anal Chem 59:1417–1423.

Kim ND, Fergusson JF. 1991. Effectiveness of commonly used sequential extraction techniques in determining the speciation of cadmium in soils. Sci Total Environ 105:190–209.

Kukier U, Chaney RL. 2000. Remediating Ni-phytotoxicity of contaminated quarry muck soil using limestone and hydrous iron oxide. Can J Soil Sci 80:851–593.

Levesque M, Mathur SP. 1988. Soil tests for Cu, Fe, Mn and Zn in Histosols. I. A comparison of eight extractants for measuring active and reserve forms of the elements. Soil Sci 145:215–221.

Li Z, Shuman LM. 1996. Redistribution of forms of zinc, cadmium and nickel in soils treated with EDTA. Sci Total Environ 19:95–107.

Lindsay WL, Norvell WA. 1978. Development of a DTPA soil test for zinc, iron, manganese and copper. Soil Sci Soc Am J 42:421–428.

Ma LQ, Rao GN. 1997. Chemical fractionation of cadmium, copper, nickel, and zinc in contaminated soils. J Environ Qual 26:259–264.

McLaughlin MJ, Zarcinas BA, Stevens DP, Cook N. 2000. Soil testing for heavy metals. Commun Soil Sci Plant Anal 31:1661–1700.

Mellum HK, Arnesen AKM, Singh BR. 1998. Extractability and plant uptake of heavy metals in alum shale soils. Commun Soil Sci Plant Anal 29:1183–1198.

Mench M, Vangronsveld J, Didier V, Clijsters H. 1994. Evaluation of metal mobility, plant availability and immobilization by chemical agents in a limed silty soil. Environ Pollut 86:279–286.

Miller WP, Marteus DC, Zeluzny LW. 1986. Effect of sequences in extraction of trace metals from soils. Soil Sci Soc Am J 50:598–601.

Narwal RP, Singh BR. 1998. Effect of organic materials on partitioning, extractability and plant uptake of metals in an alum shale soil. Water Air Soil Pollut 103:405–421.

Neilsen D, Høyt PB, MacKenzie AF. 1987. Measurement of plant-available zinc in British Columbia orchard soils. Commun Soil Sci Plant Anal 18:161–186.

Oughten DH, Salbu B, Rise G, Lein H, Østby G, Nøren A. 1992. Radionuclide mobility and bioavailability in Norwegian and Soviet soils. Analyst 117:481–486.

Oughten DH, B Salbu.1994. Influence of physio-chemical forms on transfer. In: Dahlsgård H, editor. Nordic radioecology—the transfer of radionuclides through Nordic ecosystems to man. Amsterdam: Elsevier. p 165–184.

Peterson RG, Calvin LD. 1986. Sampling. In: Klute A, editor. Methods of soil analysis. Part 1—physical and mineralogical methods. Madison (WI): American Society of Agronomy. p 33–52.

Pickering WF. 1986. Metal ion speciation-soils and sediments (a review). Ore Geol Rev 1:83–146.

Rendell PS, Batley GE, Cameron AJ. 1980 Adsorption as a control of metal concentration in sediment extracts. Environ Sci Technol 14:314–318.

Rubio R, Ure AM. 1993. Approaches to sampling and sample pretreatments for metal speciation in soils and sediments. Int J Environ Anal Chem 51:205–217.

Salbu B, Krekling T, Oughton DH. 1998. Characterisation of radioactive particles in the environment. Analyst 123:843–849.

Sauerbeck DR, Stypereck B. 1985. Evaluations of chemical methods for assessing the Cd and Zn availability from different soils and sources. In: Leschber R, Davis RD, L'Hermite P, editors. Chemical methods for assessing bio-available metals in sludges and soils. Amsterdam: Elsevier. p 49–61.

Singh BR, Narwal RP, Jeng AS, Almås Å. 1995. Crop uptake and extractability of Cd in soils naturally high in metals at different pH levels. Commun Soil Sci Plant Anal 26:2123–2142.

Symeonides C, McRae SG. 1977. The assessment of plant available Cd in soils. J Environ Qual 6:120–123.

Tessier A, Campbell PGC, Bisson M. 1979 Sequential extraction procedure for the speciation of particulate trace metals. Anal Chem 51:844–851.

2 Techniques for Measuring Attenuation: Isotopic Dilution Methods

Scott Young, Neil Crout, Julian Hutchinson, Andy Tye, Susan Tandy, and Lenah Nakhone

2.1 INTRODUCTION

Isotopic dilution (ID) has been used to discriminate between "labile" and "nonlabile" pools of various elements in soils for several decades. The earliest applications of the general technique focused on major nutrient availability and dynamics (^{32}P, ^{42}K, ^{45}Ca), and gave rise to the terminology and methods still widely used today (Larsen 1952; Russell et al. 1954; Larsen and Cooke 1961; Deist and Talibudeen 1967). They also exposed many of the methodological and conceptual problems discussed in contemporary studies. In the 1970s, there was a marked increase in the application of ID methods to determining micronutrient metal availability (Lopez and Graham 1972; Tiller et al. 1972a,b; Graham 1973), using isotopes such as ^{60}Co, ^{64}Cu, ^{54}Mn, ^{59}Fe, ^{63}Ni, and ^{65}Zn. However, in more recent years the emphasis within soil chemistry has shifted in response to greater concern over environmental contamination. Applications of ID to soils over the past 2 decades have increasingly reflected this change, and recent studies have utilized ^{73}As, ^{109}Cd, ^{111}Cd, ^{115}Cd, ^{203}Hg, ^{63}Ni, and ^{65}Zn to assess the reactivity and bioavailability of metal and metalloid contaminants (Fujii and Corey 1986; Nakhone 1989; Nakhone and Young 1993; Hamon et al. 1997, 1998; Echevarria et al. 1998; Pandeya et al. 1998; Tandy 1998; Sinaj et al. 1999; Smolders et al. 1999; Gerard et al. 2000; Hutchinson et al. 2000; Stanhope et al. 2000; Sun et al. 2000; Young et al. 2000, 2001; Ahnstrom and Parker 2001; Collins et al. 2001; Tye et al. 2002).

This chapter will outline some of the ID methods currently under development, and their applications to the study of trace metal attenuation in soils.

2.2 METHODOLOGY

The application of ID methods to the study of soil metal lability may take several forms. However, the objective is often identification of a "chemically reactive" or "bioavailable" pool of metal. Typically, an isotope of the analyte metal is physically mixed into soil and becomes partitioned between metal pools that are kinetically accessible (to the isotope) following the lapse of a specified contact time. The distribution of the isotope will then reflect that of the labile or chemically reactive metal pool in the soil. For example, a metal and its isotopic analogue in a labeled soil suspension will have the following distribution:

$$\frac{A^*_{Soln}}{M_{Soln}} = \frac{A^*_{Total}}{M_{Labile}}, \tag{2.1}$$

where

$[A^*_{Total}]$ and $[A^*_{Soln}]$ are the total activities of the isotope added to the soil (Bq kg^{-1}) and measured in the solution phase (Bq L^{-1}), and
$[M_{Labile}]$ and $[M_{Soln}]$ are the concentrations of the labile metal in the soil ($mg \cdot kg^{-1}$) and in solution (mg L^{-1}), respectively.

Equation 2.1 is only valid provided 1) all metal species included in the assay of M_{Soln} are radiolabile, 2) the isotope has physically mixed with the entire labile metal pool, and 3) the isotope has not accessed the nonlabile metal pool through secondary reaction processes such as coprecipitation.

Thus, where radioactive tracers are used, the total labile metal content can be determined from the specific activity of the isotope in any compartment that is accessible only to labile metal and can be conveniently sampled. Typically, the phases used to assay the specific activity of the labile pools have included the solution phase of a soil suspended in electrolyte, ion exchange resins equilibrated with the soil, or plants grown in labeled soil. If the soil solution phase is used for this purpose, the radiolabile metal content is traditionally called the "*E*-value." Where plants are used to sample the labile pool, the isotopically exchangeable analyte is called the "*L*-value" (e.g., Larsen and Cooke 1961).

For example, the metal *E*-value (M_E, $mg \cdot kg^{-1}$) can be determined from working expressions such as Equation 1.2 (Smolders et al. 1999):

$$M_{Labile} = M_E = M_{Soln} \times \left(k_d^* + \frac{V}{W} \right), \tag{2.2}$$

where k_d^* is the distribution coefficient (L kg^{-1}) describing the partitioning of isotope between a weight of soil W (kg) and volume of liquid V (L).

Similarly, where a plant is used to sample the labile pool, the *L*-value (M_L, $mg \cdot kg^{-1}$) is effectively determined from the specific activity in plant tissue (A^*_{Plant}/M_{Plant}) grown on labeled soil (Equation 2.3):

$$M_{Labile} = M_L = M_{Plant} \times \left(\frac{A^*_{Total}}{A^*_{Plant}} \right) \tag{2.3}$$

The recent increase in the use of ID methods has also seen renewed examination of methodology, and perhaps a greater appreciation of the conceptual and practical problems that limit the application of such techniques to soils.

2.2.1 Suspending Electrolyte: Composition and Preequilibration Time

Measurements of *E*-values may be more reliable if there is preestablishment of equilibrium between soil and an electrolyte prior to isotopic labeling. This approach removes the need for extended contact between the isotope and soil while waiting for the native metal to reach solubility equilibrium. In various studies, preequilibration times (before addition of the isotope) have included 7 d (Tiller et al. 1972a), 5 d (Young et al. 2000), 24 h (Smolders et al. 1999), 18 h (Gerard et al. 2001), 30 min (Fujii and Corey 1986), and zero (Pandeya et al. 1998). Generally, establishment of equilibrium during the preequilibration period has been reported to be fairly rapid (Gerard et al. 2000). The possibility of redox changes associated with longer equilibration times must limit the period of preequilibration in the absence of sample aeration.

Preequilibration also ensures that studies which specifically include determination of isotopic dilution kinetics are not compromised by a counter-flux of metal, released from the soil solid phase. Furthermore, if metal is added to the soil as part of an *E*-value protocol, preequilibration with the soil minimizes the possibility of coprecipitation of isotope. Thus, Tiller et al. (1972a) measured larger *E*-values for Zn in a calcareous soil if the isotope was added with, rather than after, carrier Zn. This was thought to be due to initial fixation or precipitation of ^{65}Zn along with the unlabeled Zn added with the radioisotope, where they were added together.

The suspending electrolyte must dissolve enough metal to achieve reliable measurements of the distribution of the isotope (k_d^*) and the metal concentration in solution (M_{Soln}) in Equation 2.2. However, the electrolyte should not dissolve any nonlabile metal from inorganic matrices. Clearly, this requires a compromise in terms of the "extracting power" of the electrolyte. Prior to the widespread availability of inductively coupled plasma atomic mass spectrometer (ICPMS) and graphite furnace atomic absorption spectrophotometer (GFAAS), complexing agents were used to increase the solution–solid ratio of labile metal in order to bring M_{Soln} into the analytical range of the day. For example, Lopez and Graham (1972), Graham (1973), and Dyanand and Sinha (1985) used a mix of 0.005 M DTPA in 0.01 M $CaCl_2$ and 0.1 M Na-acetate with 48 h equilibration to measure *E*- and *L*-values for a wide range of metals (Mn, Fe, Zn, Co, Cu, and Ca). Lopez and Graham (1972) found that 80% to 100% of the added isotope remained in solution, using this suspending electrolyte. Pandeya et al. (1998) used the same approach more recently to determine Cd_E at a range of soil pH values. By contrast, Tiller et al. (1973a, b) determined Zn *E*-values in 0.05 M $CaCl_2$ and relied on solvent extraction (APDC-MIBK) to determine Zn_{Soln}. In a recent study to measure radiolabile arsenate, using ^{73}As, Tye et al. (2002) achieved greater analytical precision by including a dilute phosphate solution (0.005 M $NH_4H_2PO_4$) in the suspending solution to increase the solubility of labile As.

Neutral salts may be preferable to complexing (or acidifying) agents when considering the risk of dissolving nonlabile metal. Young et al. (2000) investigated the effects of solid–solution ratio (0.02 to 0.08 g ml^{-1}), submicron filtration (<2 μm) and supporting electrolyte composition (0.1 M $Ca(NO_3)_2$ and $CaCl_2$) on the determination of Cd_E. Consistent *E*-values were found for the 2 electrolytes, even though the chloride electrolyte produced a fivefold greater value of Cd_{Soln}. Similar suspending electrolytes have included 0.01 M $CaCl_2$ (Cd and Zn: Nakhone and Young 1993 and Smolders et al. 1999); 0.05 M $CaCl_2$ (Zn: Tiller et al. 1972a); water (Ni: Echevarria et al. 1998); 0.01 M $Ca(NO_3)_2$ with 3×10^{-5} M EDTA (Cd and Zn: Fujii and Corey 1986); 0.1 M $Sr(NO_3)_2$ (Cd: Ahnstrom and Parker 2001). However, the lower concentrations of metal and isotope in solution also aggravate errors related to the presence of (metal-containing) suspended solids. Thus, Gerard et al. (2001) used water to suspend soil during the measurement of Cd *E*-values and found that submicron filtration (0.45, 0.2, and 0.025 μm), following centrifugation at 10,000 g, made a significant difference to Cd_{Soln}. A larger salt concentration may reduce the magnitude of such errors by greater flocculation of colloids.

It is more difficult to guard against the effects of soluble nonlabile metal complexes being included in the measurement of M_{Soln}. The effect of including such species in the assay of M_{Soln} is the same as that arising from suspended solids — the apparent *E*-value is increased (Equation 2.2). Given sufficient care over the removal of suspended solids, the problem of including nonlabile metal in M_{Soln} may be most likely to affect metals which can form kinetically stable complexes with dissolved organic matter. Figure 2.1 shows an attempt to measure radiolabile Hg following extended equilibration of $Hg(NO_3)_2$ with 15 contrasting soils (Tandy 1998). It is clear that the apparent *E*-values for Hg exceeded the amount of metal extractable either by 0.05 M EDTA or 1 M $CaCl_2$. Yet, either extractant would be expected to dissolve (at least) the labile Hg pool given the high affinity of both ligands (chloride and EDTA) for Hg^{2+} ions. One possible explanation may be the presence of kinetically stable organic-Hg complexes, with which the ^{209}Hg does not mix during the period of the measurement. In such circumstances, the *E*-value would be exaggerated by a factor equivalent to the ratio of nonlabile–labile Hg within $[Hg_{Soln}]$ (Equation 2.2).

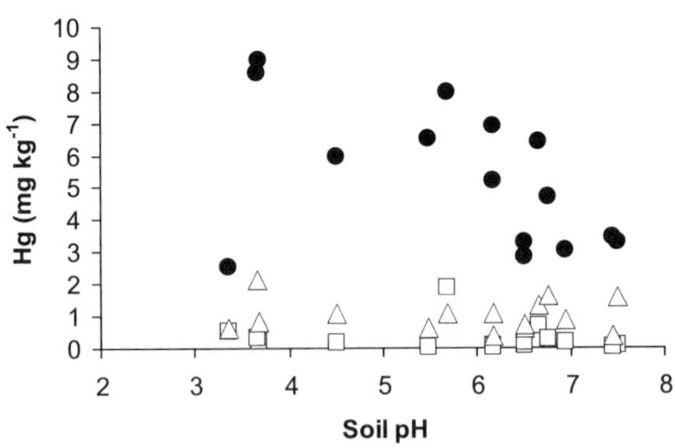

FIGURE 2.1 A comparison of radio-labile Hg (●) with extraction by 1 M $CaCl_2$ (□) and 0.05 M EDTA (△); 15 soils were incubated for 811 d at 16 °C and 80% field capacity with 20 mg Hg kg^{-1} added as $Hg(NO_3)_2$.

One approach which may resolve the problem of nonlabile analyte, in colloidal or dissolved forms, is the use of "resin purification." This involves the addition of an ion-exchange resin strip to a labeled soil suspension or separated supernatant solution to sample the specific activity of the analyte isotope in solution. The method assumes:

1. The resin can be cleaned of all colloidal material.
2. Nonlabile species in solution are not adsorbed.
3. Nonlabile metal is not dissolved by the presence of the resin.

This approach was recently developed and tested for the determination of phosphate *E*-values (Hamon and McLaughlin 2002). The same technique was also successfully applied to trace metals using a cation exchange resin (Lombi et al. 2003).

2.2.2 Isotope Equilibration Time

Slow fixation of metals added to soil, following an initial rapid reaction, is widely reported. It is therefore reasonable to expect that metal isotopes will progressively migrate into nonlabile sites following addition to soil. Clearly, this raises concern over the pursuit of a simple binary classification of metal reactivity into labile and nonlabile pools.

Such reactions may be more significant where there is a solid phase composed of the analyte, as in the cases of Mn_E or Fe_E determinations (Lamm et al. 1963). Incorporation of constituent metal isotopes into solids has been described as a series of kinetic reactions and shown to depend upon the structure or degree of hydration of the solid phase (Imre 1937 and 1939; Pullmann and Haissinsky 1947). Similar processes are likely where a foreign metallic isotope can form a new solid-solution deposit on the surface of an adsorbent such as Cd-$CaCO_3$ (Papadopoulos and Rowell 1988), or become incorporated through solid phase diffusion into the structure of a preexisting adsorbent such as Fe oxides (Barrow 1986 and 1987).

Thus, the isotopically exchangeable pool is strictly an operationally defined assay, whose value must depend to some extent upon equilibration time. This is of particular concern in the measurement of *L*-values, where contact between the isotope and soil may extend to several weeks or months (Hamon et al. 1997; Echevarria et al. 1998; Smolders et al. 1999; Hutchinson et al. 2000). There is division within the literature regarding the extent to which the slow reaction between the isotope and soil compromises the interpretation of measured *L*-values in terms of a simple "bioavailable" metal fraction.

Some investigations of possible methodological artifacts suggest that "simple" *E*- and *L*-values provide a reasonably meaningful separation of metal pools in soils. Reports of progressive change in isotopic distribution between the solid and solution phases vary between studies which have measured *E*-values. Young et al. (2000) tested a range of isotopic equilibration times (1–6 d) on Cd *E*-values in a range of contaminated soils. The method was reasonably robust over the range of conditions tested, although there was a small increase in *E*-value (adsorption of isotope) over the first 3 d. A preequilibration time of 5 d and isotopic contact time of 2 d were adopted in further studies. Similarly, Smolders et al. (1999) found little change in either adsorption of added ^{109}Cd or desorption of native soil Cd in soil suspensions when comparing isotopic equilibration periods ranging from 1–14 d. They used 1 d for preequilibration followed by 7 d contact with the isotope as a standard procedure. However, Tiller et al. (1972a) observed a rapid change in isotope adsorption over the first 2 to 3 d contact with soil, followed by a more gradual but almost linear decrease in solution activity (in 0.05 M $CaCl_2$) from 3–15 d. Hutchinson et al. (2000) used periodic extraction of ^{109}Cd with 1 M $CaCl_2$, following addition to 2 soils, to demonstrate that there was an average isotopic fixation rate of just 0.03 and 0.07% per d following a small initial drop in lability. At worst, this represented around 6% "loss" of labile isotope during the growth period of their pot experiments to determine *L*-values. Hamon et al. (1997) found similar results by monitoring the soil pore water of pot experiments labeled with ^{65}Zn and ^{109}Cd; activity fell slightly during the first week after potting and then remained fairly constant.

Other workers have attempted to describe more rigorously the progressive mixing of isotopes with the whole soil metal pool in order to predict apparent *E*-values following extended contact times (Tiller et al. 1972a). Echevarria et al. (1998), Sinaj et al. (1999), and Gerard et al. (2000) used an expression based on an infinite series of exponential terms to describe the degree of mixing of ^{63}Ni, ^{65}Zn, and ^{109}Cd, respectively, with the total soil metal pools (Equation 2.4).

$$P^*_{(t)} = P^*_{(1)}\,[t + P^*_{(1)}{}^{1/n}]^{-n} + P_M \tag{2.4}$$

P* is the proportion of isotope in the solution phase at a specified time (t) or after 1 min (1), and n is an empirical constant. The proportion of soil: metal in solution (P_M) is the value of $P^*_{(t)}$ at infinite time and assumes that the measured value of metal concentration in solution reflects an equilibrium state. This enabled prediction of apparent *E*-values for up to several weeks contact between the isotope and the soil from measurements over just a few minutes (Sinaj et al. 1999). It was therefore possible to compare *E*- and *L*-values for similar isotopic contact times. Such a comparison assumes an *L*-value dominated by metal uptake in the latter stages of plant growth but compensates for the early fixation of isotope reported by several authors.

Currently, the literature remains divided on both the validity and the interpretation of comparisons between short-term *E*-values (the classical approach) and *L*-values derived from labeled pot experiments.

2.2.3 Use of Stable Isotopes

Stable isotopes may be used in place of radioisotopes for ID studies and offer some important advantages in terms of safety and experimental longevity. Thus, Ahnstrom and Parker (2001) used ^{111}Cd as the tracer isotope in a comparison of apparent *E*-values in 4 soils. They were able to follow isotopic penetration of Cd pools, defined by a sequential extraction scheme, over 59 weeks. A similar study could have been undertaken with ^{109}Cd ($t_{0.5}$ = 463 d), but this would have involved a reduction in activity of nearly 50% in the incubated soil during the lifetime of the experiment. Studies of metal fixation rates extending over several years would clearly be impractical with some isotopes which have been used to measure *E*-values, for example: ^{65}Zn ($t_{0.5}$ = 244 d), ^{73}As ($t_{0.5}$ = 76 d), ^{203}Hg ($t_{0.5}$ = 47 d), or ^{64}Cu ($t_{0.5}$ = 12.7 h). In addition, the extended storage of soil contaminated with gamma-emitting isotopes is obviously undesirable for reasons of radiological safety.

The analytical requirements for assay of stable isotopes (e.g., ICP-MS) are more demanding than for radioisotopes. However, the greater flexibility offered by stable isotopes and the pace of development in surface analysis techniques (e.g., laser ablation ICP-MS) suggests that this may be an approach which will eventually replace the use of radioisotopes in ID studies.

2.2.4 Comparison of ID Methods and Soil Extractants

The use of chemical extraction agents to fractionate soil metals is an attractive alternative to isotopic dilution for reasons of greater simplicity, analytical accuracy, and radiological safety. However, developing a protocol for an extraction procedure which accesses only the radiolabile pool of metal presents a number of problems. There have been a limited number of studies which directly compare the 2 approaches.

Nakhone and Young (1983) compared 4 chemical extractants with Cd_E measured on 33 soils contaminated by a wide range of sources. They found that the pools of Cd resolved generally followed the sequence: "total" (HNO_3 digestion) >0.05 M EDTA > *E*-value >0.01 M $CaCl_2$ >1.0 M KNO_3. Thus, it was suggested that radiolabile Cd was chemisorbed rather than exchangeable and that EDTA may partially dissolve solid phase adsorbents such as carbonates and oxides which could harbor nonlabile Cd. Stanhope et al. (2000) confirmed the latter finding as the amount of Cd extracted by 0.05 M EDTA (250 mmol kg^{-1} soil) from a sewage sludge-amended soil exceeded Cd_E. However, in the same study, extraction at concentrations of EDTA equivalent to "phytoremediation treatments" (up to 10 mmol EDTA kg^{-1}) dissolved substantially less than the Cd *E*-value.

Echevarria et al. (1998) compared extraction with 0.005 M DTPA and *E*-values for Ni. The latter assay was extrapolated to allow for 122 d contact between ^{63}Ni and the soil prior to extraction. They found broad similarity between both assays, although DTPA may have extracted slightly more than the ID pool. Sinaj et al. (1999) found that the value of Zn_E for 15 d contact (Equation 2.4) was close to the amount of Zn extracted with a mixture of 0.005 M DTPA and 0.1 M TEA in 0.01 M $CaCl_2$. Fujii and Corey (1986) found that 0.005 M DTPA underestimated Cd_E (73%) and Zn_E (69%), whereas 0.005 M EDTA removed around 100% of isotopically exchangeable Cd and Zn, but with some variability between soils. Fujii and Corey (1986) list several studies from the 1970s where isotopically exchangeable metal pools have been compared with dilute chelate extractants. For example, Tiller et al. (1972a) found that a single extraction with 0.01 M EDTA removed 30 to 50% of isotopically exchangeable Zn, measured as Zn_L, on 25 soils using clover as the test plant. In contrast, extraction with 0.1 M HCl removed more Zn than could be accounted for by Zn_L.

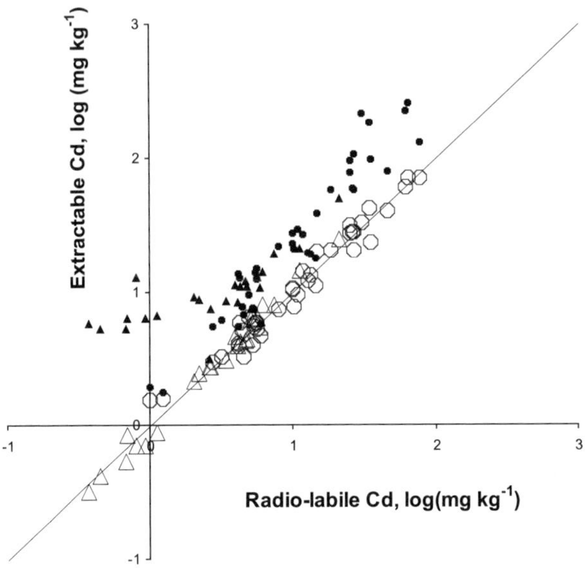

FIGURE 2.2 Comparison of the radio-labile Cd assay with extraction of Cd by 1 M $CaCl_2$ (open symbols) and nitric acid digestion (closed symbols). Soils were historically contaminated with Pb/Zn minespoil (○●, *n* = 41) or sewage sludge (△▲, *n* = 25).

Young et al. (2000) showed that extraction of Cd using 1 M $CaCl_2$ provided a good alternative to determination of radiolabile Cd for 25 soils contaminated with sewage sludge and 41 mine-spoil soils. Figure 2.2 shows their data replotted on a single log scale and a comparison with the total soil metal content. It was postulated that the large concentration of chloride ions has the effect of dissolving all labile Cd by chloro-complexation ($CdCl^+$, $CdCl_2^0$, etc.) while leaving mineral sorption surfaces (FeOOH, MnO_2, $CaCO_3$, etc.) relatively intact. The comparability of the 2 methods was confirmed by Hutchinson et al. (2000) for a limited number of contaminated soils.

Tye et al. (2002) tested several concentrations of phosphate as possible analogues for radiolabile arsenate, measured using ^{73}As. They found that the minimum concentration of phosphate required to extract a level of As which was independent of P concentration was around 0.5 M $NH_4H_2PO_4$. However, when compared with radiolabile arsenate ($As^V{}_E$), it was found that the phosphate-extractable fraction consistently exceeded the soil As content accessible to isotopic dilution with $^{73}As^V$ (Figure 2.3).

Ahnstrom and Parker (2001) tested a 5-step sequential extraction procedure (SEP) against apparent *E*-values for Cd after allowing the tracer isotope (^{111}Cd) to access "less labile pools" through extended contact with soil (up to 59 weeks). There was some correspondence between the apparent value of Cd_E for the whole soil and the first 2 fractions of the SEP (nominally "exchangeable" and "sorbed"). However, Ahnstrom and Parker noted that this was largely fortuitous, because mixing of the isotope with the second fraction was not complete and there was also some penetration of the third SEP fraction ("oxidizable" Cd). Åsgeir and Singh (2001) undertook a

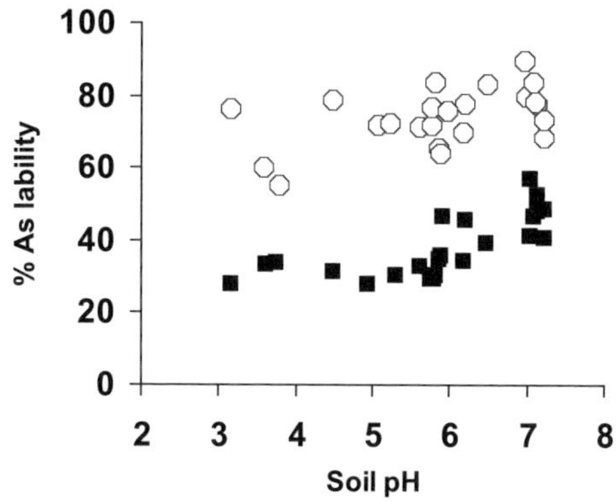

FIGURE 2.3 Radio-labile As (■) and phosphate-extractable As (O), as a proportion of total (%), in 23 diverse soils incubated (16 °C; 80% field capacity) for 818 d after addition of 50 mg As kg^{-1} as $Na_2HAsO_4.7H_2O$. (Reproduced with permission from *Environ. Sci. Technol.* 2002, **36**, 982–988. ©2002 American Chemical Society.)

similar study after adding ^{109}Cd and ^{65}Zn to an alum-shale soil. They stored the labeled soil at 60% field capacity and extracted both isotopes at regular intervals using a 7-step SEP. There was a decline in magnitude of the "reversible" pools and an increase in the "irreversible" fractions. For example, reversibly adsorbed ^{65}Zn fell from 20% at 0.5 h to 10% after 8760 h. However, given that reversibly adsorbed Zn was extracted with a relatively weak extractant (1 M NH_4OAc at pH 5), it may be that a considerable proportion of Zn_E was actually still adsorbed following the extraction.

It is unlikely that any extractant can match exactly the radiolabile metal pool. Use of indifferent electrolytes to remove electrostatically bound (exchangeable) metal will underestimate radiolabile metal because most labile metal is specifically adsorbed in soil. Chelating agents will dissolve specifically adsorbed metal but, partly through dissolution of adsorption surfaces, they may also access nonlabile pools. It is possible that the range of findings reported in the literature simply reflects the range of chelate concentrations and operating conditions used (e.g., solid: solution ratio, time, etc.). In some circumstances, such as fairly acidic organic soils, extraction of the labile pool with competing metals (Lofts et al. 2001) or with dilute acid (Tipping et al. 2000) may provide a useful estimate of labile metal content. However, these, and other simple extraction schemes, remain untested as alternatives to isotopic dilution.

2.3 APPLICATIONS

2.3.1 Source of Contaminant

The degree of attenuation of trace elements, as determined by isotopic dilution, may reflect the origins of the metal as much as the prevailing soil conditions or contact time. In uncontaminated soils a proportion of the trace element complement will reside within primary mineral matrices and will not be accessible to isotopic dilution.

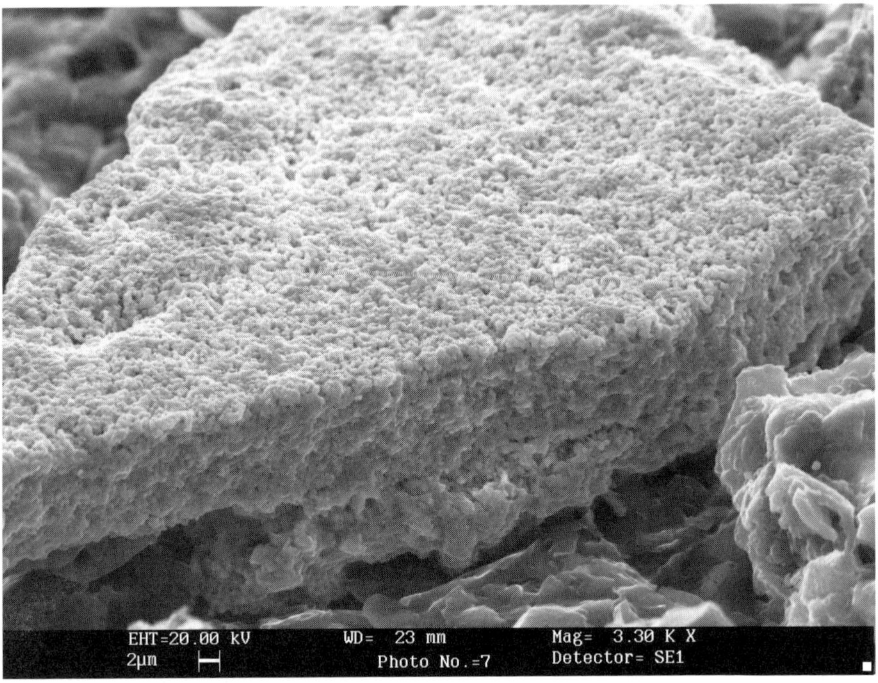

FIGURE 2.4 A Pb- and Ti-rich paint flake isolated in the heavy fraction (10–63 μm) from a garden soil adjacent to railway sidings in Nottingham, U.K. IPR/23-25C British Geological Survey.

In addition, the persistence of nonlabile contaminant materials in soils is well documented, for example, where mine-spoil materials (Young et al. 2000) or anthropogenic artifacts are the original source. In soils subject to industrial contamination, for example, recognizable particulates can be isolated, following soil physical fractionation. Figures 2.4 and 2.5 are electron micrographs showing examples of anthropogenic metal-rich particulates in the silt-sized fraction of a contaminated urban soil. It is probably safe to assume that the majority of the metal within such particles would not be accessible to an added metal isotope without extremely extended equilibration times. Even in studies where soils are amended with metal salts or other contaminants, the original form of the metal, if not completely soluble initially, will affect the measured lability even after extended equilibration (Smolders et al. 2000).

Tye et al. (2002) measured arsenate lability (As_E) in soils contaminated by mine-spoil and historical applications of sewage sludge. They found that the original source of soil As largely dictated the pattern and extent of arsenate lability, whereas soil conditions were of relatively minor importance. Figure 2.6 shows the proportion (%) of arsenate radiolability for field soils from 3 sources as a function of soil pH. The trend shown by soils artificially amended with $Na_2HAsO_4.7H_2O$ in Figure 2.3 is included as a fitted line. An extremely contaminated soil, adjacent to a prospective gold mine in Malaysia, contained 4900 to 17200 mg As kg^{-1}, although As lability varied from just 0.44% to 1.01% due to the presence of As-containing primary minerals

FIGURE 2.5 A segment of Sn/Pb solder isolated in the light fraction (10–63 μm) from a garden soil adjacent to railway sidings in Nottingham, U.K. IPR/23-25C British Geological Survey. ©NERC. All rights reserved.

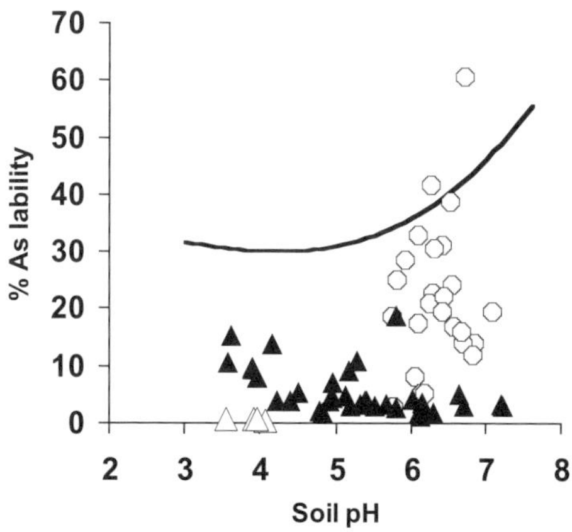

FIGURE 2.6 Radio-labile As, as a proportion of total (%), in 1) 23 soils incubated (16 °C; 80% field capacity) for 818 d after addition of 50 mg As kg^{-1} (fitted solid line from data in Figure 2.3), 2) a sewage sludge disposal facility (O) and soils contaminated with minespoil containing arsenopyrite from Cumbria, U.K. (▲) and Malaysia (△). (Reproduced with permission from *Environ. Sci. Technol.* 2002, **36**, 982–988. ©2002 American Chemical Society.)

such as arsenopyrite. Arsenic in the soils around a 19th century mine site in Cumbria, U.K., was slightly more labile (1.4% to 19% of total soil As). The trend with pH in the latter soils was apparently the reverse of that shown by the artificially amended soils in Figure 2.3, where 50 mg As kg^{-1} had been added originally as a soluble salt. Again, this comparison is a reflection of the origins of the As rather than soil conditions. The more acidic, humus-rich soil samples were found adjacent to the Cumbrian mine-spoil area and had relatively low concentrations of As. Thus, particles of mine-spoil deposited on such soils would be expected to weather more rapidly and oxidize to release (labile) arsenate. Radio labile As in the sewage-sludged soils showed no discernible trend with pH, but the lability was, on average, half that found in the As-amended soils. It is thought that much of the nonlabile As may be tied up with Ca-phosphate minerals in sludge-treated soils due to the rigorous liming regime practiced on such land in the U.K. and large P contents arising from sludge application (Tye et al. 2002).

2.3.2 Effect of Soil pH on Lability

Soil properties which affect adsorption strength, such as pH, are also likely to influence measured *E*- and *L*-values. There cannot be a single consistent relation between soil pH and radiolability, partly because of the determining role played by the original form of the metal contaminant. However, in the special case of studies where metals are added to soil as salt ions in solution, there is at least a common starting point from which to examine influences such as soil conditions and time. The results from a study of 23 diverse soils incubated (80% field capacity, 16 °C) for 811 d with 3.0 mg Cd kg^{-1} and 300 mg Zn kg^{-1} are shown in Figure 2.7. There

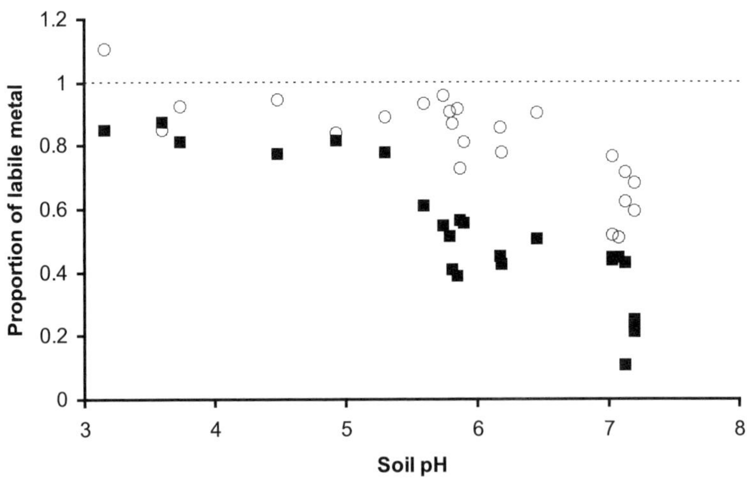

FIGURE 2.7 Variation in radio-labile Cd (○) and Zn (■) with soil pH, in 23 soils. Labile metal concentration is expressed as a proportion of added metal concentration. Both metals (3 mg Cd kg^{-1} and 300 mg Zn kg^{-1}) were added as nitrate salts; soils were equilibrated (16 °C; 80% field capacity) for 818 d.

was a clear decline in relative lability with pH for both metals, which broadly reflects the trend in adsorption. There were also clear differences between Cd and Zn, with the latter metal apparently subject to a greater degree of "fixation." In the case of Cd, lability exceeded 70% for all soils <pH 7. Collins et al. (2003) found a similar trend in Cd_E with pH when they adjusted the pH of a contaminated calcareous soil with HNO_3. There was approximately a threefold reduction in Cd_E from pH 5.85 to pH 7.5. Sun et al. (2000) measured *E*-values for Cd and Zn in 2 soils from a long-term liming trial (pH range 4.3 to 7.7) amended with 3.0 mg Cd kg^{-1} and 300 mg Zn kg^{-1} as metal sulfate salts. The soils, a silty clay loam and a sandy loam, had been used in a single ryegrass pot trial shortly after the addition of metal and then stored for 6 years in an air-dry condition. The trend in *E*-values with pH was broadly similar to that seen in Figure 2.7, but there was evidence of greater fixation in the fine-textured soil compared to the sandy loam. Similarly, the trend in As lability of Tye et al. (2002) (Figure 2.3) suggests a relationship with adsorption. Lower lability is associated with stronger adsorption — at low pH in the case of arsenate.

Immediate changes in *E*-values following short-term adjustment of pH have also been recorded. Several studies have used buffered electrolytes to determine *E*-values across a range of (adjusted) pH values. Thus, Lopez and Graham (1972) found that *E*-values for a range of trace metals were greater at pH 5 compared to pH 7. Pandeya et al. (1998) measured Cd_E in buffered suspensions at pH 5, 6, and 7, and found that Cd_E at pH 5 was greater than at pH 7 in virtually all cases, for the same soil. Using a similar approach, Dyanand and Sinha (1985) found that iron lability (Fe_E) was also greater at low pH.

E-values are expected to differ between soils which have undergone extended contact with metal at different pH values. If surface adsorption is a prelude to time-dependent fixation reactions, then lower *E*-values for cationic species would be expected following equilibration at low pH in comparison to high pH soils. However, it is apparent that changes in *E*-value may also occur shortly after adjustment of soil pH. One can speculate that such responses may arise either through dissolution of surfaces which harbor nonlabile metal (Hamon et al. 2002) or simply because of a change in the desorption kinetics of surface-adsorbed metal complexes.

2.3.3 Effect of Time on Lability

Time-dependent changes in sorption, solubility, bioavailability, and toxicity of metals added to soil have been studied over time periods varying from minutes to years (Boawn et al. 1960; Armour et al. 1989; Brennan 1990; Brennan and Gartrell 1990; Smit et al. 1997). Slow attenuation of metals has clear implications for studies of risks associated with bioavailability and mobility (see Chapter 9). The mechanisms for such changes in soil are only understood qualitatively (e.g., McLaughlin 2001). Processes of fixation may include solid-phase diffusion into mineral lattices (Barrow 1986 and 1987; Bruemmer et al. 1988; Trivedi and Axe 2000); slow entrapment in mineral deformities or cavities, within clay interlayer sites (Ma and Uren 1998); occlusion by coprecipitation (Tiller et al. 1972a); and the formation of superficial solid solutions (Papadopoulos and Rowell 1988). Isotopic dilution provides a useful way to quantify changes associated with progressive attenuation but cannot reveal specific mechanisms without either inference or further measurement.

Hamon et al. (1998) measured *L*-values for Cd on soils contaminated by various rates of superphosphate addition, for either 27 or 47 years, using 28-d-old wheat as the test crop. They were able to describe Cd fixation using a model based on annual additions of labile Cd subject to an irreversible first-order fixation rate. The apparent fixation rate was approximately 1% to 1.5% of labile Cd per annum.

Young et al. (2001) followed the change in Zn and Cd radiolability in 23 soils over 811 d (Figure 2.8) to determine the rate of fixation from an assumed initial condition of 100% lability. Figure 2.8 shows the proportional reduction in radiolability for Zn with the soils grouped into 3 pH classes. It was possible to fit reversible first-order kinetic equations to the data (solid lines) but, clearly, these were empirically dependent upon pH. The data for soil pH >6.5 appear to gravitate to a mean value of around 30% which is similar to several studies of field soils in this pH range (Sun et al. 2000; Young et al. 2000). Figure 2.8 also shows the mean ±1 standard deviation for 25 soils historically amended with sewage sludge (Young et al. 2000).

It is worth noting the apparent reversibility of "fixed" metal as determined by the isotopic dilution method. The data in Figure 2.8 was described using a reversible kinetic model, which requires a final equilibrium position with less than 100% fixation of metal. Furthermore, field soils, which may have been contaminated for decades prior to analysis, also show a substantial degree of metal lability. This strongly suggests that "fixation" reactions resolved by ID methods are effectively reversible.

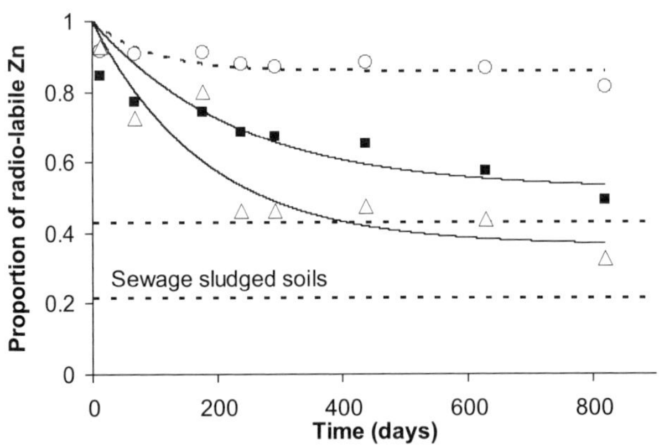

FIGURE 2.8 Time-dependent reduction in radio-lability of Zn added as $Zn(NO_3)_2$ to 23 soils and incubated (16 °C; 80% field capacity) for over 800 d. The soils are grouped into 3 pH ranges: 6 soils pH <5.5 (○); 10 soils pH 5.5–6.5 (■); 7 soils pH >6.5 (△). Solid lines are the fit of a reversible first-order kinetic equation to each grouped dataset. Broken lines represent the mean ± 1 standard deviation for a survey of soils historically amended with sewage sludge.

2.3.4 Describing the Solubility of Metals

The solid⇔solution equilibria of metals added to soils have traditionally been described using empirical adsorption isotherm equations. Recently, Elzinga et al. (1999) presented a compilation of such adsorption studies. More mechanistic "assemblage" models, which combine descriptions of ion adsorption on well characterized adsorbents to simulate whole-soil behavior, are currently under development. These include extensions of the *NICA* model (Gooddy et al. 1995) and the "assemblage" model SCAMP, developed by Lofts and Tipping (1998).

A complicating factor in any equilibrium adsorption model is discriminating between metal in the solid phase which is in dynamic equilibrium with the solution phase and that which has become fixed. For soils sampled from the field, it is therefore necessary to either make assumptions within the model structure regarding the apportionment of the metal or to determine labile metal content by extraction. Thus, Tipping et al. (2000) used the WHAM speciation model to predict metal ion activities in moorland soils and found that the pool of active soil metal was well represented by extraction with 0.1 M HNO_3.

For metal that has been freshly added to soil, it is possible to demonstrate both time-dependent adsorption and desorption hysteresis (Barrow 1986; Nakhone 1989; Filius et al. 1998), suggesting a progressive change in the energy status of the adsorbed metal species. To resolve a coherent description of metal ion sorption under such circumstances it is therefore necessary to describe both (equilibrium) adsorption and the kinetics of metal transfer to nonlabile forms. Thus, Barrow (1986) adapted a mechanistic description of specific ion adsorption to include diffusive penetration of adsorption surfaces as a fixation mechanism. This combined model closely described time-dependent adsorption of Zn on soil.

Isotopic dilution cannot resolve the chemical nature of adsorbed or fixed metal without further assumptions. However, it may offer a useful partition between reactive and nonreactive forms of metal which defines the fractions involved in the solid⇔solution equilibrium being described by the adsorption model. The strength of this approach has been appreciated by several workers who have employed ID methods to resolve "true" adsorption distribution coefficients. Thus, Graham (1973) measured k_d values of radiolabile Mn, Fe, Zn, Co, Cu, and Ca using appropriate isotopes and 0.1 M $CaCl_2$ as the background electrolyte. Tiller et al. (1972b) derived quantity–intensity (Q/I) relations for Zn from the distribution of the isotope ^{65}Zn in solution. More recently, Pandeya et al. (1998) measured Q/I relations for Cd using ^{115}Cd; again, the k_d value for radiolabile Cd was measured in 0.1 M $CaCl_2$.

Nakhone (1989) demonstrated the ability of ID methods to resolve a consistent adsorption behaviour by labile metal in the presence of fixation reactions. Adsorption and desorption of Cd on a calcareous clay subsoil was measured using several equilibration times. Radiolabile Cd was determined following adsorption and also at the end of desorption sequences. Figure 2.9 shows progressive sorption of Cd with time when "total sorbed Cd" is presented as the adsorbed species. By contrast, radiolabile Cd, measured after 1, 5, 20, and 60 d apparently follows a single adsorption trend with Cd concentration in solution (Freundlich isotherm). Nakhone also

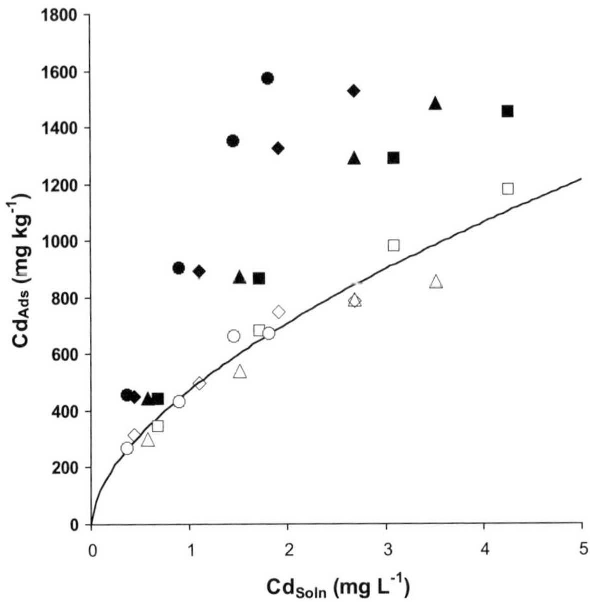

FIGURE 2.9 Cd adsorption on a calcareous clay measured after equilibration for 1 d (■), 5 d (▲), 20 d (◆), and 60 d (●). The open symbols are equivalent data using radiolabile adsorbed Cd. The solid line is the fit of a Freundlich equation to all the radiolabile points treated as a single dataset.

found that the same adsorption isotherm for labile Cd could be resolved when including radiolabile Cd data measured at the end of desorption sequences (data not shown). This was despite the marked hysteresis shown by sorbed Cd when subject to desorption.

2.3.5 Bioavailability: Comparison of *E*- and *L*-Values

Growing plants on isotopically labeled soil enables the specific activity in the plant tissue to be used as a means of estimating bioavailable metal (Equation 2.3); this measure is often called the *L*-value. Both *E*- and *L*-values represent the labile metal pool in soils and therefore might be expected to be equal. As described previously, specific assumptions regarding the extent of mixing of the isotope with indigenous soil metal underpin this expectation (Stanhope et al. 2000). In addition, identical *E*- and *L*-values will only be found if the presence of the plant has no effect on transfers between labile and nonlabile forms of metal. Therefore, a comparison of *E*- and *L*-values may be a useful tool with which to investigate possible chemical or biological mobilization of fixed forms of metal in the soil (Smolders et al. 1999).

Dissolution of nonlabile forms of metal may occur in the rhizosphere through processes such as alteration of pH or the production of metal-solubilizing root exudates or microbial metabolites (Grinsted et al. 1982; Bernal et al. 1994). In soil spiked with a metal radioisotope, uptake of nonlabile (unlabeled) metal would cause a decrease in the expected specific activity of the isotope in the plant and an increase in the apparent *L*-value over the *E*-value. Thus, differences between the *L*- and *E*-values may provide quantitative estimates of the mobilization of nonlabile metal within the rhizosphere. Reports of such differences vary depending upon methodology and choice of plant and soil. One of the largest reported differences in *E*- and *L*-values was found for Fe by Dyanand and Sinha (1985). They measured Fe_L using *Sorghum vulgare* with ^{59}Fe and used the electrolyte mix 0.005 M DTPE + 0.01 M $CaCl_2$ + 0.1 M Na-acetate to measure Fe_E. The ratio of *L:E* varied from 1.3 to 17.8 and approached unity as labile Fe increased. It was postulated that when Fe is in short supply, the plant roots reduce pH locally and release citric acid to promote nonlabile Fe dissolution.

Smolders et al. (1999) measured Cd *E*- and *L*-values for wheat grown in 10 Belgian soils (total Cd = 0.33 to 6.5 mg·kg^{-1}). They found that *L*-values slightly exceeded the corresponding *E*-values, by a factor of between 1.05 and 1.4, suggesting that rhizosphere conditions caused solubilization of the nonlabile soil Cd. Tiller et al. (1972a) reported only small differences between *E* - and *L*-values for Zn uptake by subterranean clover below pH 7, while Echevarria et al. (1998) reported similar results for ^{63}Ni uptake by red clover.

Hutchinson et al. (2000) compared 6 populations of the Zn and Cd hyperaccumulator *Thlaspi caerulescens* grown on 2 contrasting soils in which there was a clear distinction between total and labile Cd. They found very little difference between *E*- and *L*-values for Cd in most cases (e.g., Figure 2.10): the *L*-values generally followed the soil *E*-value, even for a population of the hyperaccumulator with a markedly greater affinity for Cd. However, for 1 population of *Thlaspi caerulescens* grown on a soil contaminated with minespoil, the ratios of Cd_L:Cd_E

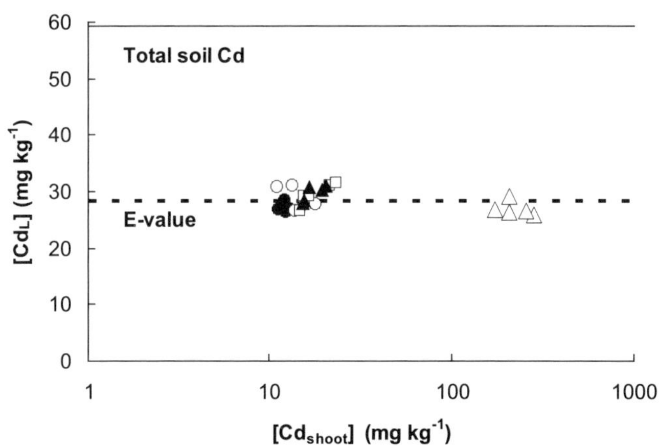

FIGURE 2.10 Variation in radiolabile Cd (the *L*-value, [Cd_L]), with the concentration of Cd in plant shoots, [Cd_{shoot}] grown in a soil historically contaminated with sewage sludge. Plants include *Lepidium heterophyllum* (▲) and *Thlaspi caerulescens* from Darley Dale, U.K. (●), Whitesike, U.K. (○), Prayon, Belgium (❑) and Ganges, France (△). The solid lines represent total and radiolabile Cd measured in vitro as the *E*-value. (Reproduced with permission from *New Phytologist* 2000, **146**, 453–460.)

were in the range for wheat found by Smolders et al. (1999). Similarly, Gerard et al. (2000, 2001) found that Cd *E*- and *L*-values were fairly similar for ryegrass, lettuce, and *Thlaspi caerulescens*. There were small differences which may have been due to inadequate penetration of the isotope in the potted soils ($Cd_L < Cd_E$) and rapid uptake of Cd *by Thlaspi caerulescens* possibly leading to local desorption of nonlabile Cd ($Cd_L > Cd_E$).

Stanhope et al. (2000) tested the ability of chelate-assisted phytoextraction with *Brassica juncea* to access nonlabile soil metal (Cd) using applications of EDTA in the range commonly reported in phytoremediation studies (0 to 10 mmol EDTA kg^{-1}). There was no evidence of phytomobilization of fixed forms of Cd as *E*- and *L*-values were broadly similar, and no trend in the ratio Cd_L:Cd_E was seen with the level of EDTA application (Figure 2.11).

Collins et al. (2001) measured Cd_E in a soil contaminated with smelter fallout after equilibration with several organic ligands known to be present in rhizospheres. They found that Cd_E was essentially unaffected, except where pH was changed by the organic acids, and concluded that it is unlikely that nonlabile Cd could be solubilized at organic ligand concentrations normally found in the rhizosphere. Hamon et al. (1997) compared 7 species of plants for possible indications of rhizosphere effects in relation to mobilization of nonlabile Cd or Zn. However, all species produced similar values of Zn_L (around 12% of total soil Zn) and only 1 of the 7 gave an anomalous result for Cd. Similarly, Hamon and McLaughlin (1999) com-

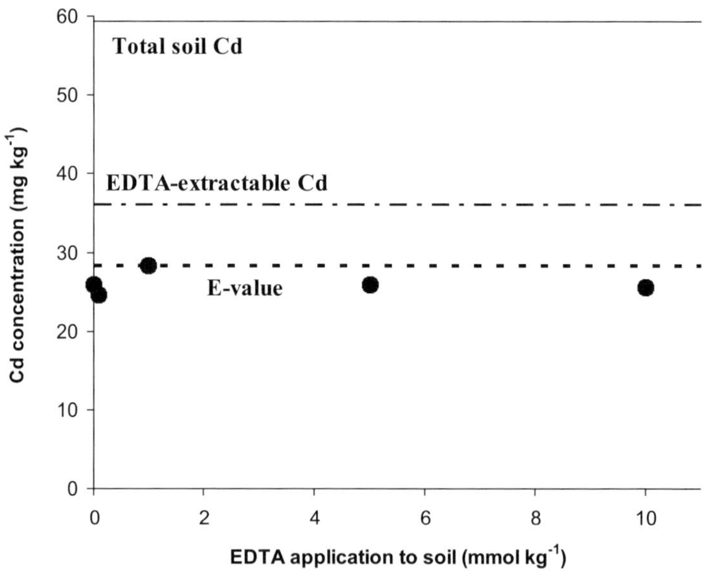

FIGURE 2.11 Effect of EDTA amendment of soil on radiolabile Cd (the *L*-value; ●), using *Brassica juncea* as the test plant. Three chemical assays of soil Cd are shown, including "total" soil Cd (HNO_3 digestion), EDTA-extractable Cd (0.05 M EDTA) and radiolabile Cd measured *in vitro* as the *E*-value. (Reproduced with permission from *Environ. Sci. Technol.* 2000, **34**, 4123-4127. ©2000 American Chemical Society.)

pared specific activities in the Zn hyperaccumulator *Thlaspi caerulescens* and wheat, planted sequentially in soil spiked with ^{65}Zn and ^{109}Cd. For both metals, there were no significant differences in *L*-values between the cereal and hyperaccumulator plants.

2.4 CONCLUSIONS AND FUTURE DIRECTIONS

The literature dealing with isotopic dilution methods over the past 50 years suggests that many of the uncertainties associated with both methodology and interpretation remain stubbornly intact. This is partly a reflection of the limitations inherent in ID methods when these are applied in isolation to soil chemical studies. Nevertheless, it is clear that progress is being made and ID methods have already contributed substantially to our understanding of metal attenuation by soils. Recent work has explored some long-standing deficiencies in methodology, and it seems likely that more standardized procedures will emerge shortly.

For obvious reasons, there is a long-standing preference in soil chemistry for simple metal-extraction schemes. Thus, future work will undoubtedly continue the search for extractants which can serve as analogues for ID-based estimates of labile metal pools. The justification for this effort is clear, provided the degree of compromise required to accept chemical extraction methods is sufficiently small.

For more mechanistic studies, exploration of the use of stable isotopes appears particularly attractive in view of the technological improvements in mass spectroscopy. It is likely that ID methods will be increasingly used as a complementary tool, alongside more precise speciation techniques (e.g., XAFS; see Chapter 4 and Chapter 5) and surface examination tools such as LA-ICP-MS. It is also expected that the ongoing development of mechanistic models of metal solubility and speciation may take greater advantage of the kinetic information which can be gained from ID methods. The full understanding of metal dynamics in soils is a clear objective, but realization of such a goal will rely upon the marriage of a number of complementary approaches.

REFERENCES

Ahnstrom ZAS, Parker DR. 2001. Cadmium reactivity in metal contaminated soils using a coupled stable isotope dilution-sequential extraction procedure. Environ Sci Technol 35:121–126.

Armour JD, Ritchie GSP, Robson AD. 1989. Changes with time in the availability of soil applied zinc to navy beans and in the chemical extraction of zinc from soils. Aust J Soil Res 27:699–710.

Åsgeir A, Singh BR. 2001. Partitioning and reaction kinetics of Cd-109 and Zn-65 in an alum shale soil as influenced by organic matter at different temperatures. In: Iskandar IK, Kirkham MB, editors. Trace elements in soil: bioavailability flux and transfer. Boca Raton, FL: Lewis Publishers. p 127–143.

Barrow NJ. 1986. Testing a mechanistic model. 2. The effects of time and temperature on the reaction of zinc with a soil. J Soil Sci 37:277–286.

Barrow NJ. 1987. Reactions with variable-charge soils. Developments in plant and soil sciences. Dordrecht: Martinus Nijhoff Publishers.

Bernal MP, McGrath SP, Miller AJ, Baker AJM. 1994. Comparison of the chemical changes in the rhizosphere of the nickel hyperaccumulator *Alyssum murale* with the non-accumulator *Raphanus sativus*. Plant Soil 164:251–259.

Boawn LC, Viets FG, Crawford CL, Nelson JL. 1960. Effect of nitrogen carrier, nitrogen rate, zinc rate, and soil pH on zinc uptake by sorghum, potatoes, and sugar beets. Soil Sci 90:329–337.

Brennan RF. 1990. Reaction of zinc with soil affecting its availability to subterranean clover. II. Effect of soil properties on the relative effectiveness of applied zinc. Aust J Soil Res 28:303–310.

Brennan RF, Gartrell JW. 1990. Reaction of zinc with soil affecting its availability to subterranean clover. I. The relationship between critical concentrations of extractable zinc and properties of Australian soils responsive to zinc. Aust J Soil Res 28:293–302.

Bruemmer GH, Gerth J, Tiller KG. 1988. Reaction kinetics of the adsorption and desorption of nickel, zinc and cadmium by goethite. I. Adsorption and diffusion of metals. J Soil Sci 39:37–52.

Collins R, McLaughlin MJ, Morel J-L, Gerard E, Merrington G. 2001. Effect of organic ligands on isotopically exchangeable Cd (E-value) in and desorption from two heavy metal contaminated soils. In: Evans L, editor. Proc 6th International Conference on the Biogeochemistry of Trace Elements (6th ICOBTE); July 2001; University of Guelph, Ontario, Canada.

Collins R, Merrington G, McLaughlin MJ, Morel J-L. 2003. Organic ligand and pH effects on isotopically exchangeable cadmium in polluted soils. Soil Sci Soc Am J 67:112–121.

Deist J, Talibudeen O. 1967. Ion exchange in soils from the ion pairs K-Ca, K-Rb and K-Na. J Soil Sci 18:125–137.

Dyanand S, Sinha MK. 1985. Labile pool and selective distribution of iron in calcareous and sodic soils. Plant Soil 88:11–21.

Echevarria G, Morel JL, Fardeau JC, Leclerc-Cessac E. 1998. Assessment of phytoavailability of nickel in soils. J Environ Qual 27:1064–1070.

Elzinga EJ, Van Grinsven JJM, Swartjes FA. 1999. General purpose Freundlich isotherms for cadmium, copper and zinc in soils. Eur J Soil Sci 50:139–50.

Filius A, Streck T, Richter J. 1998. Cadmium sorption and desorption in limed topsoils as influenced by pH: Isotherms and simulated leaching. J Environ Qual 27:12–18.

Fujii R, Corey RB. 1986. Estimation of isotopically exchangeable cadmium and zinc in soils. Soil Sci Soc Am J 50:306–308.

Gérard E, Echevarria G, Morel C, Sterckeman T, Morel JL. 2001. Isotopic exchange kinetics method for assessing cadmium availability in soils. In: Iskandar IK, Kirkham MB, editors. Trace elements in soil: bioavailability flux and transfer. Boca Raton, FL: Lewis Publishers. p 127–143.

Gérard E, Echevarria G, Sterckeman T, Morel JL. 2000. Cadmium availability to three plant species varying in cadmium accumulation pattern. J Envirol Qual 29:1117–1123.

Gooddy DC, Shand P, Kinniburgh DG, Van Riemsdijk WH. 1995. Field-based partition coefficients for trace elements in soil solutions. Eur J Soil Sci 46:265–286.

Graham ER. 1973. Selective distribution and labile pools of micronutrient elements as factors affecting plant uptake. Proc Soil Sci Soc Am J 37:70–74.

Grinsted MJ, Hedley MJ, White RE, Nye PH. 1982. Plant induced changes in the rhizosphere of rape (*Brassica napus* var. Emerald) seedlings. I. change and the increase in P concentration in the soil solution. New Phyt 91:19–29.

Hamon RE, Wundke J, McLaughlin MJ, Naidu R. 1997. Availability of zinc and cadmium to different plant species. Aust J Soil Res 35:1267–1277.

Hamon RE, McLaughlin MJ, Naidu R, Correll R. 1998. Long-term changes in cadmium bioavailability in soil. Environ Sci Technol 32:3699–3703.

Hamon RE, McLaughlin MJ. 1999. Use of the hyperaccumulator *Thlaspi caerulescens* for bioavailable contaminant stripping. In: Wenzel WW, Adriano DC, Alloway B, Doner HE, Keller C, Lepp NW, Mench M, Naidu R, Pierzynski GM, editors. Proc 5th International Conference on the Biogeochemistry of Trace Elements (5th ICOBTE), Vienna. p 908.

Hamon RE, McLaughlin MJ. 2002. Interferences in the determination of isotopically exchangeable P in soils and a method to minimise them. Aust J Soil Res 8:1383–1397.

Hamon RE, McLaughlin MJ, Cozens G. 2002. Mechanisms of attenuation of metal availability in in situ remediation treatments. Environ Sci Technol 36:3991–3999.

Hutchinson JJ, Young SD, McGrath SP, West HW, Black CR, Baker AJ. 2000. Determining uptake of 'non-labile' soil cadmium by *Thlaspi caerulescens* using isotopic dilution techniques. New Phytol 146:453–460.

Imre L. 1937. Kinetic-radioactive investigations on the active surface of crystalline powders. Trans Farad Soc 33:571–583.

Imre L. 1939. Kinetic-radioactive investigations on the active surface of crystalline powders, II. Trans Farad Soc 35, 751–758.

Lamm CG, Hansen EH, Jørgensen JA. 1963. Isotopic exchange in soil fertility studies. Soil Sci 95:16–23.

Larsen S. 1952. The use of P^{32} in studies on the uptake of phosphorus by plants. Plant Soil 4:1–10.

Larsen S, Cooke IJ. 1961. The influence of radioactive phosphate level on the adsorption of phosphate by plants and on the determination of labile soil phosphate. Plant Soil 14:43–48.

Lofts S, Woof C, Tipping E, Clarke N, Mulder J. 2001. Modelling pH buffering and aluminium solubility in European forest soils. Eur J Soil Sci 52:189–204.

Lofts S, Tipping E. 1998. An assemblage model for cation binding by natural particulate matter. Geochem Cosmochim Acta 62:2069–2625.

Lombi E, Hamon RE, McGrath SP, McLaughlin MJ. 2003. Lability of Cd, Cu and Zn in polluted soils treated with lime, beringite and red mud and identification of non-labile colloidal fraction of metals. Environ Sci Technol 37:979–984.

Lopez PL, Graham ER. 1972. Labile pool and plant uptake of micronutrients: 1. Determination of labile pool of Mn, Fe, Zn, Co, and Cu in deficient soils by isotopic exchange. Soil Sci 114:295–299.

Ma YB, Uren NC. 1998. Dehydration, diffusion and entrapment of zinc in bentonite. Clays Clay Min 46:132–138.

McLaughlin MJ. 2001. Ageing of metals in soils changes bioavailability. Fact sheet on environmental risk assessment. International Council on Metals and the Environment (ICME), September 2001.

Nakhone LN. 1989. Factors affecting the lability of cadmium in soil [PhD thesis]. University of Nottingham, UK.

Nakhone LN, Young SD. 1993. The significance of (radio) labile Cadmium pools in soil. Environ Pollut 82:73–77.

Pandeya SB, Singh AK, Jha P. 1998. Labile pool of cadmium in sludge-treated soils. Plant Soil 203:1–13.

Papadopoulos P, Rowell DL. 1988. The reactions of cadmium with calcium carbonate surfaces. J Soil Sci 39:23–36.

Pullmann B, Haissinsky M. 1947. Echange isotopique dans les systemes MnO_2/Mn^{2+} et PbO_2/Pb^{2+}. J Phys Radium 8:36–38.

Russell RS, Rickson JB, Adams SN. 1954. Isotopic equilibria between phosphates in soil and their significance in the assessment of fertility by tracer methods. J Soil Sci 5:85–105.

Sinaj S, Machler F, Frossard E. 1999. Assessment of isotopically exchangeable zinc in polluted and nonpolluted soils. Soil Sci Soc Am J 63:1618–1625.

Smit CE, Van Beelen P, Van Gestel CAM. 1997. Development of zinc bioavailability and toxicity for the springtail *Folsomia candida* in an experimentally contaminated field plot. Environ Pollut 98:73–80.

Smolders E, Brans K, Foldi A, Merkx R. 1999. Cadmium fixation in soils measured by isotopic dilution. Soil Sci Soc Am J 63:78–85.

Smolders E, Kortekaas A, Van Den Brande K. 2000. The fate of Zn from tyre debris in soil. Report of a project sponsored by the Rubber Industry of the EU (BLIC) and the Zinc Oxide Producers Association (ZOPA), co-ordinated by the International Lead and Zinc Research Organisation (ILZRO), Research Triangle Park, NC.

Stanhope KG, Young SD, Hutchinson J, Kamath R. 2000. Use of isotopic dilution techniques to assess the mobilization of non-labile Cd by chelating agents in phytoremediation. Environ Sci Technol 34:4123–4127.

Sun B, Zhao FJ, Young SD, Tye A, McGrath SP. 2000. Availability and fixation of Zn and Cd in soils amended with metal sulphate. In: Luo YM, McGrath SP, Cao ZH, Zhao FJ, Chen YX, Xu JM, editors. Proceedings of the International Conference on Soil Remediation (SoilRem2000), October 15–19, Hangzhou, China. p 354–365.

Tandy S. 1998. The behaviour of mercury in contaminated soils [Masters Thesis]. University of Nottingham, UK.

Tiller KG, Honeysett JL, deVries MPC. 1972a. Soil zinc and its uptake by plants. I. Isotopic exchange equilibria and the application of tracer techniques. Aust J Soil Res 10:151–164.

Tiller KG, Honeysett JL, deVries MPC. 1972b. Soil zinc and its uptake by plants. II. Soil chemistry in relation to prediction of availability. Aust J Soil Res 10:165–182.

Tipping E, Lawlor AJ, Lofts S, Vincent CD. 2000. Development of critical level methodologies for toxic metals in soils and surface waters-dynamic modelling. Environmental Diagnostics GST/04/1709 Final Report. Centre for Ecology & Hydrology, NERC, UK.

Trivedi P, Axe L. 2000. Modeling Cd and Zn sorption to hydrous metal oxides. Environ Sci Technol 34:2215–2223.

Tye AM, Young SD, Crout NMJ, Zhang H, Preston S, Bailey EA, Davison W, McGrath SP, Paton GI, Kilham K. 2002. Predicting arsenic solubility in contaminated soils using isotopic dilution techniques. Environ Sci Technol 36:982–988.

Young SD, Tye A, Carstensen A, Resende L, Crout N. 2000. Methods for determining labile cadmium and zinc in soil. Eur J Soil Sci 51:129–136.

Young SD, Tye A, Crout NMJ. 2001. Rates of metal ion fixation in soils determined by isotopic dilution. In: Evans L, editor. Proceedings of the 6th International Conference on the Biogeochemistry of Trace Elements (6th ICOBTE); July 2001; University of Guelph, Ontario, Canada. p 105.

3 Biological Assessment of Natural Attenuation of Metals in Soil

Enzo Lombi, Daryl P. Stevens, Rebecca E. Hamon, and Mike J. McLaughlin

3.1 INTRODUCTION

Natural attenuation, also known as aging, of metals in soil refers to the processes by which the mobility, bioavailability, and toxicity of soluble metals in soils decline with time. These processes are important in terms of both risk assessment and crop production because of their effects on metal toxicity and micronutrient availability, respectively. However, most of the research related to natural attenuation of metals in soil has been conducted using a chemocentric approach. Adsorption and desorption methods were used over 30 years ago by Tiller et al. (1972) to investigate long-term binding of Co in soil. Single or sequential extractions have also been widely used to assess aging of a variety of metals and metalloids in soil (see Chapter 1). More recently, isotopic dilution techniques have been employed to assess changes in metal lability over time (see Chapter 2). These chemical techniques provide information regarding changes in soil-solution partitioning, extractability, and lability of metals in soil, but do not directly measure changes in biological availability or toxicity of metals (Stevens and McLaughlin 2001). Therefore, biological and chemical assessment of aging must be integrated to adequately understand both the mechanisms and the effects of natural attenuation of metals in soil.

In this chapter, we will also provide a biological viewpoint of natural attenuation.

Growth response curves can be used to examine attenuation from the perspective of micronutrient deficiency. In this case, the relationship between yield, or any other plant growth parameter (y), and the amount of micronutrient metal applied to the soil (x) is generally fitted using a Mitscherlich equation (Barrow and Mendoza 1990):

$$y = a - b\exp(-cx) \tag{3.1}$$

where

a provides an estimate of the maximum yield plateau,

b estimates the difference between the asymptote and the intercept on the y-axis, and

c describes the shape of the relationship and governs the rate at which y increases as x increases.

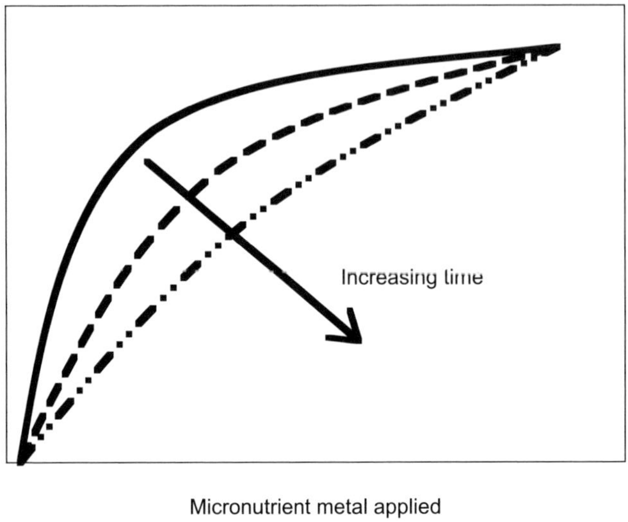

FIGURE 3.1 Effect of different incubation times of added essential nutrient metal in micronutrient-deficient soil on crop yield response.

Using this equation, a set of curves (outlined in Figure 3.1) can be obtained, relating a crop growth parameter to the metal applied at different times. It should be noted that the slope of the curves decreases with time, indicating that the effectiveness of the micronutrient metal in terms of plant growth declines over time (Figure 3.1). Using the parameters of the Mitscherlich equation (3.1), it is possible to calculate the relative effectiveness (RE) of the micronutrient metal applied at different times. RE is calculated by dividing the *c* parameter of Equation 3.1 calculated for each year, by the *c* parameter obtained with the fresh addition of micronutrient metal (Brennan et al. 2001). Therefore, the RE of a fresh addition is equal to 1; RE declines with time as shown in Figure 3.2.

The biological approaches to estimating the natural attenuation rates of metals, when present in the toxicity range in soils, are quite different from those used in agronomic studies on micronutrient metals. One way to characterize the reversion of metals to nontoxic forms is to determine toxicity-evaluated abatement rates. A toxic response to a metal can be defined (see Figure 3.3) according to a response curve (sigmoidal or other appropriate function) at time zero. This toxic response curve can be defined for any soil using a variety of biological end points. Following a certain period of incubation of metal with soil (t_1), the relevant end point (e.g., plant growth or nitrification) is measured again and related to the original toxic response curve for that soil at the start of the experiment (t0). The ratio of end points measured as EC50 values at different times can be used to calculate the rate of abatement of toxicity for each soil.

A number of biological end points can be used to assess the natural attenuation of metals. Plants, invertebrates, and the soil microbial community represent important classes of soil-dwelling organisms and differ in terms of sensitivity, pathways of exposure, and trophic levels. Therefore, these 3 classes of organisms will be

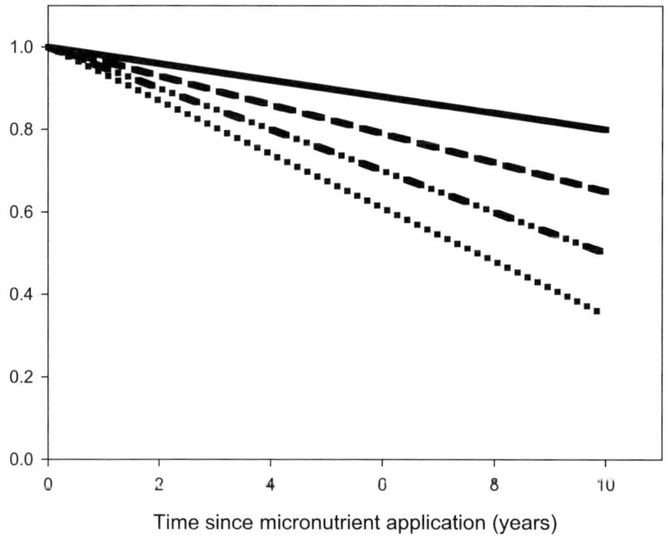

FIGURE 3.2 Relationship between relative effectiveness of micronutrient metals and time since micronutrient application. The different slopes are representative of curves expected for either different metals with different attenuation rates in the same soil or, for the same metal, differences in aging rate in different soils.

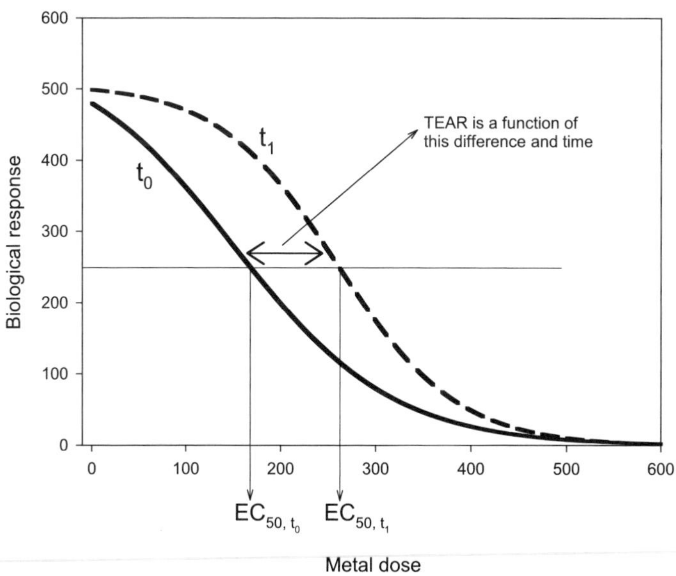

FIGURE 3.3 Hypothetical change in toxicity response to added metal as a function of time of soil–metal contact, where TEAR indicates toxicity evaluated abatement rate (t_0 is the time at which metal is freshly applied, and t_1 the test conducted some time later).

reviewed individually in relation to their use in biological assessment of natural attenuation. Furthermore, their relative limitations and advantages will also be discussed.

3.2 PLANTS AS BIOLOGICAL INDICATORS OF NATURAL ATTENUATION OF METALS IN SOIL

Plant bioassays are sensitive to a number of environmental conditions that can interfere with the plant's response to the factor being investigated, in this case, metal availability or its subsequent toxicity. Such interferences are particularly pertinent to aging studies, in which the time variable must also be considered. The optimum method to account for the variability of environmental conditions affecting plant responses is to add the metal to be investigated to soils at different times (creating different ages or times of incubation), maintain the soils at a controlled moisture and temperature, and then perform the bioassay at the same time for all of the treated soils. However, for practical or logistic reasons, this is not always possible. For instance, in field agronomical trials, micronutrient metals are often added at the same time, and the attenuation monitored by progressive bioassays (e.g., yield or uptake) conducted over a number of years. The problem with this type of assessment is that variability in environmental factors such as temperature, humidity, and soil moisture will generally give rise to variability in the biotic response, thereby potentially confounding the assessment of metal availability. In this case, increasing the number of replicates and, especially, the number of time points, may help to improve the clarity of relationships between environmental influences and attenuation processes on biological responses. For field trials, comparing the effectiveness of a micronutrient metal added in the past with a fresh metal addition is a preferable option, because the plants in the assay are grown at the same time and hence experience the same growth conditions. Obviously, pot trials under controlled conditions offer the possibility to minimize any confounding environmental factors. However, it can be argued that this type of standardized experiment is less environmentally relevant than field trials.

Various plant tests can be used to assess natural attenuation. In agronomic trials investigating changes in micronutrient availability over time, yield of crops or uptake of micronutrients is usually measured. Toxicity studies may focus on chronic or acute plant responses. In chronic toxicity tests, plant parameters such as biomass are generally measured after a few weeks or months of growth in contaminated soils. On the other hand, acute toxicity can be measured using much shorter tests such as root elongation or germination assays, which can be completed in a few days. This implies that the time frame for investigation of attenuation will be different depending on whether chronic or acute tests are conducted. Obviously, chronic tests cannot be used to investigate attenuation processes over a few days. Also, when plants are grown for several months in a soil, the difference in metal bioavailability between the beginning and the end of the period of plant growth could be significant because of attenuation processes. Therefore, the use of chronic tests is limited to investigation of natural attenuation over a long period of time, in the order of years. Acute toxicity

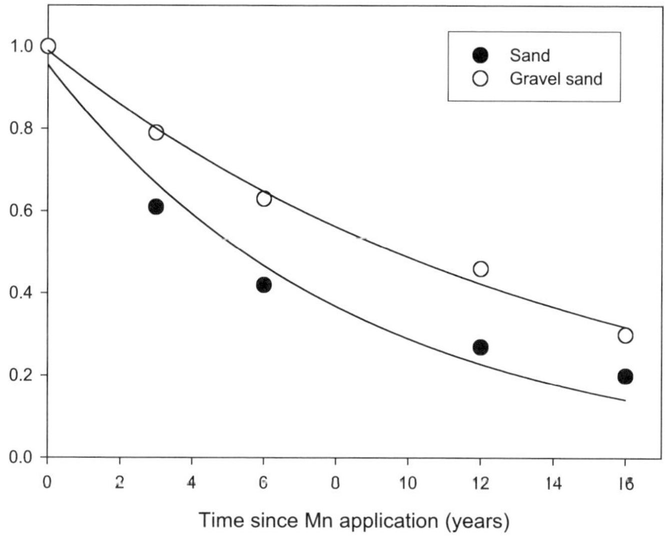

FIGURE 3.4 Relationship between the relative effectiveness of Mn, as measured by lupin grain yield, and the length of time after Mn application to 2 Mn-deficient soils. (Data from Brennan, RF, Gartrell, JW, Adcock, KG. 2001. Residual value of manganese fertilizer for lupin grain production. *Aust J Exp Agr* 41:1187–1197. With permission.)

tests based on short-term parameters such as germination, emergence, and root elongation offer a better time resolution for aging processes.

Some examples in which biological end points (agronomic trials) have been used to provide experimental evidence of changes in micronutrient availability over time are reported in the following paragraphs. A more comprehensive review of these experiments is given in Chapter 10. Experiments assessing changes in micronutrients' RE have been used to compare attenuation rates: for a single metal in different soils and for different metals in the same soil, as well as to identify soil parameters that influence the rate of natural attenuation.

Brennan et al. (2001) compared the relationship between RE of Mn, measured by grain yield of lupin (*Lupinus angustifolius*), in 2 different soils in Western Australia. The results indicated that the rate of Mn aging depended on soil properties (Figure 3.4). The decline in effectiveness was larger in a gravel-sandy soil than in a sandy soil. The gravel sand had a greater content of clay, silt, and organic C and a larger cation exchange capacity in comparison to the sandy soil.

However, the study conducted by Brennan et al. (2001) considered only 2 soils, and therefore did not allow assessment of soil characteristics affecting natural attenuation of metals. This kind of information is only attainable if a large number of soils, differing in chemical and physical characteristics, are used. Brennan (1990) assessed the effect of 30-d aging at 30 °C on the effectiveness of Zn measured by the growth of subterranean clover *(Trifolium subterraneum)*. A simple regression between soil pH and RE explained 75% of the variation (Figure 3.5). This result is similar to that reported by Young et al. (Chapter 2), who showed that soil pH enhances natural attenuation of metals and therefore decreases their effectiveness.

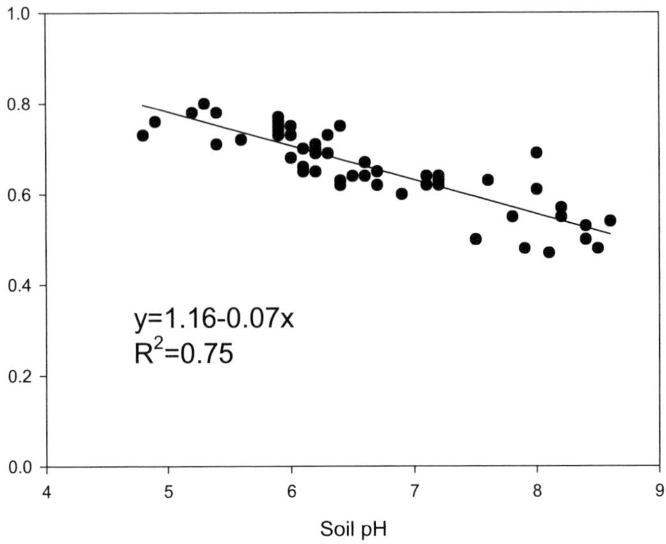

FIGURE 3.5 Relative effectiveness, measured as Zn uptake by clover, of aged vs. freshly added Zn as a function of soil pH. (Data from Brennan, RF. 1990. Reaction of zinc with soil affecting its availability to subterranean clover: II. Effect of soil properties on the relative effectiveness of applied Zn. *Aust J Soil Res* 28:303–310. With permission.)

The last example of changes in micronutrient effectiveness over time relates to the aging comparison between 2 micronutrients, Mo and Zn, in the same soil (Brennan 2001, 2002). RE of Mo appears to decline much more rapidly than that of Zn (Figure 3.6).

The following examples describe studies in which plant bioassays have been used to indicate natural attenuation of metals from a toxicological point of view. Bruus Pedersen et al. (2000) investigated the natural attenuation of Cu to black bindweed (*Fallopia convolvulus*) over a period of 12 weeks. In their study, Cu was added to an uncontaminated soil and dose-response curves were reported between root or shoot biomass and total, DTPA- or $CaCl_2$-extractable soil Cu. Results showed that the EC50 (concentration at which growth was reduced by 50%) for Cu was not significantly different when soil was spiked 1, 5, and 12 weeks before the plant bioassay. As suggested by the authors, the time span used in this study may have been too narrow to reveal any clear effects of Cu aging. This is corroborated by the fact that concentrations of extractable metals did not change between 1 and 12 weeks of incubation. Interestingly, Bruus Pedersen et al. (2000) also compared their results to those of Kjær et al. (1998), who used the same plant to test Cu toxicity in the same soil that was contaminated in 1918 by a timber preservation plant. In their study on the field-contaminated soil, Kjær et al. (1998) did not report any toxicity up to total Cu concentrations of 928 $mg \cdot kg^{-1}$. This concentration is much larger than EC50 (270 mg Cu kg^{-1}) reported by Bruus Pedersen et al. (2000) for the soil spiked in the laboratory. Similarly, extractability of Cu from the aged field-contaminated soil was much less than in the laboratory-spiked soil at similar total soil Cu

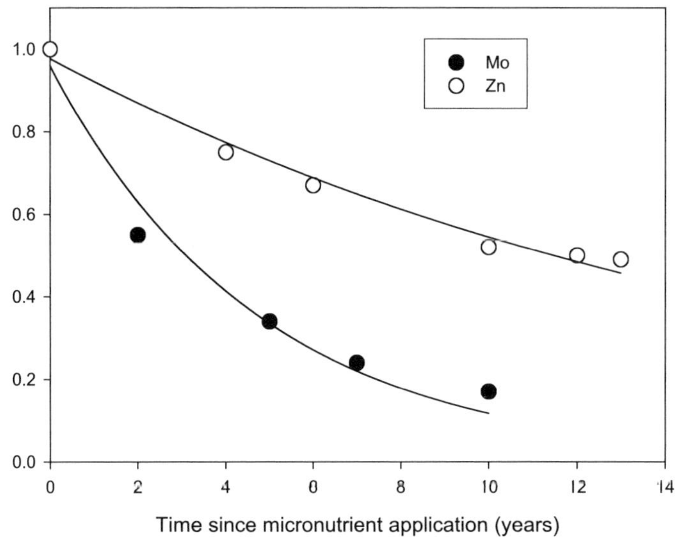

FIGURE 3.6 Relationship between the relative effectiveness of Mo and Zn and the length of time after fertilizer application to the same soil. (Data from Brennan, RF. 2001. Residual value of zinc fertilizer for production of wheat. *Aust J Exp Agr* 41: 541–547; Brennan, RF. 2002. Residual value of molybdenum trioxide for clover production on an acidic sandy podsol. *Aust J Exp Agr* 42: 565–570. With permission.)

concentrations, confirming that, under field conditions, natural attenuation of Cu had occurred over the 70-y period.

Lock and Janssen (2003) studied the influence of aging on Cu bioavailability and toxicity to red clover (*Trifolium pratense*). In this study, the bioavailability of Cu in 25 contaminated soils collected by Cu runoff from bronze statues was compared to Cu bioavailability in control soils freshly spiked to the same Cu concentration of the corresponding field soils. Their results indicated that growth of clover was less in the freshly spiked soils and that this was related to the Cu concentration in soil-pore water. However, the pH of the soil after spiking decreased in comparison to the historically contaminated soils. Therefore, it was not possible to unequivocally relate the decrease in Cu bioavailability to attenuation of Cu by the soil.

Hamon et al. (1998) measured changes in Cd lability to wheat (*Triticum turgidum*), using an isotopic dilution technique (*L*-value). With this method, they developed a model that estimated Cd attenuation in the soil to be occurring at a rate of 1% to 1.5% of the total Cd applied per year. This method, as well as other isotopic dilution techniques used to investigate natural attenuation of metals in soil, are reviewed in Chapter 2.

3.3 INVERTEBRATES AS BIOLOGICAL INDICATORS OF NATURAL ATTENUATION OF METALS IN SOIL

Among soil invertebrates, a number of feeding guilds, such as detritivores, herbivores, bacterial and fungal feeders, omnivores, and predators, are represented. Nutrient cycling in the soil is heavily dependent on the activity of soil invertebrates. Therefore, invertebrates are an important class of organism for soil toxicity testing and could be used to assess natural attenuation of metals in soil. Similarly to plants, a variety of invertebrate tests are available.

Short-term survival tests can provide a snapshot of acute toxicity. However, if the goal of the study is the protection of soil invertebrate populations, then information regarding the effects of soil contaminants on reproduction of species is required (Løkke et al. 2002). As discussed for plants, the time frame for aging investigations will vary with the test chosen. For instance, survival studies that can be completed in a few days may provide a higher time resolution than reproduction tests that require several weeks to be completed.

One aspect that needs to be considered when soil invertebrates are used to assess attenuation processes relates to the pathways of exposure to metals. For invertebrates, the major exposure routes can include both the soil-pore water as well as direct ingestion of soil particles. In the case of soil ingestion, metal speciation may be changed in the gut of the organism. Therefore, it is possible that the attenuated metal could be remobilized in the gut environment. If this occurs, use of invertebrates to assess metal attenuation may provide different results in comparison to other biological or chemical assessments. However, recent results using earthworms and isotopic dilution techniques suggest that Zn uptake by *Eisenia andrei* is predominantly via the exchangeable pools (possibly the soil-pore water) rather than by any gut-induced dissolution of Zn held in nonexchangeable pools in the soil (Scott-Fordsmand et al. 2004).

A limited number of studies have attempted to assess natural attenuation of metals using soil invertebrates. Smit and Van Gestel (1998) investigated the effect of aging on bioaccumulation and toxicity of Zn for the springtail *Folsomia candida*. Using the weight of the springtail as the biological end point, they determined that the effect of Zn was larger by a factor of 5 to 8 in freshly spiked soils in comparison to test soils subjected to aging under field conditions for 1.5 years. However, these results were complicated by a significant difference in pH between freshly spiked and field-aged soils, which has a strong effect in terms of Zn solubility and availability. Also, Smit and Van Gestel (1998) suggested that the presence of relatively large amounts of Cl (introduced with the spiking of $ZnCl_2$) may have resulted in an enhanced toxicity in freshly contaminated soils through a synergistic toxic effect of Zn and Cl, or through Cl altering the bioavailability of Zn.

In a similar experiment, Smit et al. (1997) studied the effect of aging on bioavailability and toxicity of Zn in a field trial. In this case, the end point investigated was mortality and reproduction of *F. candida.* Although a relationship between mortality and Zn concentration was not observed, production of juveniles appeared to be a sensitive and reproducible parameter to assess Zn toxicity and to study Zn aging. Similar to other studies, comparison between freshly spiked soils and soils incubated in the field for several months was difficult because of a lower pH of the freshly spiked soils. However, after a few months, the pH of treatments stabilized, and a comparison between EC50s at different times was possible. Even though variability in the field trial was large, EC50s calculated on total Zn concentrations increased significantly with time. This data indicated that an increase in Zn sorption and a decrease in bioavailability over time lead to reduced toxicity.

Smit and Van Gestel (1996) also compared the difference in Zn toxicity between soil collected around a Zn smelter and control soil spiked with Zn salts. In this case, the reproduction of *F. candida* was used as the end point to calculate EC50s, and the soil pH was adjusted to lessen the importance of pH as a confounding factor. In the freshly spiked soils, EC50 for reproduction was 210 mg Zn kg^{-1}. In contrast, no effects on reproduction were found in the transect field soils up to the highest Zn concentration (approximately 1500 $mg \cdot kg^{-1}$).

Lock and Janssen (2001) used a central composite experimental design to develop a model to predict Zn toxicity to *F. candida*. This model was able to predict chronic Zn toxicity to the springtail in freshly spiked soils as a function of soil pH, cation exchange capacity, and total Zn in soil. However, when applied to 19 soils historically contaminated by previous smelting activities or application of dredge-contaminated sediments, the model failed to predict Zn toxicity. Lock and Janssen (2001) suggested that this was probably due to changes in Zn availability caused by aging processes. In subsequent work, the same authors investigated the influence of aging on acute Cu toxicity to the potworm *Enchytraeus albidus* and on chronic toxicity to *F. candida*. The toxicity of historically contaminated soils collected around bronze statues and of freshly spiked control soils was compared. Toxicity in the historically contaminated soils was lower than in the laboratory-spiked soils. However, pH differences between the 2 sets of soil treatments made it impossible to ascertain whether aging, or simply a difference in Cu partitioning due to pH, was responsible for the decrease in Cu toxicity observed in the field-contaminated soils.

3.4 MICROBIAL END POINTS AS BIOLOGICAL INDICATORS OF NATURAL ATTENUATION

The same microbial tests to assess metal toxicity in soil could potentially be used to investigate attenuation processes. A large number of tests are available, and most relate to some function or characteristic of the indigenous soil microbial community or of a specific group of this community. This is in contrast to laboratory tests conducted with invertebrates or plants in which these organisms are usually placed in the soil at the time of testing. Consequently, in the case of microbial tests, aging generally occurs in the presence of the organisms used in the test. As a result of this, and because of the rapid life cycle of a number of microorganisms, there is a factor in addition to attenuation that may influence the toxic response. This factor is adaptation. Adaptation of microorganisms to metals can be achieved by means of intrinsic properties (tolerance) or by developing specific mechanisms of detoxification (resistance) as a result of exposure to metals. Adaptation is therefore a factor that, similar to aging, will develop with time after the application of metals to soil. Also, adaptation, as with aging, results in a decrease of the detrimental effect when assessed with microbial toxicity tests. Because adaptation not only depends on time but also causes an effect on the bioassay similar to aging, it represents a potentially

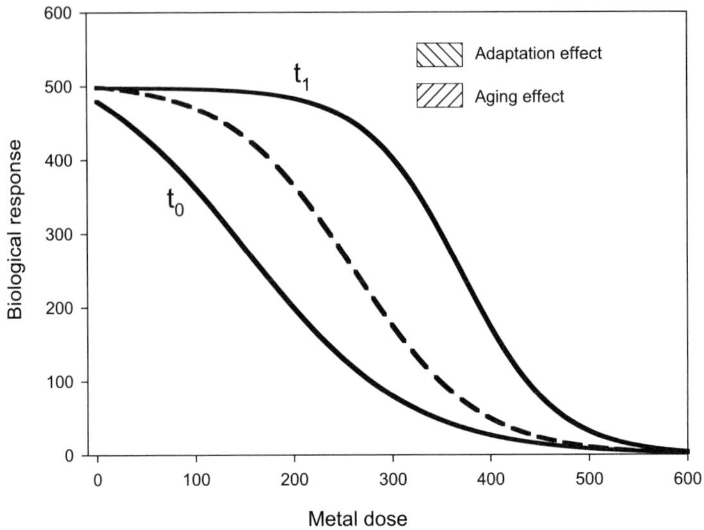

FIGURE 3.7 Diagram showing potential contributions of adaptation and aging to the change in toxicity response to added metal as a function of time of soil–metal contact (t_0 is the time at which metal is freshly applied, and t_1 the test conducted some time later).

large confounding factor in the assessment of natural attenuation using microbial tests (Figure 3.7).

As discussed in Chapter 2, there is generally a very rapid decrease in availability immediately after the addition of a metal to a soil, but this decrease in availability slows down with time. Similarly, adaptation will proceed more rapidly immediately after metals are added to the soil because their availability, and therefore the selective pressure to which the microbial community is exposed, is larger immediately after metal addition and decreases with time. Another potential confounding factor is related to the possibility that a "shock" effect on the indigenous microbial population, when large concentrations of metals are added in a single application, could influence the short-term results of a bioassay. Such large additions of metals with a single application are very rare in the environment and would be restricted to accidental spills.

Various studies have demonstrated induction of microbial community tolerance to metals when soils have been contaminated. Rutgers et al. (1998) used a method involving Biolog plates to demonstrate increased tolerance of the soil microbial community because of Zn exposure. Other methods, such as ^{3}H-thymidine incorporation and phospholipid-ester-linked fatty acid (PLFA) analyses have also measured increases in the tolerance of soil microorganisms upon exposure to metal (Baath 1992; Frostegard et al. 1993; Diaz-Ravina et al. 1994).

Recently, Hamon et al. (2003) proposed a method based on the sterilization and reinoculation of soils with either an exposed or unexposed soil microbial community to investigate tolerance. This method can be used for the investigation of metal aging avoiding the problem of tolerance. In this case, soils spiked and aged for different periods of time are first sterilized and then reinoculated by adding a small amount of the original uncontaminated soil. In this way, the microbial community is not exposed to metals during the aging process, thus avoiding the confounding factor of adaptation (excluding adaptation processes taking place during the testing itself). However, care would need to be taken to account for any effects that the sterilization process may have on influencing metal availability and toxicity. Another way to avoid the problem of metal adaptation of the indigenous microbial community is based on the use of added, or nonindigenous, microorganisms. An example is the use of microbial biosensors to test metal bioavailability in soil-pore water (Paton et al. 1997). However, so far these methods have not been employed in natural attenuation studies, and the interpretation of published results must take into account the possibility of microbial adaptation as a confounding factor.

Kelly et al. (1999) investigated changes in a microbial community over time after a single application of Zn. Significant changes in the community structure were observed immediately after spiking (15 d) but decreased with time (420-d incubation). However, the effect of changes in Zn availability over time was not separated from adaptation of the microbial community. Doelman and Haanstra (1984) studied the changes in basal soil respiration in 5 soils spiked with Zn and incubated for up to 43 weeks. A decline of Zn toxicity with time was found in some soils but not in all. Smolders et al. (2003) compared the Zn toxicity for soil microbial processes between laboratory-contaminated and polluted-field soils collected around galvanized structures. Nitrification, respiration, and N-mineralization rates were significantly reduced in the laboratory-spiked soils in comparison to the control but were unaffected, or even increased with Zn concentration, in the field transect soils (Figure 3.8). Even though adaptation of the soil microbial population may have been a reason for the difference in Zn toxicity between field- and laboratory-contaminated soils, attenuation of Zn bioavailability has probably been a contributing factor. In fact, large differences in Zn solubility were observed between aged and freshly spiked soils. The highest soil-solution Zn concentrations in the field soils were always lower than the soil-solution EC50s of spiked soils.

Not all the evidence obtained with microbial tests suggests that natural attenuation of metals in soil occurs. For instance, Chaudri et al. (1992) followed the survival of the rhizobium population for a period of 18 months after spiking soil with Zn at 6 different concentrations. No significant effects were reported up to 12 months. However, after 18 months, the population decreased at higher Zn concentrations. This is opposite to what should be expected if bioavailability of metals was the only factor controlling the survival of rhizobium in the soil. These results indicate that the effect of time on metal aging must be related to the life cycle and any lag-time response at community or population levels.

3.5 LIMITATIONS OF BIOLOGICAL APPROACHES TO INVESTIGATE NATURAL ATTENUATION

Some of the limitations of biological approaches to investigate natural attenuation of metals in soil are specific to particular groups of biological tests and have been

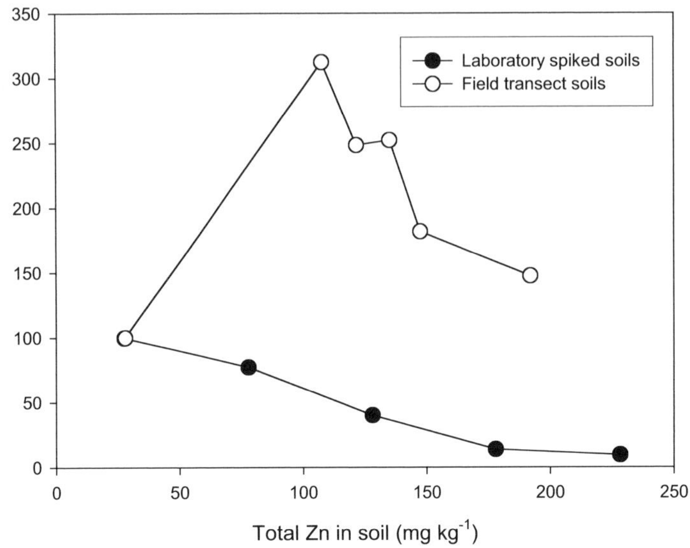

FIGURE 3.8 Potential nitrification rate in soils of Zn-contaminated transects and laboratory-spiked control soils. (From Smolders E, McGrath SP, Lombi E, Bernhard R, Karman CC, Cools D, Van den Brande K, Van Os B, Walrave N. 2003. Comparison of toxicity of zinc for soil microbial processes between laboratory-contaminated and polluted field soils. *Environ Toxicol Chem* 22:2592–2598. With permission.)

discussed in the previous sections. For instance, adaptation is a confounding factor, which is important to consider when indigenous microbial assays are used, but is not applicable for short-term tests in which organisms (such as plants or invertebrates) are added to the soil. However, other limitations are more generic in character and have implications for every biological approach, independent of the test organism considered.

It is widely accepted that biological variability, and consequently measurement of biological parameters, is larger than analytical errors in chemical tests. This must be taken into account when biological methods are used to assess natural attenuation. In fact, natural attenuation usually produces small changes in bioavailability over periods that can be considerably long. If these changes are smaller than the error associated with the measurement of the biological parameters, aging responses cannot be revealed using these techniques. Variability in biological assays is caused by the intrinsic biological variability in the bioassay used, the response of the bioassay to environmental conditions, and the analytical or physical measurement of the particular biological parameter used. Each of these sources of variability can be minimized by using, for instance, clonal biological material, conducting the testing under controlled environmental conditions, and following standard analytical procedures supported by adequate quality controls. However, the introduction of some of these measures to decrease variability may also decrease the environmental or ecological significance of the assessment. For instance, results from a laboratory assessment using a specific microbial biosensor, under very defined environmental conditions, is probably less environmentally relevant than field measurements of changes in soil microbial functions or community structure over time.

A large number of toxicological studies reviewed in the previous sections compared freshly spiked control soils with field-contaminated soils in which metals were aged. This is a useful shortcut because it allows the assessment of natural attenuation in a short time avoiding the need to age soils for long periods of time. However, it is not always easy to find field soils contaminated by a single metal and with a known history of contamination. Moreover, spiking of uncontaminated soils with large amounts of metal salts can cause several artifacts that confound the interpretation of the results. For instance, soil pH in the spiked soils can significantly decrease in comparison to field soils that have been contaminated slowly with a metal and have undergone leaching through natural rainfall events. The decrease in soil pH after addition of metal salts is due to metal hydrolysis and metal adsorption by soil, and consequent protons released in the soil solution (Basta and Tabatabai 1992). Stevens et al. (2003) reported that counterions of metals present in the salts may also be toxic to the biological end point investigated. Furthermore, metal spiking can significantly increase the salinity in the soil-pore water with consequences for the solid-solution partitioning of metals.

Biological assessment of natural attenuation using the RE approach does not suffer the same confounding factors discussed for toxicological studies. In fact, micronutrient metals are usually added at small rates that do not cause changes in pH or salinity. However, RE approaches are limited to deficiency situations and to elements for which there is a positive response in the organism tested in comparison to an untreated control.

3.6 FUTURE USES AND CHALLENGES

One of the key aspects of natural attenuation of metals in soil relates to the decrease in bioavailability and toxicity of micronutrients and pollutants to biological end points. Consequently, a biological assessment of aging processes appears to be extremely appropriate. On the other hand, because of the natural variability of biological systems and a series of confounding factors previously discussed, at present it is difficult to assess natural attenuation of metals, at least when the interest is focused on the toxicity range. Attenuation processes taking place at low metal loading may be quite different from those occurring under toxic conditions. More research is needed to understand the implications of natural attenuation for metals in the toxic range. It should be emphasized that a biological approach to the assessment of natural attenuation of metal toxicity is quite recent, and most of the methods employed are the same as those used in ecotoxicological studies. More work is therefore needed to fine-tune these methods to obtain reliable information regarding processes resulting in attenuation of trace element bioavailability.

REFERENCES

Baath E. 1992. Measurement of heavy metal tolerance of soil bacteria using thymidine incorporationn into bacteria extracted after homogenisation-centrifugation. Soil Biol Biochem 24:1167–1172.

Barrow NJ, Mendoza RE. 1990. Equations for describing sigmoid yield responses and their application to some phosphate responses by lupins and by subterranean clover. Fert Res 22:181–188.

Basta NT, Tabatabai MA. 1992. Effect of cropping systems on adsorption of metals by soils. 2. Effect of pH. Soil Sci 153:195 204.

Brennan RF. 1990. Reaction of zinc with soil affecting its availability to subterranean clover. II. Effect of soil properties on the relative effectiveness of applied zinc. Aust J Soil Res 28:303–310.

Brennan RF. 2001. Residual value of zinc fertiliser for production of wheat. Aust J Exp Agr 41:541–547.

Brennan RF. 2002. Residual value of molybdenum trioxide for clover production on an acidic sandy podsol. Aust J Exp Agr 42:565–570.

Brennan RF, Gartrell JW, Adcock KG. 2001. Residual value of manganese fertiliser for lupin grain production. Aust J Exp Agr 41:1187–1197.

Bruus Pedersen M, Kjær C, Elmegaard N. 2000. Toxicity and bioaccumulation of copper to black bindweed (*Fallopian convolvolus*) in relation to bioavailability and age of soil contamination. Arch Environ Contam Toxicol 39:431–439.

Chaudri AM, McGrath SP, Giller KE. 1992. Survival of the indigenous population of *Rhizobium leguminosarum* biovar *trifolii* in soil spiked with Cd, Zn, Cu and Ni salts. Soil Biol Biochem 24:625–632.

Diaz-Ravina M, Baath E, Frostegard A. 1994. Multiple heavy metal tolerance of soil bacterial communities and its measurement by a thymidine incorporation technique. Appl Environ Microbiol 60:2238–2247.

Doelman P, Haanstra L. 1984. Short-term and long-term effects of cadmium, chromium, copper, nickel, lead and zinc on soil microbial respiration in relation to abiotic soil factors. Plant Soil 79:317–327.

Frostegard A, Baath E, Tunlid A. 1993. Shifts in the structure of soil microbial communities in limed forests as revealed by phospholipid fatty-acid analysis. Soil Biol Biochem 25:723–730.

Hamon RE, McLaughlin MJ, Naidu R, Correll R. 1998. Long-term changes in cadmium bioavailability in soil. Environ Sci Technol 32:3699–703.

Hamon RE, Rusk J, McLaughlin MJ. 2003. Metals and soil microbes: how much adaptation is OK? In: Gobran G. editor. Proceedings of the 7th International Conference on the Biogeochemistry of Trace Elements (7th ICOBTE), Uppsala, Sweden. p 258–259.

Kelly JJ, Haggblom M, Tate RL, 1999. Changes in soil microbial communities over time resulting from one time application of zinc: a laboratory microcosm study. Soil Biol Biochem 31:1455–1465.

Kjær C, Bruus Pedersen M, Elmegaard N. 1998. Effects of soil copper on the black bindweed (*Fallopian convolvolus*) in the laboratory and in the field. Arch Environ Contam Toxicol 35:14–19.

Lock K, Janssen CR. 2001. Ecotoxicity of zinc in spiked artificial soils versus contaminated field soils. Environ Sci Technol 35:4295–4300.

Lock K, Janssen CR. 2003. Influence of aging on copper bioavailability in soils. Environ Tox Chem 22:1162–1166.

Løkke H, Janssen CR, Lanno RP, Rombke J, Rundgren S, van Straalen NM. 2002. Soil toxicity tests-Invertebrates. In: Fairbrother A, Glazebrook PW, van Straalen NM, Tarazona JV, editors. Test methods to determine hazards of sparingly soluble metal compounds in soils. Pensacola (FL): SETAC Pr. p 37–58.

Lombi E, Wenzel WW, Sletten RS. 1999 Arsenic adsorption on soil and iron oxide coated sand: changes over time. Zeit Pflanzen Bodenk 162:451–456.

Paton GI, Palmer G, Burton M, Rattray EAS, McGrath SP, Glover LA, Killham K. 1997. Development of an acute and chronic ecotoxicity assay using lux-marked *Rhizobium leguminosarum* biovar *trifolii*. Lett App Microbiol 21:296–300.

Rutgers M, Verlaat I, Wind B, Posthuma L, Breure A. 1998. Rapid method for assessing pollution-induced community tolerance in contaminated soil. Environ Toxicol Chem 17:2210–2213.

Scott-Fordsmand JJ, Stevens D, McLaughlin M. 2004. Do earthworms mobilize fixed zinc from ingested soil? Environ Sci Technol 38:3036–3039.

Sing D, McLaren RG, Cameron KC. 1997. Desorption of native and added zinc from a range of New Zealand soils in relation to soil properties. Aust J Soil Res 35:1253–1266.

Smit CE, Van Beelen P, Van Gestel CAM. 1997. Development of zinc bioavailability and toxicity for the springtail *Folsomia candida* in an experimentally contaminated field plot. Environ Poll 98:73–80.

Smit CE, Van Gestel CAM. 1996. Comparison of the toxicity of zinc for the springtail *Folsomia candida* in artificially contaminated and polluted field soils. Appl Soil Ecol 3:127–136.

Smit CE, Van Gestel CAM. 1998. Effects of soil type, prepercolation, and ageing on bioaccumulation and toxicity of zinc for the springtail *Folsomia candida*. Environ Toxicol Chem 17:1132–1141.

Smolders E, McGrath SP, Lombi E, Bernhard R, Karman CC, Cools D, Van den Brande K, Van Os B, Walrave N. 2003. Comparison of toxicity of zinc for soil microbial processes between laboratory-contaminated and polluted field soils. Environ Toxicol Chem 22:2592–2598.

Stevens DP, McLaughlin MJ. 2001. Bioindicators for assessing natural attenuation of metals in soil. In: Evans L, editor. Proceedings of the 6th International Conference on the Biogeochemistry of Trace Elements (6th ICOBTE); July 2001; University of Guelph, Ontario, Canada. p 106.

Stevens DP, McLaughlin MJ, Heinrich T. 2003. Determining toxicity of runoff lead and zinc in soils-salinity effects on metal partitioning and phytotoxicity. Environ Toxicol Chem 22:3017–3024.

Tiller KG, Honeysett JL, De Vries MPC. 1972. Soil zinc and its uptake by plants. I. Isotopic exchange equilibria and the application of tracer techniques. Aust J Soil Res 10:151–164.

4 Long-Term Fate of Metal Contaminants in Soils and Sediments: Role of Intraparticle Diffusion in Hydrous Metal Oxides

Paras Trivedi and Lisa Axe

4.1 INTRODUCTION TO SORPTION KINETICS

Hydrous oxides of Al, Fe, and Mn are ubiquitous components of soils and sediments, which play an important role in controlling the mobility and bioavailability of metal contaminants released into the environment from various anthropogenic sources (Hesterberg 1998; Jenne 1998; Martinez and McBride 1999). These oxides occur as discrete particles or as coatings on other mineral surfaces; they have large surface areas, microporous structures, and high affinities for metal ions (Jenne 1998; Sparks 1998). The sorption of contaminants to these microporous oxides is recognized as a 2-step process: an initial rapid adsorption that reaches a pseudo equilibrium at the mineral–water interface, followed by a much slower uptake that may continue for a period of days to years (Hodges and Johnson 1987; Brümmer et al. 1988; Fuller et al. 1993; Waychunas et al. 1993; Axe and Anderson 1995, 1997; Papelis et al. 1995a; Raven et al. 1998; Scheidegger et al. 1998; Sparks 1998; Strawn et al. 1998; Scheinost et al. 2001).

Many mechanistic models have been invoked to explain transient sorption, including diffusion into micropores (Hodges and Johnson 1987; Barrow et al. 1989; Fuller et al. 1993; Waychunas et al. 1993; Axe and Anderson 1995, 1997; Liang and Tsai 1995; Papelis 1995; Papelis et al. 1995a, 1995b; Misak et al. 1996; Strawn et al. 1998; Trivedi and Axe 1999, 2000, 2001a,b; Cole et al. 2000; Scheinost et al. 2001), surface precipitation (Farley et al. 1985; Fitts et al. 2000), and the formation of layered double hydroxides (Scheidegger et al. 1996, 1998; Ford et al. 1999).

In modeling intraparticle diffusion, adsorbent characteristics that significantly affect the intra-aggregate mass transport of the contaminants include particle size and geometry, porosity, and pore size distribution (Papelis et al. 1995a; Axe and Anderson 1997). For example, Papelis et al. (1995a) observed that the uptake of Cd and Se in smaller alumina particles was much faster than in larger ones. Surface area also serves as another tool in assessing the physical properties of an adsorbent (Gregg and Sing 1982). Strauss et al. (1997) compared the properties of goethite with varying crystallinity; they found that as the degree of crystallinity of goethite increased, surface area also increased, whereas the mean particle size and the porosity decreased. Such structural differences significantly affect sorption behavior. In addition, Papelis et al. (1995a) noted that Cd and Se uptake in microporous alumina was much slower than in the mesoporous one.

Recently, Trivedi (2001) demonstrated the role of porosity in assessing the long-term sorption of contaminants. Accordingly, amorphous oxides of aluminum (HAO), iron (HFO), and manganese (HMO), whose individual porosities (based on freeze-dried particles) ranged from 35 to 50%, exhibited significant diffusivity. Likewise, Cole et al. (2000) also demonstrated that in clays with a porosity of 55 to 58%, surface diffusion affected contaminant mobility in soils and sediments. In contrast, for the crystalline iron oxide goethite, with porosity measured on freeze-dried particles less than 1%, the slow sorption process was insignificant (Trivedi and Axe 2001b). These results emphasize the need to account for the microporosity of the different components of soils and sediments to appraise their role in the long-term fate of the contaminants. Furthermore, to identify the type of diffusion mechanism, it is important to evaluate the pore size distribution in porous sorbents.

When a metal ion is transported through the internal pores, pore size and effective radius of the cation will determine the type of intraparticle diffusion (Froment and Bischoff 1990; Kärger and Ruthven 1992). If the effective radius of the pore is much greater than the radius of the diffusing species, then bulk diffusion is expected. When adsorption is significant and the cation and effective pore radii are approximately equivalent, then the transport occurs along the pore walls and is referred to as "configurational" or "surface diffusion." In hydrated oxides, layers of water are sorbed to the pore surface, resulting in micropores. In such cases, surface diffusion dominates the intraparticle transport of the metal contaminants (Froment and Bischoff 1990; Kärger and Ruthven 1992; Axe and Anderson 1995, 1997; Trivedi and Axe 1999, 2000, 2001a; Trivedi 2001).

4.2 EXPERIMENTAL METHODS

The adsorbents considered in this research include HAO, iron (HFO and goethite), and HMO. Methods of preparation and characterization have been detailed elsewhere (Axe and Anderson 1995; Trivedi and Axe 1999, 2000). Thermodynamic assessment, based on the short-term adsorption studies with Ca, Sr, Cd, Ni, and Zn, demonstrated that these metal ions sorb, via intraparticle diffusion (physical sorption), to HAO, HFO, and HMO, but are chemisorbed to goethite (Axe and Anderson 1997; Trivedi and Axe 1999, 2000, 2001a,b).

Diffusion kinetics can take years to reach equilibrium, hence methods are needed to understand the sorption in a convenient time frame. In traditional sorption studies, the total amount of the adsorbate is fixed, and, consequently, the driving force for diffusion decreases with time. To overcome these limitations, semibatch studies were

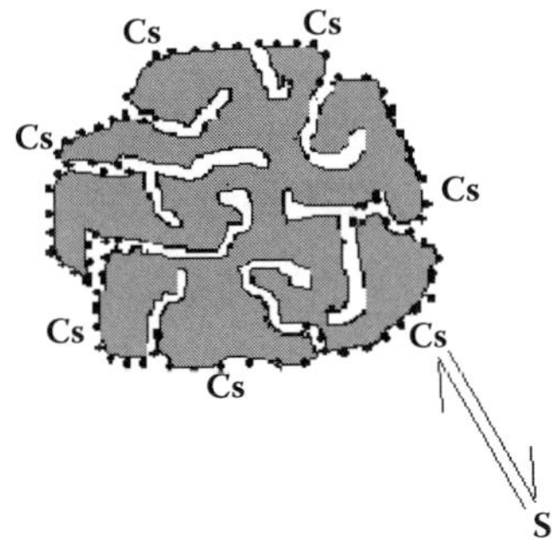

FIGURE 4.1 Constant boundary condition (CBC) experiments: Intraparticle diffusion of metal contaminants along the micropore walls of amorphous oxides. The grey spherical area represents the amorphous oxide particle with micropores and dark points represent the metal adsorbate. (Reproduced from Trivedi P. 2001. Predicting Thermodynamic and Transport Parameters for Metal Contaminant Sorption to Hydrated Metal Oxides in Aquatic Systems, Ph.D. thesis. New Jersey Institute of Technology, NJ. With permission.)

conducted under constant boundary conditions, wherein the bulk adsorbate concentration was monitored and maintained constant by adding the adsorbate as needed (Figure 4.1) (Axe and Anderson 1995, 1997; Trivedi and Axe 1999, 2000, 2001a; Cole et al. 2000). The relative importance of external mass transfer vs. intraparticle diffusion can be assessed by evaluating dimensionless numbers: the Biot number (Bi) and the Reynolds' number (Re) (Bird et al. 2001; Traegner and Suidan 1989). For systems with a $Bi<<1$, the external mass transfer resistance controls the reaction, whereas for $Bi>>100$, intraparticle diffusion is the rate-determining step. Therefore, for experimental studies, the system must be maintained in a turbulent hydraulic regime to minimize the external mass transfer resistance (Fogler 1992).

4.3 MODELING APPROACH

Diffusion-based sorption kinetics have been described through various approaches, including the Elovich equation, the parabolic diffusion equation, and the fractional power equation. Although they have been employed to model the sorption kinetics, they have serious limitations where their use in fitting often does not provide any mechanistic insight (Sparks 1998). In contrast, using a mass balance with well-defined boundary conditions and sorbent characteristics has resulted in 1 fitting parameter, surface diffusivity (Axe and Anderson 1995, 1997; Trivedi and Axe 1999, 2000, 2001a). With negligible external mass transfer resistance and dilute adsorbate concentrations, and assuming the internal sites are similar to the external ones, the mass balance for spherical aggregates yields the following partial differential equation (Crank 1975):

$$\frac{\partial C}{\partial t} = \frac{D_S}{\left[1 + \frac{\varepsilon}{(\rho K_i)}\right]} \frac{1}{r^2} \frac{\partial \left(r^2 \frac{\partial C}{\partial r} \right)}{\partial r} \tag{4.1}$$

where

C = the moles of adsorbate sorbed g^{-1} of hydrous oxide,
D_s = the surface diffusivity in $cm^2\ s^{-1}$,
ε = the porosity,
ρ = the bulk density of the oxide in $g\ cm^{-3}$,
R = the radial position within the sphere, and
K_i = the distribution coefficient for sorption to internal sites.

Given the boundary conditions, including a constant surface concentration, integration of the analytical solution to Equation 4.1 over the volume of the particle yields the mass sorbed per particle at a specific time:

$$\frac{M}{M_\infty} = 1 - \frac{6}{\pi^2} \sum_{n=1}^{\infty} \frac{1}{n^2} \exp\left[\frac{-D_s n^2 \pi^2 t}{R^2 \left(1 + \varepsilon / \rho K_i \right)} \right] \tag{4.2}$$

For a given time, the mass sorbed internally for each particle size times the number of particles from the particle size distribution is summed with the amount sorbed to the external surface to obtain the total metal sorbed. All the parameters, except the surface diffusivity, are determined from the characterization and the isotherm studies. By minimizing the variance, the only fitting parameter — surface diffusivity — is obtained.

4.4 RESULTS AND DISCUSSION

For Cd, Ni, Sr, and Zn, the experimental surface diffusivities ranged from 10^{-16} to 10^{-10} $cm^2\ s^{-1}$ (Axe and Anderson 1995, 1997; Trivedi and Axe 1999, 2000, 2001a). These modeling results show that internal sorption accounts for approximately 50% of the total sorption to HAO, 40% to HFO, and 90% to HMO. Given the particle size distribution of each oxide, sorption reaches equilibrium in approximately 3 to 4 months for Sr, 2 to 5 years for Cd, and as much as 5 to 8 years for Zn (Figure 4.2). Nearly 30% of the total selenite sorption occurred because of diffusion in Fe_2O_3, where the fraction of metal sorbed internally was independent of pH (Hamdy and Gissel-Nilsen 1977). Brümmer et al. (1988) observed that Ni sorption in silica-associated goethite increased from 12 to 70% after a contact time of 42 d, whereas Cd and Zn sorption increased up to 21 and 33%, respectively, over the same time. They attributed this slow sorption to intraparticle diffusion. Barrow et al. (1989)

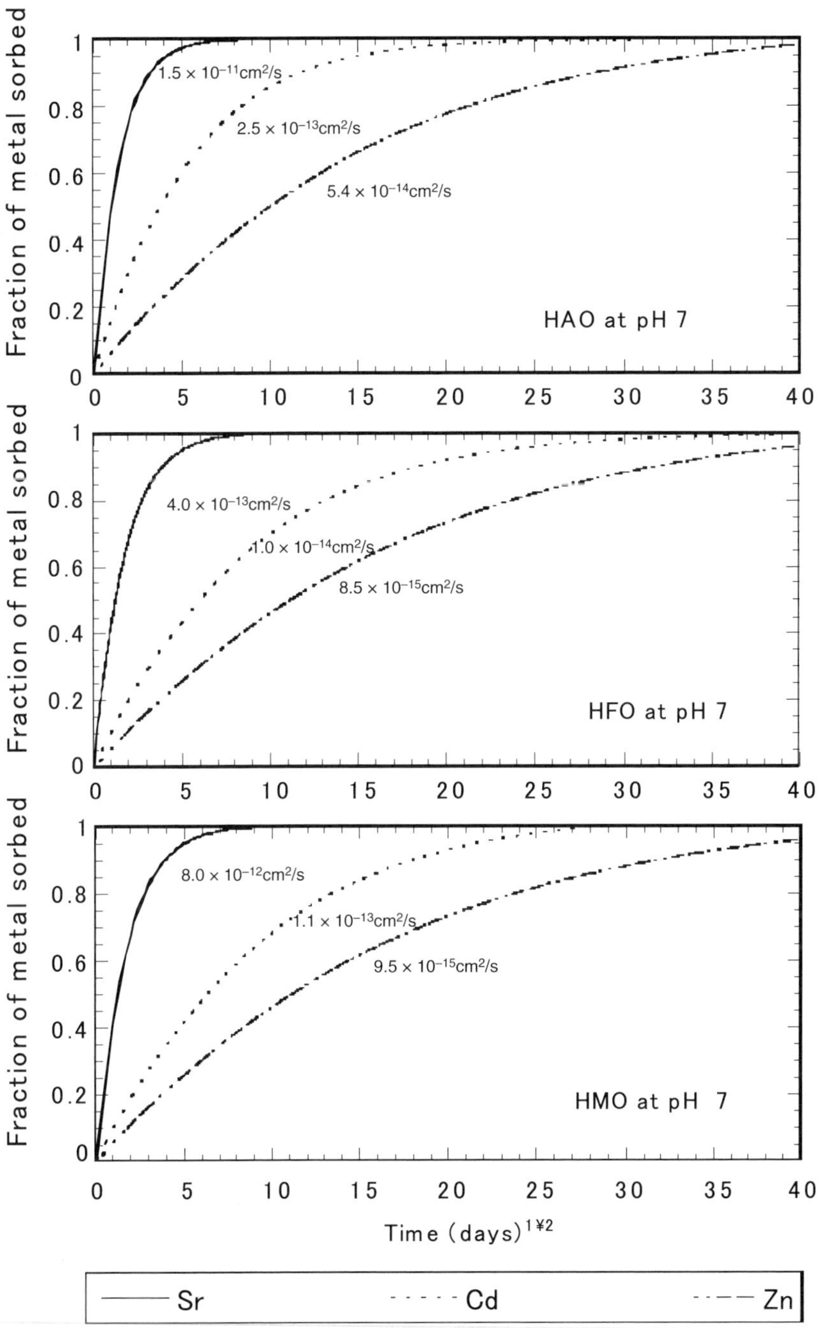

FIGURE 4.2 Predicting equilibrium at pH 7 and 25 °C for Sr, Cd, and Zn sorption to hydrous oxides of aluminum (HAO), iron (HFO), and manganese (HMO) given their respective particle size distributions. (Reproduced from *Environ. Sci. Technol.*, 2000, **35**, 2215–2223, ©2000 American Chemical Society.)

modeled these studies and estimated the diffusivity for Zn and Cd to be 6.5×10^{-19} and 3.0×10^{-20} cm^2 s^{-1}, respectively. Fuller et al. (1993) estimated effective diffusivities for As(V) in HFO to be on the order of 10^{-11} cm^2 s^{-1}. Liang and Tsai (1995) estimated the diffusivity of Cs in natural modernite in the order of 10^{-14} cm^2 s^{-1}. Misak et al. (1996) reported diffusivities of Co and Zn in hydrous ferric and stannic oxide to be approximately 10^{-11} cm^2 s^{-1}. Papelis, Roberts et al. (1995) found that Cd and Se(IV) diffusivities in porous alumina ranged between 10^{-12} and 10^{-10} cm^2 s^{-1}. All these studies indicate that in soils and aquatic environments where microporous oxides and clay minerals are present, these sorbents potentially act as sinks for metal contaminants.

4.5 INTRAPARTICLE DIFFUSION AND SITE ACTIVATION THEORY

Transient studies are time consuming, and therefore predictive methods would be useful for determining the surface diffusivities. Based on site activation theory (Kärger and Ruthven 1992; Axe and Anderson 1997), the surface diffusivity is a function of the jump frequency, which, in turn, is a product of the vibrational frequency and the Boltzmann factor. The vibrational frequency is a function of the force constant k (d^2U/dx^2), where U is the surface potential energy. This potential can be described by a sinusoidal function whose minima represent the sorption sites and maxima signify the energy barrier or the activation energy, E_A. Accordingly, the surface diffusivity can be estimated (Kärger and Ruthven 1992; Axe and Anderson 1997; Trivedi and Axe 1999, 2000, 2001a):

$$D_s = \lambda[E_A/(2m)]^{1/2} \exp[-E_A/(RT)] \tag{4.3}$$

where m is the molecular mass of the diffusing species. As the pH increases, the site density increases for cations based on isotherm studies (Trivedi and Axe 1999), and, in turn, the mean distance between the sites (λ) decreases; consequently, the surface diffusivity decreases. The random walk model has also been applied, for example, in predicting the diffusivities of volatile organic compounds in adsorbents such as activated carbon (Lordgooei et al. 2001). In Equation 4.3, based on the Polanyi relation, the activation energy (E_A) is equal to the product of the adsorption enthalpy (ΔH°) and proportionality constant (α) (Boudart 1968). In amorphous oxide systems, the activation energy was found to be unique for a specific metal ion (Trivedi 2001; Trivedi and Axe 2001a). Additionally, metals from the same group of the Periodic Table form similar sorption complexes with amorphous oxides and, hence, exhibit an equivalent α. Furthermore, this activation energy, which is simply the energy barrier between sites, was correlated to the hydrated radius (R_H) of the cation with charge Z and its primary hydration number (N) (Figure 4.3) (Trivedi 2001; Trivedi and Axe 2001a):

$$E_A = f\,[N \times (Z/R_H)^2] \tag{4.4}$$

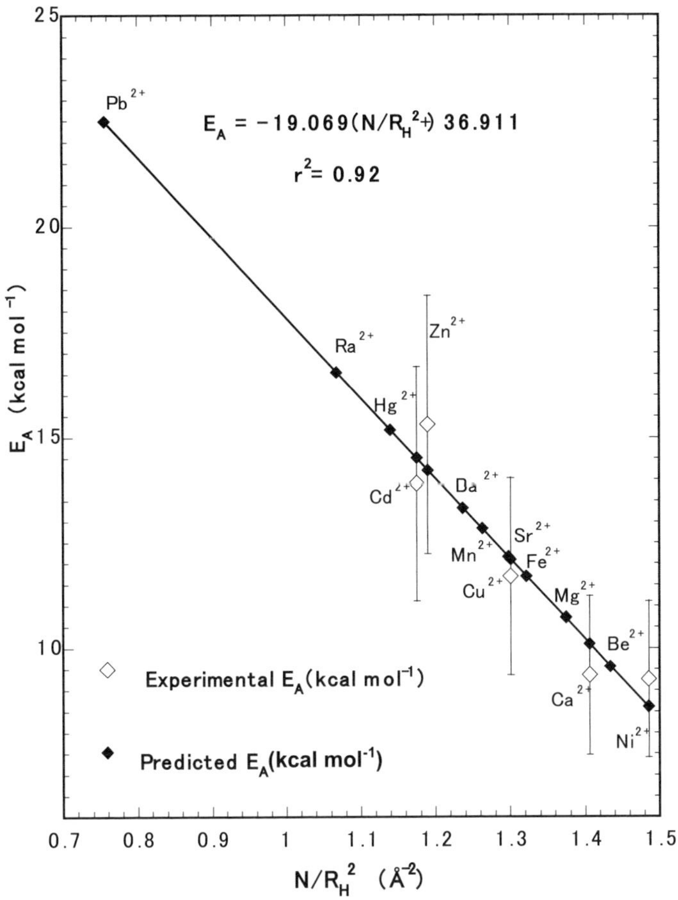

FIGURE 4.3 Correlation between activation energy and the structural parameters (R_H is the hydrated radius (Å) and N is the hydration number of the hydrated metal ions). The correlation is based on the experimental activation energies of Sr, Cd, and Zn (with ± 2 S.D.). The experimental activation energies of Ca and Ni (with ± 2 S.D.) are compared with the ones predicted from this model. (Reproduced from Trivedi P. 2001. Predicting Thermodynamic and Transport Parameters for Metal Contaminant Sorption to Hydrated Metal Oxides in Aquatic Systems, Ph.D. thesis. New Jersey Institute of Technology, NJ. With permission.)

Evangelou (1998) noted that reactions with activation energies less than 42 kJ mol^{-1} are indicative of diffusion-controlled processes, where as those with greater activation energies are surface controlled. The diffusivities predicted using Equation 4.3 and Equation 4.4, as seen in Figure 4.4 and Figure 4.5, are comparable with experimental ones, thus demonstrating the predictive capability of these equations. Brümmer et al. (1988) reported a diffusivity trend for Zn, Cd, and Ni sorption to silica-contaminated goethite, which is consistent with the 1 shown for these metals in Figure 4.5.

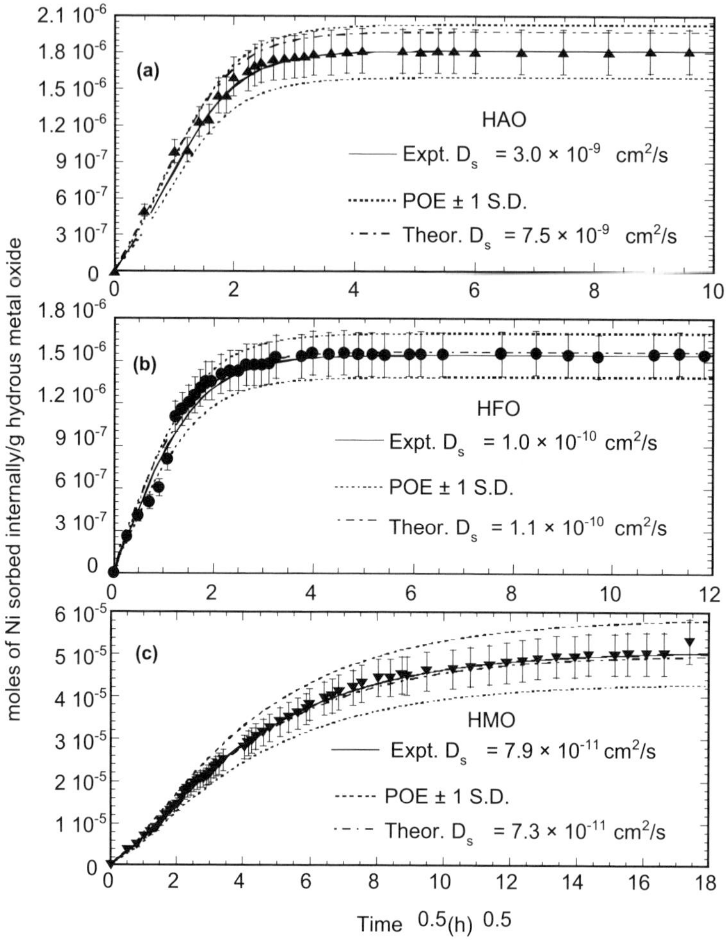

FIGURE 4.4 CBC studies at pH 7 and 25 °C — modeled (solid lines) Ni internal sorption to (a) 1 g L^{-1} HAO with $[Ni]_{bulk} = 2.5 \times 10^{-8}$ M, (b) 1 g l^{-1} HFO with $[Ni]_{bulk} = 1.1 \times 10^{-7}$ M, and (c) 0.1 g l^{-1} HMO with $[Ni]_{bulk} = 3.2 \times 10^{-9}$ M. Solid lines represent the best fit of the experimental diffusivity. Long dashed lines represent predictions of theoretical diffusivities based on site activation theory. Short dashed lines are modeling errors for the experimental diffusivity. (Reproduced from *Environ. Sci. Technol.*, 2001, **35**, 1779–1784. ©2001 American Chemical Society.)

4.6 SPECTROSCOPIC EVIDENCES OF INTRAPARTICLE DIFFUSION

In order to glean mechanistic information, macroscopic studies need to be complemented with spectroscopic ones that assist in resolving the local chemical environment of sorption complexes (Sposito 1986; Brown 1990; Sparks 1995). X-ray absorption spectroscopy (XAS) is 1 such advanced tool that can provide in situ structural information about the sorbate and its surroundings (Stern 1976; Brown

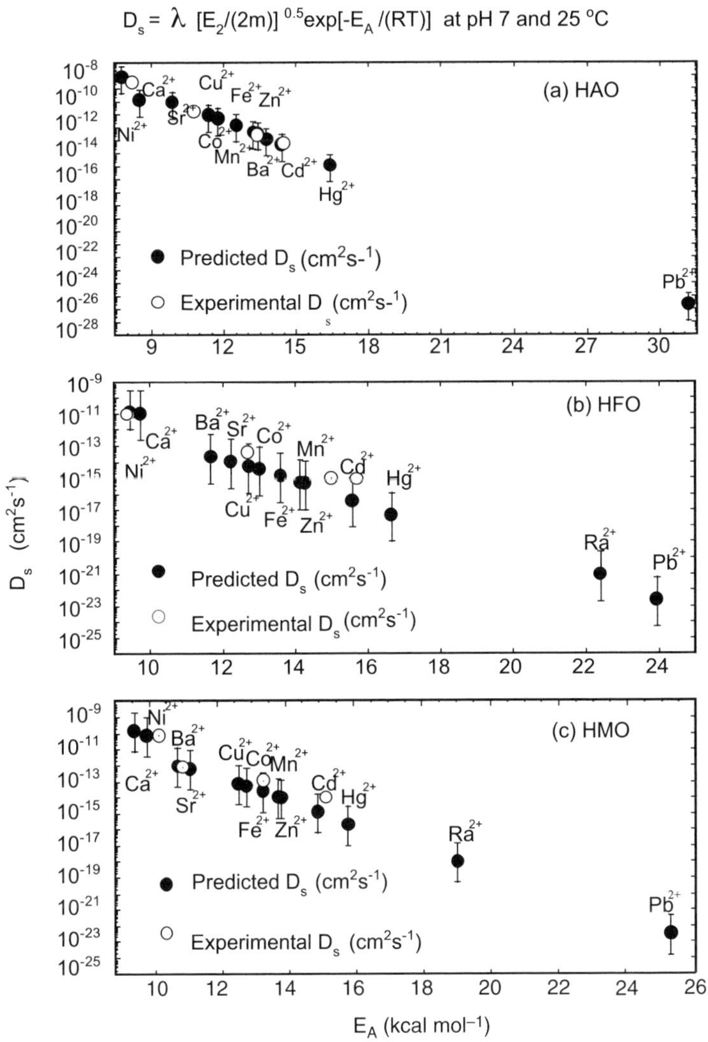

FIGURE 4.5 Theoretical surface diffusivities, at pH 7, based on site activation theory. Solid symbols with error bars represent the theoretical D_s, and open symbols represent experimental D_s (for Sr, Cd, Ni, and Zn). (Reproduced from *Environ. Sci. Technol.*, 2001, **35**, 1779–1784. ©2001 American Chemical Society.)

1990; Bunker 1999). A number of researchers have employed XAS to elucidate the sorption mechanism of environmentally relevant metal contaminants to important soil components (Charlet and Manceau 1992; Bargar et al. 1997; Manning et al. 1998; Scheidegger et al. 1998; Spadini et al. 1994; Sparks 1998). However, the application of XAS to elucidate diffusion kinetics is limited. Waychunas et al. (1993) studied As(V) sorption to ferrihydrite with the help of XAS studies; in their analyses of long-term sorption samples, there was no evidence of surface precipitation or solid-solution formation; thus, they attributed the limiting mechanism to intraparticle diffusion. Eggleston and Stumm (1993) also observed Cr(III) surface diffusion and dimer formation on the (001) surface of α-Fe_2O_3, using scanning tunnel microscopy. In contrast, in Cr-γ-Al_2O_3 sorption systems, Fitts et al. (2000) found from their XAS analyses an increase in Cr-Cr interactions and a decrease in Cr-Al interactions with time. Hence, they proposed that the slow uptake of Cr(III) by γ-Al_2O_3 is a result of the Cr(III) polymerization. In another study with aluminum oxides, Papelis and coworkers employed XAS (Papelis et al. 1995b) and x-ray photoelectron spectroscopy (XPS) (Papelis 1995) to complement their macroscopic results (Papelis et al. 1995a), where Cd and selenite were found to diffuse into the micropores of the alumina and sorb as mononuclear complexes. Strawn et al. (1998) supplemented their macroscopic work with XAS in studying Pb sorption to aluminum oxide. They found that Pb sorbed to aluminum oxide via an inner sphere bonding mechanism and that, as sorption increased as a function of time, the coordination environment was invariant, thus suggesting diffusion of Pb along the micropores of the oxide.

Axe et al. (1998, 2000) found that Sr sorption to HFO and HMO is a physical reaction and its overall uptake by these oxides is limited by intraparticle diffusion. Furthermore, the combined results of the XRD, XAFS, and XANES analyses demonstrated that the local structure of Fe in HFO and Mn in HMO did not change in the presence or absence of Sr for up to several months. Recently, similar XAS studies were conducted to assess the local structure of Zn (Figure 4.6) and Ni (Figure 4.7) sorbed to HMO at pH 7 as a function of the contact time (Trivedi et al. 2001). In the long-term samples aged for as long as 110 d, metal ions were found to have similar local structure to that of their respective short-term (4 h) samples. These results support the conclusion that the slow uptake of metal ions in HMO results from intraparticle diffusion wherein the sorption sites located along the micropore walls are similar to those available on the external surfaces and macropores. Likewise, Scheinost et al. (2001) employed XAS and found that Cu and Pb ions formed mononuclear monodentate types of sorption complexes with freshly precipitated, as well as with resuspended, freeze-dried ferrihydrite particles. Interestingly, the local structure of Pb and Cu sorbed to these iron oxides was invariant with time up to 8 weeks in the absence or presence of competing ions or fulvic acid; thus, they attributed the slow uptake of metal ions by ferrihydrite to intraparticle diffusion. The spectroscopic evidence asserts the need to include contributions from diffusion for modeling the long-term fate and bioavailability of metal pollutants in soil and aquatic environments.

4.7 OXIDE COATINGS

For accurate modeling of the mobility and bioavailability of the metal contaminant, a thorough understanding of metal sorption onto microporous surfaces of oxide-coated minerals is as important as understanding sorption onto discrete particles. Limited research has been conducted to address this issue. For example, Dong et al. (2000) demonstrated that, for natural surface coatings, sorption of Cd and Pb ions onto the oxides of Al, Fe, and Mn dominates over the interactions of these metals with organic materials and biofilms. Stahl and James (1991) studied interactions between Fe hydroxide-coated sands and Zn, and concluded that the surface coatings could induce

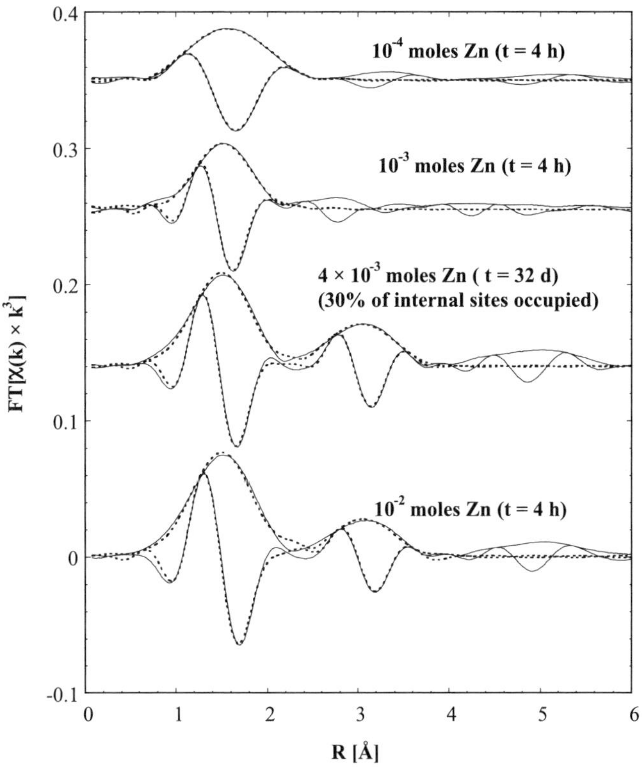

FIGURE 4.6 Fourier transforms (solid lines) of Zn K-edge XAS spectra of Zn-HMO adsorption samples at pH 7 and 298K, presented as a function of zinc concentration, each filtered over k-range 2.4–9.4 $Å^1$ and fitted with chalcophanite (dashed lines) from 0.5 to 3.78 Å (except single shells, which were fitted between 0.5 and 2.2 Å). Fitting parameters: N_1 = 6 O, R_1 = 2.18 Å; N_2 = 8 O, R_2 = 3.50 Å. (Reproduced from *Environ. Sci. Technol.*, 2001, **35**, 4515–4521. ©2001 American Chemical Society.)

surface hydrolysis of sorbed Zn, effectively fixing it at the surface. Anderson and Benjamin (1985), as well as Meng and Letterman (1993), have developed models of adsorption at multicomponent oxide surfaces. These studies also indicate that coatings change the surface chemistry of the substrate, and thus significantly influence the mobility of the metal contaminants. Most of the studies conducted on oxide coatings are based on short-term macroscopic experiments, suggesting the need for long-term experiments on kinetics of metal sorption onto the coatings. Lai et al. (2000) demonstrated through BET and EDAX analyses that the iron oxide-coated sands had relatively more micropores and mesopores, and a larger surface area, than the discrete sand particles; their studies suggest that these coatings must significantly elongate the equilibration periods of sorption reactions. In light of the development of composite adsorbents for wastewater treatment, a number of studies have demonstrated intraparticle diffusion of aqueous metal contaminants onto composite adsorbents as the predominant mode of sorption. These adsorbents range from cemented goethite (Theis et al. 1992) and fulvic acid-coated ferrihydrite (Scheinost

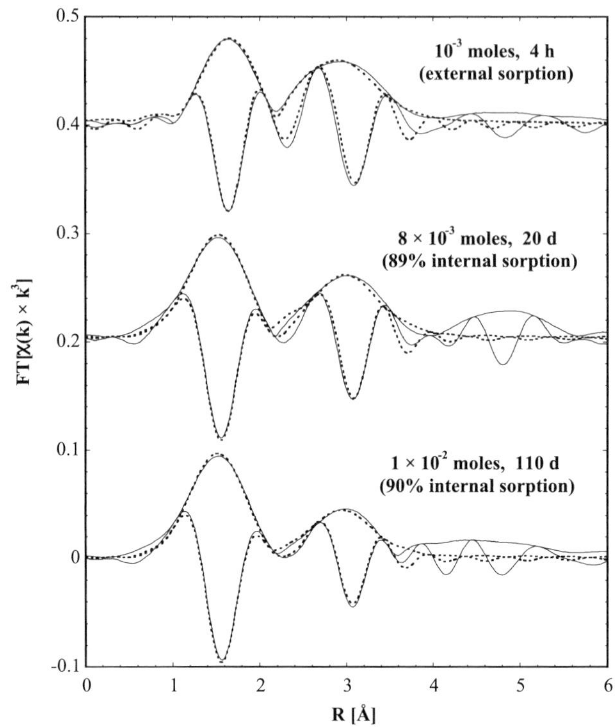

FIGURE 4.7 Fourier transforms (solid lines) of Ni K-edge XAS spectra of Ni sorbed to HMO (1 g l^{-1}) studied at pH 7 and 298 K, presented as a function of contact time, each filtered over k-range 2.45–9.21 $Å^{-1}$ and fitted with Ni_6MnO_8 (dashed lines) over 0.41–4.00 Å. Fitting parameters: $N_1 = 6$ O, $R_1 = 2.07$ Å; $N_2 = 8$ O, $R_2 = 3.38$ Å. (Reproduced from *Environ. Sci. Technol.*, 2001, **35**, 4515–4521. ©2001 American Chemical Society.)

et al. 2001) to poorly crystalline iron oxide coated onto activated carbon (Wang 1995) and manganese oxide-coated carbon (Fan 1996). Thus, there is a strong need to understand metal sorption onto oxide-coated mineral and organic surfaces to effectively predict the fate of metals in soils and sediments.

4.8 CONCLUSIONS

This research is largely focused on identification of the intraparticle diffusion mechanism that can contribute to a decrease in long-term mobility and bioavailability of contaminants in soils. Macroscopic modeling efforts provide the basis for evaluating diffusivities. XAS studies have further supported the rate-limiting mechanism, intraparticle diffusion. Time-resolved in situ spectroscopic studies need to be extensively applied for better understanding of diffusion-related processes in microporous components of soils, such as hydrated metal oxides. The challenge now lies in assessing these diffusion-limited processes for multicomponent systems and incorporating them in transport models and waste management programs.

REFERENCES

Anderson PR, Benjamin MM. 1985. Effects of silicon on the crystallization and adsorption properties of ferric oxides. Environ Sci Technol 19:1048–1053.

Axe L, Anderson PR. 1995. Sr diffusion and reaction within Fe oxides: evaluation of the rate-limiting mechanism for sorption. J Colloid Interface Sci 175:157–165.

Axe L, Anderson PR. 1997. Experimental and theoretical diffusivities of Cd and Sr in hydrous ferric oxide. J Colloid Interface Sci 185: 436–448.

Axe L, Anderson PR. 1998. Intraparticle diffusion of metal contaminants in amorphous oxide minerals. In: Jenne EA, editor. Adsorption of metals by geomedia. New York: Academic Pr. p 193.

Axe L, Tyson T, Trivedi P, Morrison T. 2000. Local structure analysis of strontium sorption to hydrous manganese oxide. J Colloid Interface Sci 224:408–416.

Bargar JR, Brown GE Jr., Parks GA. 1997. Surface complexation of Pb(II) at oxide-water interfaces II. XAFS and bond-valence determination of mononuclear Pb(II) sorption products and surface functional groups on iron oxides. Geochim Cosmochim Acta 61.2639–2652.

Barrow NJ, Gerth J, Brümmer GW. 1989. Reaction kinetics of the adsorption and desorption of nickel, zinc, and cadmium by goethite: II Modeling the extent and rate of reaction. J Soil Sci 40:437–450.

Bird BR, Stewart WE, Lightfoot EN. 2001. Transport phenomena. 2nd ed. New York: J Wiley.

Boudart M. 1968. Kinetics of chemical processes. Englewood Cliffs (NJ): Prentice-Hall.

Brown GE Jr. 1990. Spectroscopic studies of chemisorption reaction mechanisms at oxide/water interfaces. In: Hochella MF, White AF, editors. Mineral-water interface geochemistry. Washington, DC: Mineralogical Society of America. p 309–363.

Brümmer GW, Gerth J, Tiller KG. 1988. Reaction kinetics of the adsorption and desorption of nickel, zinc, and cadmium, by goethite. I. Adsorption and diffusion of metals. J Soil Sci 39:37–52.

Bunker G. 1999. http://gbxafs.iit.edu/training/tutorials.html.

Charlet L, Manceau AA. 1992. X-ray absorption spectroscopic study of the sorption of Cr(III) at the oxide water interface. J Colloid Interface Sci 148:425–442.

Cole T, Bidoglio G, Soupioni M, O'Gorman M, Gibson N. 2000. Diffusion mechanism of multiple strontium species in clay. Geochim Cosmochim Acta 64:385–396.

Crank J. 1975. The mathematics of diffusion. 2nd ed. Oxford (UK): Clarendon Pr.

Dong D, Nelson YM, Lion LW, Shuiler ML, Ghiorse WC. 2000. Adsorption of Pb and Cd onto metal oxides and organic material in natural surface coatings as determined by selective extractions: new evidence for the importance of Mn and Fe oxides. Wat Res 34:427–436.

Eggleston CM, Stumm E. 1993. Scanning tunneling microscopy of chromium(III) chemisorbed on α-ferric oxide (001) surfaces from aqueous solution: direct observation of surface mobility and clustering. Geochim Cosmochim Acta 57:4843–4850.

Evangelou VP. 1998. Environmental soil and water chemistry: principles and applications. New York: J Wiley.

Fan H-J. 1996. Removal and recovery of Cu(II) and Cd(II) by Mn oxide-coated composite adsorbent [PhD thesis]. Chicago (IL): Illinois Institute of Technology.

Farley KJ, Dzombak DA, Morel FMM. 1985. A surface precipitation model for the sorption of cations on metal oxides. J Colloid Interface Sci 106:226–242.

Fitts JP, Brown GE Jr., Parks GA. 2000. Structural evolution of Cr(III) polymeric species at the γ-Al_2O_3/water interface. Environ Sci Technol 34:5122–5128.

Fogler HS. 1992. Elements of Chemical Reaction Engineering. 2nd ed. Englewood Cliffs, NJ: Prentice-Hall.

Ford RG, Scheinost AC, Scheckel KG, Sparks DL. 1999. The link between clay mineral weathering and the stabilization of Ni surface precipitates. Environ Sci Technol 33:3140–3144.

Froment GF, Bischoff KB. 1990. Chemical reactor analysis and design. New York: Wiley Interscience.

Fuller CC, Davis JA, Waychunas GA. 1993. Surface chemistry of ferrihydrite: Part 2. Kinetics of arsenate adsorption and coprecipitation. Geochim Cosmochim Acta 57:2271–2282.

Gregg SJ, Sing KSW. 1982. Adsorption, surface area and porosity. 2nd ed. Orlando, FL: Academic Pr Inc.

Hamdy AA, Gissel-Nielsen G. 1977. Fixation of selenium by clay and iron oxides. Z Pflanzeneraehr Bodenkd 140:63–70.

Hesterberg D. 1998. Biogeochemical cycles and processes leading to changes in mobility of chemicals in soils. Ecosyst Environ 67:121–133.

Hodges SC, Johnson GC. 1987. Kinetics of sulfate adsorption and desorption by Cecil soil using miscible displacement. Soil Sci Soc Am J 51:323–331.

Jenne EA. 1998. Adsorption of metals by geomedia: data analysis, modeling, controlling factors, and related issues, p.1. In: Jenne EA, editor. Adsorption of metals by geomedia. San Diego, CA: Academic Pr.

Kärger J, Ruthven DM. 1992. Diffusion in zeolites and other microporous solids. New York: J Wiley.

Lai CH, Lo SL, Chiang HL. 2000. Adsorption/desorption properties of copper ions on the surface of iron-coated sand using BET and EDAX analyses. Chemosphere 41:1249–1255.

Liang T, Tsai YJ. 1995. Sorption kinetics of cesium on natural modernite. Appl Radiat Isot 46:7.

Lordgooei M, Rood MJ, Rostam-Abadi M. 2001 Modeling effective diffusivity of volatile organic compounds in activated carbon fiber. Environ Sci Technol 35:613–619.

Manning BA, Fendorf SE, Goldberg S. 1998. Surface structures and stability of arsenic(III) on goethite, spectroscopic evidence for inner-sphere complexes. Environ Sci Technol 32:2383–2388.

Martinez CE, McBride MB. 1999. Dissolved and labile concentrations of Cd, Cu, Pb, and Zn in aged ferrihydrite-organic matter systems. Environ Sci Technol 33:745–750.

Meng XG, Letterman RD. 1993. Effect of component oxide interactions on the adsorption properties of mixed oxides. Environ Sci Technol 27:970–975.

Misak NZ, Ghoneimy HF, Morcos TN. 1996. Adsorption of Co^{2+} and Zn^{2+} ions on hydrous Fe(III), Sn(IV) and Fe(III)/Sn(IV) oxides. J Colloid Interface Sci 184:31–43.

Papelis C. 1995. X-ray photoelectron spectroscopic studies of cadmium and selenite adsorption on aluminum oxides. Environ Sci Technol 29:1526–1533.

Papelis C, Roberts PV, Leckie JO. 1995a. Modeling the rate of cadmium and selenite sorption on micro- and mesoporous transition aluminas. Environ Sci Technol 29:1099–1108.

Papelis C, Brown, GE Jr., Parks GA, Leckie JO. 1995b. X-ray absorption spectroscopic studies of cadmium and selenite adsorption on aluminum oxides. Langmuir 11:2041–2048.

Raven KP, Jain A, Loeppert HR. 1998. Arsenite and arsenate adsorption on ferrihydrite: kinetics, equilibrium, and adsorption envelopes. Environ Sci Technol 32:344–349.

Scheidegger AM, Lamble GM, Sparks DL. 1996. Investigation of Ni sorption on pyrophyllite: an XAFS study. Environ Sci Technol 30:548–554.

Scheidegger AM, Strawn DG, Lamble GM, Sparks DL. 1998. The kinetics of mixed Ni-Al hydroxide formation on clay and aluminum oxide minerals: A time-resolved study. Geochim Cosmochim Acta 62:2233–2245.

Scheinost AC, Abend S, Pandya KI, Sparks DL. 2001. Kinetics controls on Cu and Pb sorption by ferrihydrite. Environ Sci Technol 35:1090–1096.

Spadini L, Manceau A, Schindler PW, Charlet L. 1994. Structure and stability of Cd^{2+} surface complexes on ferric oxides 1. Results from EXAFS spectroscopy. J Colloid Interface Sci 168:73–86.

Sparks DL. 1998. Kinetics of sorption/release reactions on natural particles. In: Huang PM, Senesi N, Buffle J, editors. Structure and surface reactions of soil particles. New York: J Wiley p 413.

Sparks DL. 1995. Kinetics of metal sorption reactions. In: Allen HE, Huang CP, Bailey GW, Bowers AR editors. Metal speciation and contamination of soil. Boca Raton, FL: CRC Pr. p 35.

Sposito G. 1986. Distinguishing adsorption from surface precipitation. In: Geochemical processes on mineralogical surfaces. ACS Symp Ser 323:217.

Stahl RS, James RB. 1991. Zinc sorption by manganese-oxide-coated sand as a function of pH. Soil Sci Soc Am J 55:1291–1294.

Stern EA. 1976. The analysis of materials by X-ray absorption. Sci Am Appl 234:96–103.

Strauss R, Brümmer GW, Barrow NJ. 1997. Effects of crystallinity of goethite: I. Preparation and properties of goethites of differing crystallinity. Eur J Soil Sci 48:87–99.

Strawn DG, Scheidegger AM, Sparks DL. 1998. Kinetics and mechanisms of Pb(II) sorption and desorption at the aluminum oxide-water interface. Environ Sci Technol 32:2596–2601.

Theis TL, Iyer R, Ellis SK. 1992. Evaluating a new granular iron oxide for removing lead from drinking water. J Am Wat Works Assoc 84:101–105.

Traegner UK, Suidan MT. 1989. Evaluation of surface and film diffusion coefficients for carbon adsorption. Wat Res 23:267–273.

Trivedi P. 2001. Predicting thermodynamic and transport parameters for metal contaminant sorption to hydrated metal oxides in aquatic systems [PhD thesis]. New Jersey Institute of Technology, NJ.

Trivedi P, Axe L. 1999. A comparison of strontium sorption to hydrous aluminum, iron, and manganese oxides. J Colloid Interface Sci 218:554–563.

Trivedi P, Axe L. 2000. Modeling Cd and Zn sorption to hydrous metal oxides. Environ Sci Technol 34:2215–2223.

Trivedi P, Axe L. 2001a. Predicting metal sorption to hydrous Al, Fe, and Mn oxides. Environ Sci Technol 35:1779–1784.

Trivedi P, Axe L. 2001b. Ni and Zn Sorption to amorphous versus crystalline iron oxides: macroscopic studies. J Colloid Interface Sci 244:221–229.

Trivedi P, Axe L, Tyson TA. 2001. XAS studies of Ni and Zn sorbed to hydrous manganese oxide. Environ Sci Technol 35:4515–4521.

Waychunas GA, Rea, BA, Fuller CC, Davis JA. 1993. Surface chemistry of ferrihydrite: Part 1. EXAIS studies of the geometry of coprecipitated and adsorbed arsenate. Geochim Cosmochim Acta 57(10):2251–2269.

Wang T. 1995. Copper(II) and cadmium(II) removal and recovery by iron oxide-coated granular activated carbon [PhD thesis]. Illinois Institute of Technology, Chicago, IL.

5 Structural Dynamics of Metal Partitioning to Mineral Surfaces

Robert G. Ford

5.1 INTRODUCTION

Partitioning from solution to immobile solid phases in soils and sediments exerts control over the mobility and availability of inorganic contaminants. The extent of contaminant mobility will be influenced by the rates of sorption and desorption reactions. In general, ion exchange and adsorption reactions occur over relatively rapid timescales (milliseconds to days), and one might conclude that these reactions dominate partitioning in soils and sediments. However, efforts to model long-term fate and transport of contaminants are applied over much longer periods of time, more consistent with the timescale of fluid transport in natural systems (days to years). Over this timescale, solid-phase structural transformations that significantly affect the partitioning process, such as mineral nucleation and growth, can occur. The following quote excerpted from Grundl and Sparks (1998) concisely captures the significance of this issue:

> When viewed in this context, minerals should not be considered as passive solids, or even as simple sources of a reactive surface but must be considered as *bulk reactants* (emphasis added).

Solid-phase transformation processes in soils have generally been studied under the scientific field of soil genesis. Although the processes that control mineral precipitation–dissolution and, in part, soil development are a critical component of this scientific field, they are generally not evaluated under the context of contaminant partitioning. One reason for this is the difficulty of coupling the dynamics of microscopic surface phenomena to larger-scale transport processes within a quantitative analytical framework. However, the ability to reliably conceptualize and model contaminant partitioning is currently limited by chemical partitioning data primarily derived from short-term experimental systems. An assumption inherent in data interpretation for such systems is that the solid phase is a static component within the system. The lack of a basic knowledge on the types of partitioning processes and relative timescales that are active within the context of contaminant fate and transport could lead to a potentially flawed conceptualization of the system under study.

The purpose of this chapter is to highlight solid-phase transformation processes that may affect contaminant partitioning. The discussion will attempt to

1) describe relevant solid-phase transformations that may occur at the surface or within the bulk structure of soil minerals,
2) provide an overview of process rates observed in experimental systems,
3) evaluate the importance of these "slow" processes, and
4) identify current knowledge gaps that limit progress in modeling trace element fate and transport in subsurface systems.

5.2 ION PARTITIONING IN UNSATURATED AND SATURATED SOILS

In this chapter, the term "soil" will be used in a broad sense to include unconsolidated material that exists within the unsaturated and saturated zones below the ground surface. This includes unconsolidated material that exists below the plant root zone. This inclusive definition is used because of the observation of active microbial communities at depths below the root zone, and the importance of groundwater as a source of potable water in many areas of the world. In general, the chemical mechanisms controlling element partitioning are similar at all depths. However, the relative distribution and specific characteristics of the dominant processes will be influenced by climatic, hydrologic, and microbial factors that may differ between the unsaturated and saturated zones.

An underlying assumption for this review is that the soil particle is continuously bathed by an aqueous solution. This is critical for discussion of process timescales, because changes in soil moisture can influence the progress and rate of a given process. Process rates derived herein are only relevant for periodically saturated soils if the period of saturation is greater than the overall time frame of the process.

5.3 PARTITIONING PROCESSES

The partitioning of dissolved components to a solid surface is governed by the competition between sorption and desorption reactions. Sorption of a dissolved component (sorbate) to a solid surface (sorbent) may take place via several mechanisms. These mechanisms include adsorption via ion exchange or formation of a chemical complex with surface functional groups and the formation of surface precipitates. For this review, surface precipitation includes all processes that result in incorporation of the sorbate into a solid structure that could not form in the absence of the sorbent.

5.3.1 Conceptual Model of Sorbent Dynamics

The conventional representation of a mineral surface is shown in Figure 5.1a, where the sorbent is treated as a static (time-invariant) system with a complex population of surface ligating moieties. For this system, sorbate attachment to the surface is a

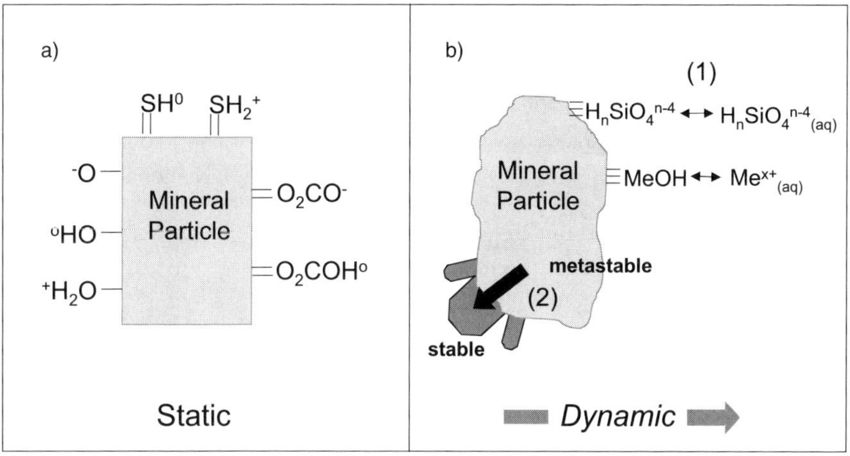

FIGURE 5.1 Conceptual models depicting mineral surface control of ion partitioning in soils. a) Static system in which the mineral surface (sorbent) has a finite, time-invariant population of surface sites. b) Dynamic system in which mineral structural transformations may alter 1) the surface site population and 2) chemical composition of the bathing solution.

function of the affinity and total population of sorption sites. The solid surface may be populated by a complex array of sorption sites, and characterization of the surface site chemical speciation can be a complex undertaking (see Davis and Kent 1990). An important assumption of this representation from a modeling perspective is the presence of a finite, time-invariant concentration of surface sites and, therefore, a finite sorption capacity. However, soils and sediments rarely achieve a state of equilibrium; hence, the minerals that constitute a given reaction volume will continue to evolve, potentially invalidating this assumption.

In contrast, Figure 5.1b depicts a dynamic system in which the mineral surface is subject to dissolution–precipitation cycles that lead to release of structural components to solution and/or reaccumulation of structural components into new mineral structures. The driving force for these reactions is the minimization of free energy at the mineral surface and/or within the bulk solid structure to reach equilibrium with respect to the bathing solution. Release of structural components at the mineral–water interface may contribute to incorporation of the sorbate into trace precipitate phases. From a macroscopic sense, the system may remain undersaturated with respect to precipitate formation. It is hypothesized that a finite volume adjacent to the mineral surface may exist that is oversaturated with respect to precipitate formation (O'Day et al. 1996; Towle et al. 1997; Ford et al. 2001). Evidence to confirm this mechanism is equivocal because of the limitations for analyzing reactions at the solid–water interface under conditions representative of the soil environment. The presence of the mineral surface may also enhance the formation of a new precipitate phase by lowering the energy barrier to nucleation of a new solid phase (Stumm 1992). Examples of the various types of surface precipitate phenomena

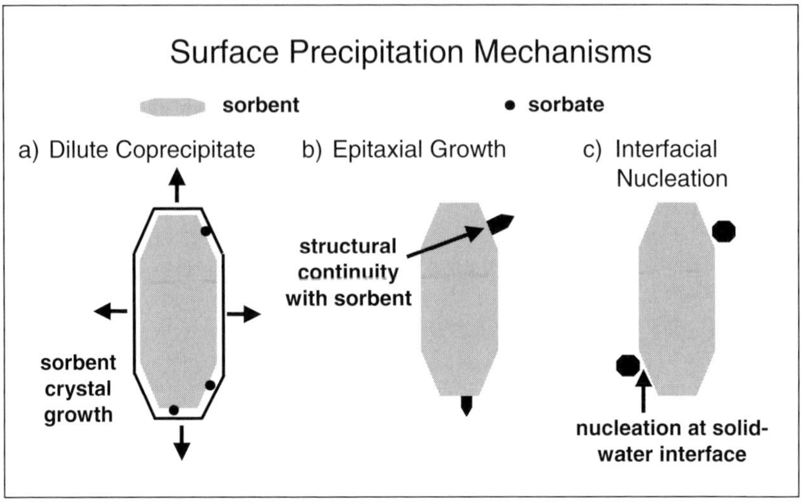

FIGURE 5.2 Conceptual model depicting possible mechanisms resulting in formation of surface precipitates incorporating the sorbate. a) Sorbate attachment occurs during the crystal growth of the sorbent. b) Sorbate attachment results in nucleation and growth of a new precipitate phase that is structurally attached to the sorbent surface. c) Nucleation of a new precipitate phase incorporating the sorbate and other soluble components at the mineral–water interface.

shown in Figure 5.2 have been documented in the literature. These include the following

1) formation of a dilute solid solution or coprecipitate with a structure that is distinct from or shared with the original sorbent,
2) epitaxial growth of a new solid phase, sharing a structural relationship with the sorbent, and
3) nucleation of a new solid phase at the solid–water interface.

5.3.2 Dilute Solid Solution

Two mechanisms resulting in the formation of a dilute solid solution have been documented in the literature. The first mechanism involves the incorporation of a sorbate concurrent with crystal growth, during which the structure of the sorbent remains essentially unchanged. The second mechanism involves incorporation of a sorbate during transformation of the sorbent from a metastable to a more thermodynamically stable phase. The dominant mechanism will depend on the stability of the sorbent with respect to the existing geochemical conditions at the time of sorption.

An example of the first solid-solution mechanism has been documented by Hazemann et al. (1992). Using polarized x-ray absorption spectroscopy, these authors characterized the structural environment of ferric iron incorporated into the structure of a natural aluminum oxyhydroxide, diaspore. Ferric iron was incorporated into the diaspore structure as polynuclear clusters possessing short-range structural order consistent with hematite. The authors proposed that the polynuclear iron fragment sorbed onto the diaspore particle via coordination with surface hydroxyl groups. The sorbate was eventually incorporated into the bulk structure during subsequent growth of the diaspore particle. Growth of diaspore leading to structural incorporation of the polynuclear iron surface species is depicted in Figure 5.3a. Although the mechanism for solid-solution formation was inferred based on structural considerations, it is plausible considering the reported stability of inner-sphere surface complexes. Given the disparity in the fundamental structure of hematite and diaspore, the formation of a solid solution via this mechanism is likely to be limited to a low sorbate surface coverage. The continued propagation of the sorbent structure would eventually be disrupted beyond some critical sorbate surface coverage.

An alternative route for solid-solution formation occurs when a sorbate is transferred to a new sorbent phase during transformation of a poorly crystalline sorbent to a more crystalline or stable phase. A model for this type of system is the modification in inorganic contaminant partitioning that may accompany transformation of the poorly crystalline iron (hydr)oxide, ferrihydrite, to more crystalline forms such as goethite or hematite (Ainsworth et al. 1994; Ford et al. 1997, 1999). During transformation, inorganic contaminants initially partitioned to the metastable sorbent phase may be structurally incorporated into the more crystalline and stable mineral. Substitution of manganese into iron structural sites during transformation of ferrihydrite into goethite is depicted in Figure 5.3b. Note that the initial association of manganese with ferrihydrite could plausibly occur through surface adsorption or solid solution. Numerous examples have been provided for incorporation of trace metals into the structures of goethite and hematite derived from ferrihydrite transformation under a range of experimental conditions (Cornell and Schwertmann 1996). The characterization of soil goethite and hematite by various techniques supports the feasibility of this process in nature (Bernstein and Waychunas 1987; Schwertmann and Pfab 1996; Manceau et al. 2000). Hence, although *in situ* observation of metal substitution during ferrihydrite-to-goethite/hematite transformation has not been demonstrated for soils or sediments, experimental studies confirm the chemical basis for this process.

The incorporation of sorbed metals during mineral transformation is limited by the structural compatibility between the sorbate and sorbent. Experimental evidence suggests that ions that are too large (Ford et al. 1999) or whose coordination environment is significantly distorted relative to that of the host structural sites (Martinez and McBride 2000) may be excluded during crystallization. This behavior has been indirectly observed for Pb partitioning to iron (hydr)oxide-rich sediments during diagenesis (Koschinsky and Halbach 1995). Thus, there are also situations in which mineral transformation may result in contaminant desorption.

5.3.3 Neoformation of Surface Precipitates

Charlet and Manceau (1994) applied the term "neoformation" to describe surface precipitate formation during dissolution of the sorbent. These neoformed phases are composed of the sorbate and components derived from the sorbent structure. This type of surface precipitate is considered a product of epitaxial growth from the sorbent surface, provided there is structural continuity between the sorbent and

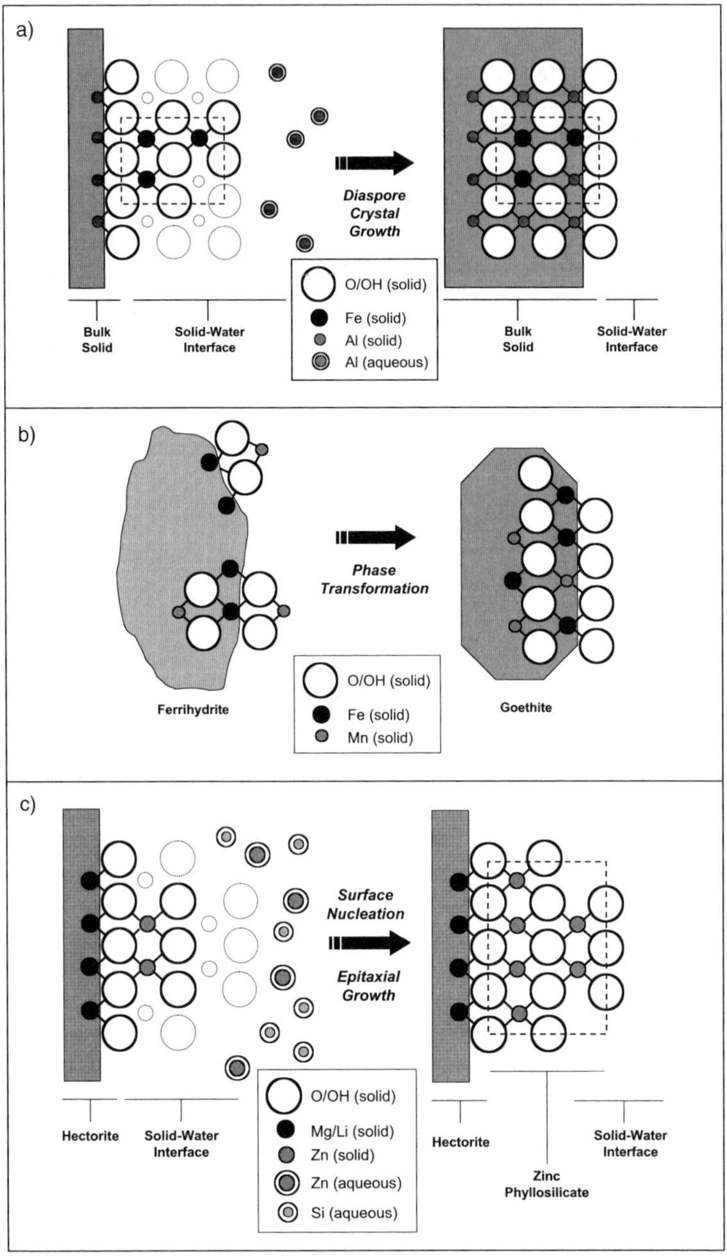

FIGURE 5.3 Two-dimensional representations demonstrating solid-solution formation and epitaxial growth for sorbed metals. a) Formation of Fe solid solution in diaspore according to Hazemann et al. (1992). Dashed rectangle highlights polynuclear Fe sorbate. b) Formation of Mn solid solution in goethite according to Ford et al. (1997). c) Formation of epitaxial Zn phyllosilicate on hectorite according to Schlegel et al. (2001). Dashed rectangle highlights newly formed Zn phyllosilicate. The silica tetrahedral layer is not shown for simplicity.

surface precipitate. Otherwise, surface precipitate formation is primarily a result of surface-facilitated nucleation of a new phase (Stumm 1992). Examples of both types of surface precipitation phenomena are documented in the literature.

5.3.3.1 Epitaxial Growth

Recent studies provide examples in which the surface-complex structure has been more closely examined with respect to the mineral surface structure (e.g., Chiarello and Sturchio 1994; O'Day et al. 1996, Schlegel et al. 2001). Schlegel et al. (2001) documented the molecular structure of Zn sorbed to the clay mineral hectorite. The molecular structure of Zn sorbed at the hectorite surface approached that of a phyllosilicate concurrent with Si release from the hectorite structure. Homogeneous precipitation of a Zn-phyllosilicate was not observed for a solution with similar total Zn and Si concentrations in the absence of hectorite. Polarized x-ray absorption spectroscopy confirmed structural continuity between the Zn surface precipitates and the hectorite surface. These experimental observations provide direct evidence for epitaxial growth of a surface precipitate under ambient conditions. The process of surface nucleation on hectorite and epitaxial growth of a zinc phyllosilicate is illustrated in Figure 5.3c. Evidence for epitaxial attachment of a Co polymeric phase on the rutile surface has also been documented (O'Day et al. 1996). In this case, the Co polymeric structure was more consistent with that of anatase, a polymorph of rutile. This suggests that structural disparities between the sorbent and surface precipitate can be accommodated during epitaxial growth.

5.3.3.2 Interfacial Nucleation

Sorption of metals onto the clay mineral, pyrophyllite, provides a model for nucleation of a surface precipitate at the mineral–water interface without structural attachment to the sorbent (depicted in Figure 5.2c). Pyrophyllite is not stable under most soil environments and will thus weather, dependent on solution conditions, into more stable clay structures (Zelazny and White 1989). Several studies have demonstrated that the sorption of Ni during aging in pyrophyllite suspensions resulted in the formation of mixed Ni-Al hydroxides (Scheidegger et al. 1998; Scheinost et al. 1999; Scheinost and Sparks 2000, and references therein). In these experimental systems, Al was derived from dissolution of the pyrophyllite structure under ambient conditions. Subsequent research has shown that long-term release of silica from pyrophyllite contributes to the transformation of the mixed Ni-Al hydroxide to a precipitate that shares structural and thermal properties common to phyllosilicates (Ford et al. 1999; Scheckel and Sparks 2000; Ford et al. 2001). Although pyrophyllite is not a commonly occurring mineral in soil environments, this system provides a model for soil/sediment systems that are partially composed of primary minerals undergoing weathering. For these systems, there is a net flux of ions derived from the primary mineral structure, which may contribute to sequestration of inorganic contaminants into trace precipitates. It should be noted that nucleation of the new solid phase may essentially proceed as a homogeneous process. However, the potential for this process is dictated by the presence of the sorbent phase and its solubility relative to the bathing solution.

5.4 RELEVANT PROCESS RATES

The evolution of mineral assemblages in soils is governed by the dynamics of pore-water chemistry from external changes to the infiltrate source (e.g., meteoric water, surface overflow, and anthropogenic inputs) and the internal dynamics of subsurface microbial communities. Processes that modify the macroscopic composition of a soil or aquifer sediment can significantly influence sorption patterns of contaminants through modification of physicochemical properties of sorbents. For example, the stability of a sorption complex formed between an inorganic contaminant and an iron (hydr)oxide depends on the ultimate stability of the sorbent. If the iron (hydr)oxide is in a metastable structural form, then the inorganic contaminant may be transferred to a more stable state during conversion to a more thermodynamically stable structure. Conversely, the sorption complex may be short-lived if changes in pore-water chemistry result in the dissolution of the metastable iron (hydr)oxide. The challenge faced in making predictions of the long-term partitioning of the inorganic contaminant involves coupling the dynamics of inorganic contaminant partitioning reactions and the overall dynamics of subsurface mineral dissolution–precipitation reactions occurring during evolution of the soil (Chadwick and Chorover 2001).

Significant effort has been made to understand the rates of ion uptake on mineral surfaces, as reflected in a number of reviews (e.g., Sparks and Grundl 1998; Sparks 1999). However, less emphasis has been placed on understanding the stability of the relevant partitioning processes with respect to potential modifications to the subsurface mineral assemblage due to abiotic and/or biotic modifications to system chemistry. The microbial cycling of redox-sensitive components drives dissolution–precipitation reactions that, in part, govern the stability and transformation of mineral components in soils. Changes in redox chemistry are particularly important for mineral components wholly or partially composed of redox-reactive elements such as carbon, iron, and sulfur. Because the evolution of trace element partitioning is intimately linked to structural changes in the soil mineral assemblage, it is important to develop a general understanding of the rates of major element cycling relative to the rates of partitioning reactions previously proposed.

5.4.1 REDOX TRANSFORMATIONS INFLUENCING MINERALOGY

A partial summary of published empirical rate constants and equivalent process half-lives for major element cycling is provided in Figure 5.4. This compilation consists of reported or derived pseudo-first-order rates for sulfate reduction, sulfide oxidation, organic matter oxidation, and iron oxidation or reduction. Preference was given to *in situ* rates derived from soil or sediment studies, but rates derived for mineral suspensions or in homogeneous aqueous systems have also been included to provide a comprehensive range of observed values. In most cases, redox transformation involved direct modification of a solid-phase component.

The majority of transformation half-lives reported in Figure 5.4 for redox transitions involving carbon, iron, or sulfur occur within a range of time from hours to a year. The slower rates reported for natural organic matter degradation are attributed

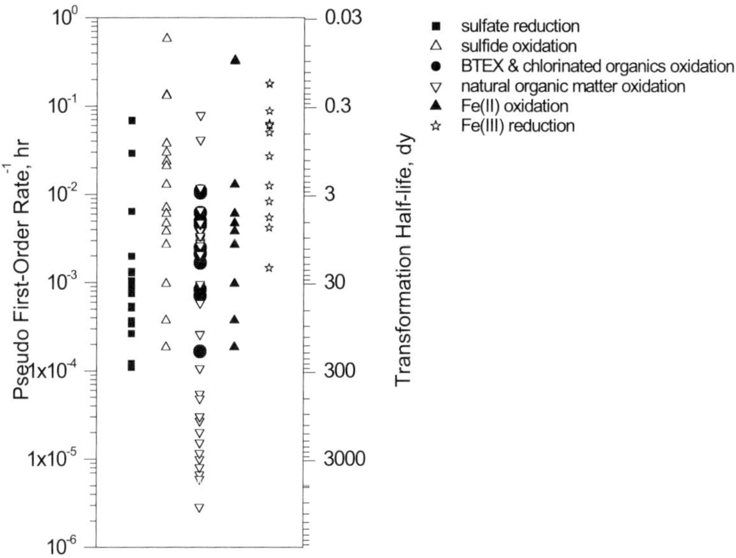

FIGURE 5.4 Relative timescales of major element oxidation-reduction reactions. Transformation half-life was calculated from published or derived pseudo-first-order rate constants. Following is a list of sources for compound-specific rates: 1) sulfate reduction (King 1983; Westrich and Berner 1984; Bak et al. 1991), 2) sulfide oxidation (Goldhaber 1983; Hassan 1994; Evangelou et al. 1998), 3) BTEX and chlorinated organics oxidation (Suarez and Rifai 1999), 4) natural organic matter oxidation (Westrich and Berner 1984; Riffaldi et al. 1996; Katterer et al. 1998), 5) Fe(II) oxidation (Goldhaber 1983; Ahmad and Nye 1990; Evangelou et al. 1998), and 6) Fe(III) reduction (dos Santos Afonso and Stumm 1992; Kukkadapu et al. 2001; Liu et al. 2001; Cocozza et al. 2002).

to the organic matter fraction in soils and sediments that displays low reactivity (Westrich and Berner 1984; Katterer et al. 1998). These redox transformations play an important role in the evolution of subsurface systems during fluid flow through porous media. They may exert a direct influence on mineral assemblages through alteration of redox-sensitive components such as Fe-bearing minerals. They may also exert an indirect influence through alteration of system pH or the production/removal of organic and inorganic aqueous ligands that influence the stability of soil minerals.

5.5 INFLUENCE ON FATE AND TRANSPORT

As discussed by Morgan and Stone (1985), reliable prediction of the fate and transport of trace elements in aqueous systems is dependent on knowledge of the relevant rates of fluid transport and chemical reactions governing element speciation. In this review, an effort has been made to establish a link between the dynamic processes that govern the composition and structure of the soil–sediment mineral assemblage and potential changes in the stability of a partitioned element. Partitioning reactions such as ion exchange and adsorption may occur within a timescale of

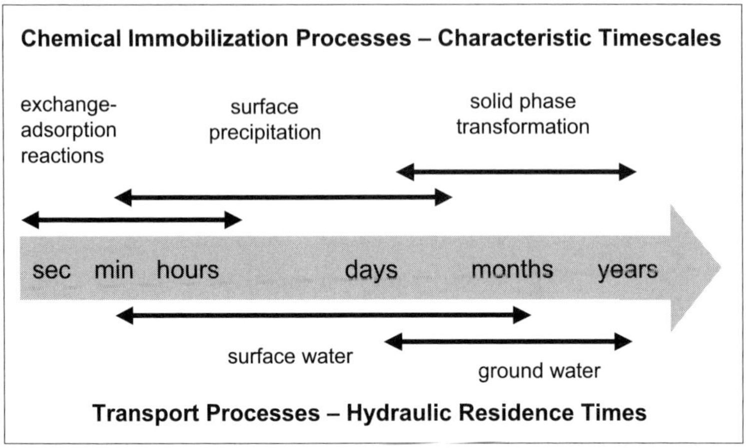

FIGURE 5.5 Relative timescales for physical transport and solid-phase partitioning processes. This diagram was developed from knowledge and concepts presented in the literature (Morgan and Stone 1985; McBride 1994; Stumm and Morgan 1996; Langmuir 1997).

milliseconds to hours (Sparks 1989). These reactions are likely to reach completion within the timescale required for significant alteration of the mineral assemblage. However, the formation, growth, and alteration of surface precipitates (as defined in this review) occur over a timescale that may be significantly impacted by alterations within the mineral assemblage.

These observations are represented in a qualitative manner in Figure 5.5. Characteristic timescales are shown for relevant processes that may lead to trace-element immobilization in soils and sediments relative to characteristic timescales for fluid transport in surface and groundwater systems. Documented timescales for surface precipitate formation range from minutes to weeks depending on sorbent identity and solution chemistry (O'Day et al. 1996; Towle et al. 1997; Scheidegger et al. 1998; Scheinost et al. 1999; Thompson et al. 1999; Thompson et al. 2000; Schlegel et al. 2001). Subsequent transformation of nucleated hydroxide surface precipitates into more stable phases with a phyllosilicate-like structure may take place in the order of weeks to years (Ford et al. 1999; Scheckel and Sparks 2000; Scheckel and Sparks 2001). This longer-term process is consistent with the observation of silication of Al-hydroxy interlayers in smectite (Lou and Huang 1993). Comparison of the characteristic timescales for surface precipitation, solid-phase transformation, and fluid transport indicate that information on the rates of potential chemical transformations is critical for the reliable prediction of the long-term mobility of trace elements.

5.6 DATA GAPS AND FUTURE DIRECTIONS

Although this review attempts to provide a rationale for considering solid-phase transformation reactions in the context of contaminant fate and transport, the practical application of this conceptualization of long-term contaminant partitioning is limited by a lack of reaction rate data for the relevant processes active in soils. Observations of the rates and structural changes in the sorbate structural environment are insufficient if the driving forces are not concurrently studied (e.g., modifications to the mineral assemblage). Disparities in the conceptualization of trace-element sorption reactions and the lack of solid-phase reaction rate data are current limitations to achieving the goal of modeling long-term trace element partitioning in soils and sediments. To address these limitations, several suggested areas for future research are highlighted in the following subsections.

5.6.1 Adsorption as a Reaction Intermediate to Precipitation

The conventional view that adsorption is a process distinct from precipitation (or crystal growth) deserves reevaluation. One of the major advances in the treatment of adsorption processes was the recognition that there are conceptual similarities between coordination reactions in solution and coordination of a sorbate to a surface functional group (Stumm 1992, and references therein). This provided the theoretical underpinning for treatment of surface coordination reactions as a result of electrostatic and chemical forces. This conceptualization has been generally accepted to the extent that we inherently view adsorption and crystal growth as distinct processes. To illustrate this point, consider the following simplified surface complexation reactions:

1. $=\text{Ni(II)-OH}^0_{(s)} + \text{NiOH}^+_{(aq)} \Leftrightarrow =\text{Ni(II)-O-Ni-OH}^0_{(s)} + \text{H}^+$
2. $=\text{Fe(III)-OH}^0_{(s)} + \text{NiOH}^+_{(aq)} \Leftrightarrow =\text{Fe(III)-O-Ni-OH}^0_{(s)} + \text{H}^+$

The first reaction depicts the addition of Ni hydrolysis species from solution to the surface of an existing nickel hydroxide, resulting in crystal growth. Throughout the process the uniformity of the crystal structure is maintained. In contrast, the second reaction depicts the addition of the same soluble Ni hydrolysis species to the surface of an existing ferric (hydr)oxide. This type of reaction is typically viewed as an adsorption reaction, because a discontinuity is introduced into the crystal structure. However, it is known that Fe and Ni are incorporated into stable mixed-cation mineral structures such as reevesite, $Ni_6Fe_2(CO_3)(OH)_{16} \cdot 4H_2O$, and trevorite, $NiFe_2O_4$ (Gaines et al. 1997).

This raises the question whether trace element adsorption onto a mineral surface might be considered simply as a form of crystal growth. From a mechanistic viewpoint, these are essentially identical reactions. Stumm (1992) has discussed precipitate nucleation and crystal growth within the context of surface complexation modeling. However, the literature suggests that the general approach, from a modeling perspective, is to view adsorption and crystal growth as 2 different phenomena. Although this may appear to be a semantic issue, this author feels that our conceptualization of "chemical" adsorption processes has allowed us to minimize the issue of the reversibility of sorbate addition to a mineral surface. A measured stability constant of large magnitude for the forward (sorption) reaction does not always indicate that the surface complex will have minimal reversibility. For example, Ford et al. (1999) demonstrated that even as the magnitude of the stability constant for sorption onto ferrihydrite is greater for Pb than for Ni, the reversibility (or lability) of the Pb sorption complex is also greater. The observed behavior for Pb is consistent both with a faster rate for water exchange in the Pb coordination sphere and the structural incompatibility of Pb with coordination sites in the crystalline iron (hydr)oxide structure. Nickel partitioning to ferrihydrite (and subsequent transformation products) may share more in common with a crystal growth process, whereas Pb partitioning is more consistent with a true adsorption process. Thus, structural considerations relative to the sorbent and sorbate appear to play an important role with respect to the energetics of solid-phase partitioning.

5.6.2 In Situ Rates of Mineral Transformation

It is recognized that precipitation of metastable phases in nature is often the first step in the formation of a more thermodynamically stable phase (Morse and Casey 1988; Steefel and Van Cappellen 1990). This observation applies to both bulk transformations in the soil or sediment mineral assemblage and surface precipitates that incorporate trace elements at the mineral–water interface. Both processes can significantly alter the reversibility of trace element partitioning. Thus, determination of the rates at which these mineral transformations occur within soils and sediments is critical toward constraining reaction-transport models for prediction of long-term partitioning processes controlling trace-element mobility. The following efforts would greatly facilitate filling this information gap

1) review and compilation of laboratory-derived mineral transformation rates for reactions involving phyllosilicates, oxyhydroxides, carbonates, and sulfides under relevant chemical conditions, and
2) comparison of laboratory- and field-derived rates coupled with identification of mechanistic factors associated with dissimilar rates (if differences exist).

The rate of solid-phase transformation is of particular importance under settings in which the chemical forces driving these reactions exceed the norm. Velde and Church (1999) provide an example of very rapid transformations (less than a few years) in the clay mineral assemblage within the sediments of a coastal salt marsh. The combination of reactive primary minerals and sharp changes in redox chemistry in this study mirrors the conditions one might anticipate in soils that are significantly affected by anthropogenic inputs. For example, dramatic chemical fronts may be generated during plume migration in contaminated groundwater. These types of chemical fronts are also observed during leachate migration from solid waste landfills (Christensen et al. 2001). Significant alterations in the subsurface mineral assemblage are anticipated as a result of plume migration and the return of the system geochemistry to background conditions. Precipitates that form as a result of perturbations to the system geochemistry will most likely be metastable and subject to Ostwald-type processes that lead to a stable mineral assemblage. These chemical changes will have a significant impact, occurring over a time frame consistent with fluid transport, on trace element partitioning.

5.6.3 Incorporation at Crystal Structure Defects

Recent experimental and field observations suggest the need to reevaluate our mechanistic understanding of crystal growth from nanophase materials (Penn and Banfield 1998; Banfield et al. 2000; Penn et al. 2001). This research demonstrates that the aggregation of nanometer-sized primary particles appears to be an important particle growth mechanism under ambient conditions. This has significant implications for trace element partitioning to nanophase precipitates such as ferrihydrite, hisengerite, and allophanc/imogolite, which are poorly crystalline with respect to characterization by x-ray diffraction (e.g., Henmi et al. 1980; Farmer 1992; Su and Harsh 1996; Schwertmann et al. 1999). These phases serve as efficient scavengers for trace elements and are precursors to more crystalline phases. Banfield et al. (2000) raise the possibility of incorporation of adsorbed ions as point defects during aggregation-based growth. The outcome of this process would be a significant change in the reversibility of ion partitioning without the rigid structural requirements for the formation of a true solid solution. If a significant population of structural defects accompanies aggregation-based crystal growth, structural changes may not be identifiable by x-ray diffraction because of structure-induced line broadening. This has implications for identification of structural vs. diffusional mechanisms resulting in decreased reversibility of trace element partitioning to metastable or poorly crystalline solids. Future research should be directed toward determining the rates and mechanisms of predominant microscale processes that generate crystalline or stable solids from poorly crystalline or metastable precursors.

In combination, these proposed research efforts suggest that a more comprehensive understanding of the link between trace element partitioning and mineral transformations at the microscopic level is required to provide the mechanistic basis for reaction-transport models. Identification of the relative rates of mineral transformations will aid model development by providing the basis to assess whether equilibrium or kinetic descriptions of ion partitioning are required with respect to fluid transport. This will involve an iterative effort incorporating both controlled laboratory investigations designed to capture reaction mechanisms and validation of this process data through application to field-scale studies. Ultimately, this approach will lead to scientifically defensible application of chemical models to describe trace element fate and transport in soils.

ACKNOWLEDGMENTS

I would like to thank John Wilson and Richard Wilkin for directing me to sources of rate data for redox reactions in subsurface systems. The U.S. Environmental Protection Agency through its Office of Research and Development partially funded and managed the research described here. It has not been subjected to agency review and therefore does not necessarily reflect the views of the agency, and no official endorsement should be inferred.

REFERENCES

Ahmad AR, Nye PH. 1990. Coupled diffusion and oxidation of ferrous iron in soils. I. Kinetics of oxygenation of ferrous iron in soil suspension. J Soil Sci 41:395–409.

Ainsworth CC, Pilon JL, Gassman PL, Van Der Sluys WG. 1994. Cobalt, cadmium, and lead sorption to hydrous iron oxide: residence time effect. Soil Sci Soc Am J 58:1615–1623.

Bak F, Scheff G, Jansen K-H. 1991. A rapid and sensitive ion chromatographic technique for the determination of sulfate and sulfate reduction rates in freshwater lake sediments. FEMS Microbiol Ecol 85:23–30.

Banfield JF, Welch SA, Zhang H, Ebert TT, Penn RL. 2000. Aggregation-based crystal growth and microstructure development in natural iron oxyhydroxide biomineralization products. Science 289:751–754.

Bernstein LR, Waychunas GA. 1987. Germanium crystal chemistry in hematite and goethite from the Apex Mine, Utah, and some new data on germanium in aqueous solution and in stottite. Geochim Cosmochim Acta 51:623–630.

Chadwick OA, Chorover J. 2001. The chemistry of pedogenic thresholds. Geoderma 100:321–353.

Charlet L, Manceau A. 1994. Evidence of the neoformation of clays upon sorption of Co(II) and Ni(II) on silicates. GeochimCosmochim Acta 58:2577–2582.

Chiarello RP, Sturchio NC. 1994. Epitaxial growth of otavite on calcite observed in situ by synchrotron X-ray scattering. GeochimCosmochim Acta 58:5633–5638.

Christensen TH, Kjeldsen P, Bjerg PL, Jensen DL, Christensen JB, Baun A, Albrechtsen H-J, Heron G. 2001. Biogeochemistry of landfill leachate plumes. Appl Geochem 16:659–718.

Cocozza C, Tsao CCG, Cheah S-F, Kraemer SM, Raymond KN, Miano TM, Sposito G. 2002. Temperature dependence of goethite dissolution promoted by trihydroxamate siderophores. Geochim Cosmochim Acta 66:431–438.

Cornell RM, Schwertmann U. 1996. The iron oxides: Structure, properties, reactions, occurrence and uses. New York: VCH.

Davis JA, Kent DA. 1990. Surface complexation modeling in aqueous geochemistry. In: Hochella MF Jr, White AF, editors. Mineral-water interface geochemistry. Washington, DC: Mineralogical Society of America. p 177–260.

Dos Santos Afonso M, Stumm W. 1992. Reductive dissolution of iron(III) (hydr)oxides by hydrogen sulfide. Langmuir 8:1671–1675.

Evangelou VP, Seta AK, Holt A. 1998. Potential role of bicarbonate during pyrite oxidation. Environ Sci Technol 32:2084–2091.

Farmer VC. 1992. Possible confusion between so-called ferrihydrites and hisingerites. Clay Minerals 27:373–378.

Ford RG, Bertsch PM, Farley KJ. 1997. Changes in transition and heavy metal partitioning during hydrous iron oxide aging. Environ Sci Technol 31:2028–2033.

Ford RG, Kemner KM, Bertsch PM. 1999a. Influence of sorbate-sorbent interactions on the crystallization kinetics of nickel- and lead-ferrihydrite coprecipitates. Geochim Cosmochim Acta 63:39–48.

Ford RG, Scheinost AC, Scheckel KG, Sparks DL. 1999b. The link between clay mineral weathering and the stabilization of Ni surface precipitates. Environ Sci Technol 33:3140–3144.

Ford RG, Scheinost AC, Sparks DL. 2001. Frontiers in metal sorption/precipitation mechanisms on soil mineral surfaces. Advan Agron 74:41–62.

Gaines RV, Skinner HCW, Foord EE, Mason B, Rosenzweig A. 1997. Dana's new mineralogy. J Wiley.

Goldhaber MB. 1983. Experimental study of metastable sulfur oxyanion formation during pyrite oxidation at pH 6-9 and 30°C. Am J Science 283:193–217.

Grundl TJ, Sparks DL. 1998. Kinetics and mechanisms of reactions at the mineral-water interface: An overview. In: Sparks DL, Grundl TJ, editors. Mineral-water interfacial reactions: kinetics and mechanisms, Vol. 715, Washington, DC: American Chemical Society. p 2–11.

Hassan SM. 1994. Sulfur speciation: methodology and application to sulfide oxidation studies at the sediment-water interface. Chemosphere 29:2555–2569.

Hazemann JL, Manceau A, Sainctavit P, Malgrange C. 1992. Structure of the α-FexAl$_{1-x}$OOH solid solution. I. Evidence by polarized EXAFS for an epitaxial growth of hematite-like clusters in Fe-diaspore. Phys Chem Min 19:25–38.

Henmi T, Wells C, Childs CW, Parfitt RL. 1980. Poorly-ordered iron-rich precipitates from springs and streams on andesitic volcanoes. Geochim Cosmochim Acta 44:365–372.

Katterer T, Reichstein M, Andren O, Lomander A. 1998. Temperature dependence of organic matter decomposition: a critical review using literature data analyzed with different models. BiolFert Soils 27:258–262.

King GM. 1983 Sulfate reduction in Georgia salt marsh soils: an evaluation of pyrite formation by use of ^{35}S and ^{55}Fe tracers. Limnol Oceanogr 28:987–995.

Koschinsky A, Halbach P. 1995. Sequential leaching of marine ferromanganese precipitates: genetic implications. Geochim Cosmochim Acta 59:5113–5132.

Kukkadapu RK, Zachara JM, Smith SC, Fredrickson JK, Liu C. 2001. Dissimilatory bacterial reduction of Al-substituted goethite in subsurface sediments. Geochim Cosmochim Acta 65:2913–2924.

Langmuir D. 1997 Aqueous environmental chemistry. New York: Prentice Hall.

Liu C, Kota S, Zachara JM, Fredrickson JK, Brinkman CK. 2001. Kinetic analysis of the bacterial reduction of goethite. Environ Sci Technol 32:2482–2490.

Lou G, Huang PM. 1993 Silication of hydroxy-Al interlayers in smectite. Clays Clay Min 41:38–44.

Manceau A, Schlegel ML, Sole VA, Gauthier C, Petit PE, Trolard F. 2000. Crystal chemistry of trace elements in natural and synthetic goethite. Geochim Cosmochim Acta 64:3643–3661.

Martinez CE, McBride MB. 2000. Aging of coprecipitated Cu in alumina: changes in structural location, chemical form, and solubility. Geochim Cosmochim Acta 64:1729–1736.

McBride MB. 1994. Environmental chemistry of soils. Oxford, UK: Oxford University Pr.

Morgan JJ, Stone AT. 1985. Kinetics of chemical processes of importance in lacustrine environments. In: Stumm W, editor. Chemical processes in lakes. New York: J Wiley. p 389–426.

Morse JW, Casey W. 1988. Ostwald processes and mineral paragenesis in sediments. Am J Science 288:537–560.

O'Day PA, Chisholm-Brause CJ, Towle SN, Parks GA, Brown GE Jr. 1996 X-ray absorption spectroscopy of Co(II) sorption complexes on quartz (α-SiO_2) and rutile (TiO_2). Geochim Cosmochim Acta 60:2515–2532.

Penn RL, Banfield JF. 1998. Imperfect oriented attachment: dislocation generation in defect-free nanocrystals. Science 281:969–971.

Penn RL, Oskam G, Strathmann TJ, Searson PC, Stone AT, Veblen DR. 2001. Epitaxial assembly in aged colloids. J Phys Chem B 105:2177–2182.

Riffaldi R, Saviozzi A, Levi-Minzi R. 1996. Carbon mineralization kinetics as influenced by soil properties. Biol Fert Soils 22:293–298.

Scheckel KG, Sparks DL. 2000. Kinetics of the formation and dissolution of Ni precipitates in a gibbsite/amorphous silica mixture. J Colloid Interface Sci 229:222–229.

Scheckel KG, Sparks DL. 2001. Temperature effects on nickel sorption kinetics at the mineral-water interface. Soil Sci Soc Am J 65:719–728.

Scheidegger AM, Strawn DG, Lamble GM, Sparks DL. 1998. The kinetics of mixed Ni-Al hydroxide formation on clays and aluminum oxides: a time-resolved XAFS study. Geochim Cosmochim Acta 62:2233–2245.

Scheinost AC, Ford RG, Sparks DL. 1999. The role of Al in the formation of secondary Ni precipitates on pyrophyllite, gibbsite, talc, and amorphous silica: a DRS study. Geochim Cosmochim Acta 63:3193–3203.

Scheinost AC, Sparks DL. 2000. Formation of layered single and double metal hydroxide precipitates at the mineral/water interface: a multiple-scattering XAFS analysis. J Colloid Interface Sci 223:167–178.

Schlegel ML, Manceau A, Charlet L, Chateigner D, Hazemann J-L. 2001. Sorption of metal ions on clay minerals. III. Nucleation and epitaxial growth of Zn phyllosilicates on the edges of hectorite. Geochim Cosmochim Acta 65:4155–4170.

Schwertmann U, Friedl J, Stanjek H. 1999. From Fe(III) ions to ferrihydrite and then to hematite. J Colloid Interface Sci 209:215–223.

Schwertmann U, Pfab G. 1996. Structural vanadium and chromium in lateritic iron oxides: genetic implications. Geochim Cosmochim Acta 60:4279–4283.

Sparks DL. 1989. Kinetics of soil chemical processes. San Diego: Academic Pr.

Sparks DL. 1999. Kinetics of reactions in pure and mixed systems. In: Sparks DL, editor. Soil physical chemistry. Boca Raton, FL: CRC Pr. p 83–178.

Sparks DL, Grundl TJ. 1998. Mineral-water interfacial reactions: kinetics and mechanisms. Washington, DC: American Chemical Society.

Steefel CI, Van Cappellen P. 1990. A new kinetic approach to modeling water-rock interaction: the role of nucleation, precursors, and Ostwald ripening. Geochim Cosmochim Acta 54:2657–2677.

Stumm W. 1992. Chemistry of the solid-water interface. New York: J Wiley.

Stumm W, Morgan JJ. 1996. Aquatic chemistry: chemical equilibria and rates in natural waters. New York: J Wiley.

Su C, Harsh JB. 1996. Alteration of imogolite, allophane, and acidic soil clays by chemical extractants. Soil Sci Soc Am J 60:77–85.

Suarez MP, Rifai HS. 1999. Biodegradation rates for fuel hydrocarbons and chlorinated solvents in groundwater. Bioremediat J 3:337–362.

Thompson HA, Parks GA, Brown Jr GE. 1999. Dynamic interactions of dissolution, surface adsorption, and precipitation in an aging cobalt(II)-clay-water system. Geochim Cosmochim Acta 63:1767–1779.

Thompson HA, Parks GA, Brown Jr GE. 2000. Formation and release of cobalt(II) sorption and precipitation products in aging kaolinite-water slurries. J Colloid Interface Sci 222:241–253.

Towle SN, Bargar JR, Brown GE, Parks GA. 1997. Surface precipitation of Co(II)(aq) on Al_2O_3. J Colloid Interface Sci 187:62–82.

Velde B, Church T. 1999. Rapid clay transformations in Delaware salt marshes. Appl Geochem 14:559–568.

Westrich JT, Berner RA. 1984. The role of sedimentary organic matter in bacterial sulfate reduction: the G model tested. LimnolOceanogr 29:36–249.

Zelazny LW, White GN. 1989. The pyrophyllite-talc group. In: Dixon JB, Weed SB, editors. Minerals in soil environments. Madison,WI:Soil Science Society of America.

6 Effects of Humic Substances on Attenuation of Metals: Bioavailability and Mobility in Soil

Christopher A. Impellitteri and Herbert E. Allen

6.1 INTRODUCTION

Soil organic matter plays a critical role in trace element behavior. It can decrease the mobility and bioavailability (hence, increase attenuation) of elements by sorption, chelation, and sequestration. Different forms of soil organic matter can also decrease trace element attenuation by soils, enhancing transport of elements into the solution phase or, by keeping them bound in solution, preventing sorption of elements onto the solid phase. The inclusion of organic substances in assessments of the environmental risk from contaminant trace elements is essential, though rarely done. This is due to a current lack of understanding of the structure of these molecules, and how this structure influences element behavior.

Perhaps no other group of substances in the environment is so ubiquitous, yet so poorly understood as organic matter. Bolin (1977) estimated that soil organic matter holds more carbon than all other surface carbon sinks (atmosphere, biomass, fresh and salt water) combined. Bohn (1976) estimated that decay of soil organic matter is the largest source of CO_2 to the atmosphere. The dynamic state of soil organic matter suggests that its effects on attenuation of trace elements may be transient. Hence, in order to understand the role organic matter plays in the attenuation of trace elements, it is necessary to understand not only the mechanisms of the interaction between the element and the organic components but also to consider the timescales for the turnover of the key organic components.

This chapter discusses the different components that constitute soil organic matter and focuses on the influence of organic matter on the natural attenuation of metal contaminants in soils.

Organic matter has been defined and redefined in the literature since the late 1700s (Stevenson 1994). Waksman (1936) suggested that all terms used to describe organic matter should be abandoned in favor of the all-encompassing term "humus." Schnitzer (1978) utilized the term humic substances to describe humic acids (HA), fulvic acids (FA), and humin. The most widely used definitions of HA, FA, and humin are based on their behavior in acidic and alkaline solutions. HA is soluble in alkaline media and insoluble in acidic solutions. FA is soluble in both, whereas humin is insoluble in both. The term "organic matter" may include carbon based detritus (leaves, twigs, animal wastes, etc.), humic substances, soot carbon, and simpler aliphatic structures (i.e., carbohydrates). Furthermore, some researchers may use the term "natural organic matter" to differentiate between humic substances and anthropogenic compounds. The terms "soil organic carbon" and "dissolved organic carbon" are often used to differentiate organic carbon in soils and solutions from inorganic carbon. The term "carbon" (as opposed to organic matter) is employed because it describes the way humic substances are measured (by carbon analyzer), often after they have been fractionated in some manner. The carbon content can be related to organic matter content (Ball 1964), as measured by oxidative procedures such as the Walkley–Black method (Walkley and Black 1934), but the terms "organic carbon" and "organic matter" are often used interchangeably. The use of these terms can lead to high degrees of confusion when attempting to compare results from different studies. For this chapter, we will use the term "humic substances" as an overall term that includes nonsoot, carbon-based compounds that have undergone the process of humification. Humification refers to the degradation of carbon-based molecules from plant, animal, and fungal tissues by biological and chemical processes. The humic substances discussed here include, but are not limited to, HA and FA.

6.2 HUMIC SUBSTANCES: DEFINITIONS AND STRUCTURE

More confusing than the words and phrases describing humic substances are the structures themselves. The science of describing humic substances can be broadly divided into 2 related paths: operational definitions and actual chemical structure. Operational definitions must be understood in order to compare and contrast studies involving trace element mobilization and attenuation by humic substances. Advances in elucidating the actual structures are important as they increase our understanding of the chemical and physical parameters that most greatly affect trace metal behavior in the environment.

Most research on humic substances invariably includes some manner of operational definition. Humic substances are frequently extracted from solids by solutions of NaOH (0.1 to 0.5 M). The substances are then characterized, based on their solubility as the extract is acidified. Stevenson (1982) noted that in addition to HA and FA, hymatomelanic acid may be defined as the alcohol soluble portion of HA, and humin as the alkali insoluble portion. The International Humic Substance Society recommends an extraction consisting of 10 mL 0.01 N NaOH/g soil (Swift 1996). Other researchers claim that increasing this ratio to 300:1 (mL 0.01 N NaOH:g C in soil) is more thorough (Kuwatsuka et al. 1992).

Another common method of characterization is separation of a solution or soil extract by resin. A good overview for resin separation procedures is given by Aiken and Leenheer (1993). Aiken et al. (1979) examined a series of nonionic, macroreticular XAD resins for the removal of fulvic acid from solution. They found that XAD 7 and XAD 8 performed best and that XAD 8 was superior, as XAD 7 exhibited bleed of C from the acrylic ester resin. Resin procedures have been used for concentration and purification of aquatic humic substances in water (Thurman and Malcolm 1981; Malcolm et al. 1994). Leenheer (1981) defined 6 separate classes of humic substances based on resin separation techniques; hydrophilic acids, bases and neutrals; and hydrophobic acids, bases and neutrals. Hydrophobic acids are retained on the resin (Leenheer and Huffman 1976) and eluted from the column using an alkaline solution (usually 0.1 N NaOH). Hydrophobic acids include HA and FA, whereas hydrophilic fractions are generally thought to contain simpler, more aliphatic type structures (Ma et al. 2001). Resins only isolate approximately 15 to 20% of the total DOC in seawater (Aiken and Leenheer 1993), but reverse phase liquid chromatography has been used for separation of humic substances in seawater (Mills et al. 1987; Zhou and Wangersky 1989). Strong and weak anion exchangers and activated carbon have also been used for separating humic substances but generally, desorption of compounds from the exchange media is problematic (Thurman 1985).

High performance size exclusion chromatography (HPSEC) has also been employed to differentiate between humic substances. Romkens and Dolfing (1998) used HPSEC to qualitatively differentiate a high molecular weight humic fraction from a low molecular weight fulvic fraction. Andersen et al. (2000) differentiated 5 different size fractions, ranging from 410 to 1170 daltons, from acidified and limed surface waters. Spectrophotometry has also been employed to quantify dissolved organic carbon in aqueous samples (Bennett and Drikas 1993; Stevenson 1994; Hongve and Akesson 1996; Hautala et al. 1999). Absorbance at different wavelengths has been used to characterize the nature of the dissolved organic carbon. Peuravuori and Pihlaja (1997) used the ratio of absorbance at 250 nm to that at 365 nm (E_2/E_3) as an estimate of the degree of aromaticity and molecular weight of aquatic humic substances. They observed that relative aromaticity of the aquatic substance increases with increasing molecular weight and that the E_2/E_3 ratio could be used as a first approximation for estimating the size and aromaticity of humic substances in bulk. Absorbance at 465 nm and 665 nm (E_4/E_6) has also been used to estimate *humification*, molecular weight, and condensation of aromatic carbon in humic substances (Chen et al. 1977; Stevenson 1982; Bloom and Leenheer 1989). E_4/E_6 ratios were used to estimate the degree of condensation of the aromatic groups in humic substances, where a low E_4/E_6 ratio indicated a high degree of condensation (Kononova 1966). Chen et al. (1977) reported that there was little evidence of a direct relationship between E_4/E_6 ratios and condensation of aromatic rings. They found that E_4/E_6 ratios best correlated with molecular size (inverse relationship) and % O_2 of the humic substances (positive relationship).

The elucidation of the actual chemical structures of humic substances has been difficult due to their complexity and the fact that they are structurally dynamic with changes in environmental master variables such as pH. Within a particular group (e.g., HA), these substances exhibit amazing similarities in bulk analyses regardless of the origin of the humic substance. For soils from a wide geographic distribution, Schnitzer (1978) reported that for HA, % C ranged from 53.8 to 57.3, % O from 32.8 to 38.3, % H from 3.2 to 6.2, % N from 0.8 to 4.3, and % S from 0.1 to 1.5. For FAs, from the same soils, % C ranged from 40.7 to 50.6 and % O from 39.7 to 49.8. It was also stated that FAs generally contained more S but less H and N when compared with HAs (Schnitzer 1978). Gjessing (1976) gave slightly lower numbers for C and N concentrations in aquatic "humus" (mean % C = 43 and mean % N = 1.1) and a higher number for % H (mean = 5.5%).

The key to the influence of humic substances on trace element behavior lies in their functional groups. In general terms, both aliphatic and aromatic compounds play very important roles as functional groups in humic substances. The primary functional groups in humic substances that interact with metals consist of carboxylic acids and phenolic compounds. Specific chemical models for humic substance structures have been proposed (Schulten and Schnitzer 1995; Schulten and Schnitzer 1997; Kubicki and Apitz 1999), and scanning and transmission electron photomicrographs have been employed to characterize humic substances (Chen and Schnitzer 1976). These portrayals of humic substances vary, but the functional groups (illustrated in the chemical models) and the presence of void spaces provide evidence to support the ability of these substances to bind and sequester a wide variety of substances in the environment. It is probable that humic substances have both the ability to chemically bind contaminants as well as to physically trap them. Improvements in analytical methodologies, including advances in nuclear magnetic resonance spectroscopy (NMR) and pyrolysis chromatography-mass spectrometry (PCMS), are breaking new ground in elucidating humic substance structure, interaction with other compounds, and formation of humic substances. Hatcher et al. (2001) have reviewed advances in analytical methodologies for the examination of humic substances. It is interesting to note that in the 70s and early 80s, researchers found NMR of little value for elucidating humic substance structure (Schnitzer 1978; Stevenson 1982), but over the course of a decade NMR has emerged as the most widely used and accepted analytical method for the study of humic substance structure (Stevenson 1994). NMR analyses depend on isotopes that exhibit magnetic spin momentum. For analyses on humic substances, the 2 isotopes are ^{13}C and ^{1}H. Additionally, ^{113}Cd has been used in the study of metal complexation by humic substances (Li et al. 2001). These isotopes resonate at a certain frequency of electromagnetic radiation, ν, when samples are placed in magnetic fields. As the isotopes resonate, the resulting voltage change is recorded. A chemical shift is based on deviation or "shift" from a standard containing C or H, such as tetramethylsilane, $Si(CH_3)_4$. Essentially, a range of chemical shift values is diagnostic for certain structures (i.e., functional groups) in a humic substance.

NMR has been employed to examine changes in humic substances as a function of pH. Chien and Bleam (1998) examined pH dependent structural changes in humic acid and found that the majority of structural differences following alteration of pH are due to alteration of aliphatic structures. NMR analyses are also providing crucial information for solving the mysteries surrounding the formation of humic substances.

For example, Chen et al. (1989) observed the increase in carboxyl and phenolic signals with a concurrent decrease in carbohydrate signals during the maturation of compost from a manure source. A similar increase in aromaticity of the organic matter in compost from a municipal waste source was observed (Chefetz et al. 1998). Chefetz et al. (2000) examined the transformation of carbon-based molecules in piles of mushroom soil. They were able to observe changes from the top of the piles, where carbohydrate concentrations rapidly decreased, and toward the bottom, where increases in aromatic groups occurred. Haiber et al. (2001) examined the actual mechanisms involved in the formation of high molecular weight, highly aromatic compounds in HA and FA reference materials. They concluded that lignin-like molecules undergo a demethylation reaction with removal of side chains. The aromatic structures then undergo a reaggregation. Simpson et al. (2001) used NMR analyses to suggest that humic substances are really aggregates of relatively simple structures, rather than cross-linked macromolecules. NMR has also been used for characterizing airborne humic substances (Subbalakshmi et al. 2000). They found that many of the organic molecules trapped in filters were similar in structure to soil HA. Ma et al. (2001) employed NMR to examine the resin-based fractionation of aquatic humic substances. They found that the hydrophilic fraction (fraction passing through XAD 8 resin) consisted primarily of simpler aliphatic structures such as carbohydrates and that the FA fraction contained more aliphatic structures than the HA fraction. Though NMR is a valuable method for characterizing humic substances, it is important to point out that many of the studies performed using NMR still depend on a macroscopic fractionation procedure, such as separation by resin. Bulk characterization of humic substances in solution is widely accepted but structural data from operationally defined fractions may be less dependable. Peuravuori (2000) examined the precision of NMR analyses on resin-separated fractions from aquatic humic substances and found that structural analyses tended to depend heavily on the fractionation itself and were imprecise.

6.3 SOLID-PHASE ORGANIC SUBSTANCES

Organic matter in, or associated with, the solid phase may be broadly characterized as parent materials (litter, humus, soot), sorbed humic materials, or humin. Soluble humic substances originate from humus and litter in soil systems, though it is not clear which of these parent materials contributes more to the formation of humic substances (Kalbitz et al. 2000). The importance of soot-based carbon for attenuation of organic contaminants in soils has received much attention in recent years. Chiou et al. (2000) examined sorption of organic contaminants of varying polarity and found that a "charcoal-like substance," or high-surface-area carbonaceous material, helped to explain nonlinearity (due to rapid, high initial sorption) in the sorption isotherms of many of the substances studied. Naes et al. (1998) illustrated the importance of soot-type carbon on the attenuation of PAHs. Compared with research in organic contaminants, the relationships between inorganic substances and soot materials in soils, sediments, and waters need further study. Other researchers have proposed the existence of a hard, glassy carbon vs. soft carbon materials (Weber and Huang 1996; Huang et al. 1997; Leboeuf and Weber 1997). The role of solid-phase organic carbon forms (litter, humus, humin, and soot) on determining the type and characteristics of potentially soluble humic substances is poorly understood.

6.4 LEACHING OF SOLID-PHASE SOIL ORGANIC MATTER

Soluble humic substances form from parent materials containing solid-phase organic carbon. These soluble humic substances are mobile and may transport metals into solution from the solid phase, thus decreasing solid-phase attenuation of metals.

System pH is the most important single factor governing the leaching of solid-phase organic matter. Erich and Trusty (1997) found that the addition of lime to forest soils caused an average of 55% more C to be released into solution vs. nonlimed soils. They also concluded that lime did not affect the functional group composition of the leached humic substances. Karlik (1995) examined dissolved organic matter outflows in limed soils during pot, lysimetric, and field experiments. Outflows from limed soils averaged 44.8% more dissolved organic matter than outflows from nonlimed soils. Karlik (1995) also found differences in the types of humic substances leached from limed vs. nonlimed soils. The limed soils tended to leach higher amounts of HA with a higher overall carboxyl group content. Andersson et al. (2000) examined the effects of pH, Ca^{2+}, and temperature on the leaching of dissolved organic carbon from a mor humus parent material. Results showed that pH had the largest effect on the leaching of dissolved organic carbon during the first 36 d of the experiment. After that period, temperature had a greater effect, where the higher temperature (15 °C vs. 4 °C) plots exhibited greater leaching of dissolved organic carbon. The temperature effect was attributed to greater microbial activity. Lastly, they showed that the presence of Ca^{2+} in solution had a moderating effect on the leaching of dissolved organic carbon from the mor humus.

6.5 SORPTION OF DISSOLVED HUMIC SUBSTANCES

Once soluble humic substances form, they are free to react with surfaces in the system (minerals, colloids, etc.). Solid-phase sorption of soluble humic substances may increase attenuation of metals by occlusion of surface-bound exchangeable forms of metals, or may limit attenuation due to competition with metals for binding on the solid phase (see Section 6.8 on Ternary Complexation). Sorption of soluble humic substances is a function of the properties of the sorbent, system pH, ionic strength, and the presence of competing ions. Shen (1999) concluded that the sorption of dissolved organic matter on 6 natural soils was largely due to ligand exchange between dissolved organic matter and hydroxyl groups on mineral surfaces and reached a maximum at pH 4 to 5. Sorption of humic substances into interlayers of 2:1 clays has been demonstrated for montmorillonite (Theng et al. 1986) but not for illite (Jardine et al. 1989). Spark et al. (1997b) studied the sorption of a coal-derived HA on goethite, alumina, kaolinite, and silica. They found that sorption of HA related to the surface charge of the mineral, with positively charged surfaces sorbing the HA to a greater extent. They also observed that the pH where dispersion of the mineral-HA complex occurred, decreased with increasing ionic strength. Huang and Yang (1995) found that kaolin had a higher sorption capacity for HA compared with FA at pH = 5.5. Sorption of a muck FA on goethite and kaolinite was compared at 3 pH values (3.7, 6, and 8) (Namjesnik–Dejanovic et al. 2000). They found that goethite had a greater affinity for FA than kaolinite at all pH values and that greater sorption of FA occurred at lower pH. They also concluded that there existed a preferential adsorption of higher molecular weight components of the FA especially at higher pH. A study on the sorption of HA on montmorillonite revealed an increase in HA sorption with decreasing pH (Liu 1999). Kaiser et al. (2000) examined the sorption of dissolved organic matter to soils containing litter-derived solid phase organic carbon and soils with highly aromatic soot-derived C. They concluded that soils containing solid phase organic carbon derived from plant or microbe sources tended to decrease sorption of dissolved organic matter, whereas the soot carbon tended to increase sorption of dissolved organic matter.

6.6 METAL ATTENUATION BY SOLID-PHASE ORGANIC MATTER

Solid phases of humic materials, such as soil organic matter (SOM) and colloidal aggregates of humic substances can sorb or chelate metals. Metals sequestered in the structure of organic molecules may not readily desorb, suggesting that this may be a mechanism for attenuation. For example, Bunzl et al. (1976) showed that the degree of desorption of Pb^{2+}, Cu^{2+}, Cd^{2+}, Zn^{2+}, and Ca^{2+} from peat was much less than the amount sorbed. Desorption of metals from organic matter will, to a certain extent, be pH dependent as the main functional groups (carboxylic and phenolic) on SOM exhibit pH-dependent charge (Sparks 1995).

The role of solid-phase organic matter in metal attenuation in biosolids-amended soils has been a subject of great debate. Most researchers agree that biosolids-applied organic matter has a function in sorbing, sequestering, and rendering toxic metals immobile in biosolids-amended soils. However, the degree of significance is arguable, and some have questioned the validity of the USEPA's regulatory loading rates for biosolids on land (USEPA 1994). The "sludge time-bomb" hypothesis speculates that, as organic matter added in the biosolids is degraded and mineralized, a sudden release of toxic metals will occur in the system (McBride 1995; McBride 1998). Other researchers strongly disagree. The "sludge protection" hypothesis argues that the metals that accumulate from repeated applications of biosolids are associated mainly with inorganic compounds in the applied biosolids, and thus their bioavailability is not influenced by behavior of organic matter in the amendments (Corey et al. 1987; Chaney and Ryan 1993). Evidence for the case of Cd was presented by Li et al. (2001). They concluded that the increased retention of Cd on biosolids-amended soils resulted from inorganic constituents in the biosolids. There is also evidence that suggests neither of the hypotheses mentioned above are entirely correct. McGrath et al. (2000) examined changes in extractability of Cd and Zn in a long-term experiment in field plots. Overall, Cd and Zn extractability (0.1 M $CaCl_2$

extractant) was higher for the sludge-amended soils. Over a 23-y period after cessation of biosolids applications, the metal extractability fluctuated, but it never consistently increased or decreased. The researchers also noted that although organic matter degradation occurred at a high rate in the initial years of sludge application, approximately 15% of the sludge organic matter persisted in the soil 23 y after the end of biosolids application.

6.7 METAL SORPTION AND CHELATION BY SOLUBLE AND POTENTIALLY SOLUBLE HUMIC SUBSTANCES

The ongoing debate regarding metal attenuation by solid-phase organic matter is important, but perhaps the more significant process is metal mobilization by soluble organic compounds. This section will review mechanisms and reactions of humic substances and metals, with emphasis on humic substances that are involved in reversing metal attenuation by moving metals from solids into solution.

Functional groups in the humic substances have an affinity for metal ions in the following order (Charberek and Martell 1959; Stevenson 1994):

$$\underset{\text{enolate}}{-O-} > \underset{\text{amine}}{-NH_2} > \underset{\text{azo}}{-N{=}N-} > \underset{\text{ring N}}{=N} > \underset{\text{carboxylate}}{-COO^-} > \underset{\text{ether}}{-O-} > \underset{\text{carbonyl}}{C{=}O}$$

Carboxyl and phenolic groups are the most significant for metal binding (Sparks 1995).

Humic substances generally have the highest affinity for Al, Cu, Fe, Hg, and Pb. Table 6.1 shows various reactivity schemes from previous studies. Conditional stability constants for metal–humic substance complexes generally increase with increasing pH as organic functional groups become more negatively charged (due to deprotonation).

Metal–humic interactions may be generally predicted by "hard" and "soft" acid or base principals (Buffle and Stumm 1994). Essentially, hard acids (Group I elements — H, Li, Na, etc.) of high positive charge and small size react with hard bases (highly electronegative, low polarizability), and soft acids (smaller positive charge, large size Group III elements Cd, Cu, Hg, Pb, etc.) with soft bases (low electronegativity, high polarizability). For humic substances, hard sites include the carboxyl and phenolic groups and soft sites include N or S containing functional groups (Sparks 1995). It should be noted that even though hard or soft and acid or base principles suggest that many of the metals would preferentially react with N and S containing functional groups, carboxylic and phenolic groups dominate metal reactions in soils due to their sheer number. These groups (phenolic, carboxyls, quinonic, etc.) contribute 55 to 80% of the cation exchange capacity in soil organic matter (Broadbent and Bradford 1952). Lu and Allen (2002) used a 4-site model to describe Cu complexation by dissolved organic matter. They concluded that Cu complexation is dominated by phenolic sites in the pH range 5–9.

TABLE 6.1
Sorption and chelation reaction sequences of metals by humic substances

Sequence	Source
$Cu^{2+} > Ni^{2+} > Pb^{2+} > Co^{2+} > Ca^{2+} > Zn^{2+} > Mn^{2+} > Mg^{2+}$	Chelation by soil organic matter (McBride 1994)
$Cu^{2+} > Ni^{2+} > Co^{2+} > Ca^{2+} > Zn^{2+} > Fe^{2+} > Mn^{2+}$	Metal-chelate complexes (Stevenson 1994)
$Hg^{2+} > Fe^{3+} > Pb^{2+} > Cu^{2+} = Al^{3+} > Ni^{2+} > Cr^{2+} = Zn^{2+} = Cd^{2+} = Co^{2+} = Mn^{2+}$ at pH = 2.4	Humic acids (Schnitzer 2000)
$Hg^{2+} > Fe^{3+} > Al^{3+} > Pb^{2+} > Cu^{2+} > Cr^{2+} > Cd^{2+} = Zn^{2+} = Ni^{2+} = Co^{2+} = Mn^{2+}$ at pH = 3.7	
$Hg^{2+} = Fe^{3+} = Al^{3+} = Pb^{2+} = Cu^{2+} = Cr^{2+} > Cd^{2+} > Zn^{2+} = Ni^{2+} > Co^{2+} > Mn^{2+}$ at pH = 4.7 and pH = 5.8	
$Pb^{2+} > Cu^{2+} > Cd^{2+}$	Humic acids (Liu and Gonzalez 2000)
$Cu^{2+} > Fe^{2+} > Pb^{2+} > Ni^{2+} > Co^{2+} > Ca^{2+} > Cd^{2+} > Zn^{2+} > Mn^{2+} > Mg^{2+}$	Humic acids (Pandey et al. 2000)
$Cu^{2+} > Cd^{2+} > Ni^{2+} > Zn^{2+} > Ca^{2+}$ for whole fulvic acid $Cu^{2+} > Cd^{2+} > Ca^{2+} > Ni^{2+} > Zn^{2+}$ for metal-binding subfraction	Fulvic acids (Brown et al. 1999)
$Fe^{3+} > Al^{3+} > Cu^{2+} > Ni^{2+} > Co^{2+} > Pb^{2+} > Ca^{2+} > Zn^{2+} > Mn^{2+} > Mg^{2+}$ at pH = 3.0	Fulvic acids (Schnitzer and Hansen 1970)
$Fe^{3+} > Al^{3+} > Ni^{2+} \approx Co^{2+} > Cu^{2+} > Pb^{2+} > Ca^{2+} > Zn^{2+} > Mn^{2+} > Mg^{2+}$ at pH = 5.0	

6.8 TERNARY COMPLEXATION

In previous sections, we discussed sorption of metals by humic substances, sorption by solid phase organic matter, and sorption of organic matter on solid-phases. Though convenient to treat these phenomena separately, there exists ample evidence that these sorption reactions occur simultaneously. These complex reactions can result in the formation of ternary complexes consisting of mineral-organic-metal constituents. Much needs to be learned about these complexes and their effects on trace element attenuation. The summaries of studies that follow describe the importance of these complexes on trace metal attenuation.

Christl and Kretzschmar (2001) examined the interactions between FA, Cu, and hematite. The presence of FA increased sorption of Cu on hematite below pH 6, and decreased sorption of Cu above pH 6 due to the formation of dissolved metal-organic complexes. They concluded that interactions between organic matter and metals must be taken into account when modeling sorption of metals at mineral surfaces. Similar effects were observed for Cd sorption in a hematite-HA system (Vermeer et al. 1999). The researchers concluded that, for the ternary system studied, the process of simply adding the sorption data for Cd from each isolated component for estimation of total Cd sorption in a ternary system was invalid. If a metal ion preferentially sorbs to the humic substance, sorption of the metal to the ternary complex will be overestimated. Conversely, if the conditions are such that the metal has a higher affinity for the oxide, overall sorption to the complex will be underestimated. This nonadditivity phenomenon was also observed for a Cu-goethite-HA system (Robertson and Leckie 1996). In contrast, Du et al. (1999) found that Cu sorption in an illite-FA system could be simulated by addition of data from 2 single-complex systems. They also suggested that FA could act to interfere with Cu sorption on illite by forming soluble complexes.

Spark et al. (1997a) concluded that metal sorption depended on the extent of sorption of the humic substance by the mineral and the solubility of the metal-humic complex. They studied the sorption of Cu, Zn, Co, and Cd on goethite, alumina, silica, and kaolinite, and found enhanced sorption of the metals on goethite and silica in the presence of HA. Metal sorption was not enhanced on the alumina and kaolinite surfaces in the presence of HA, and the authors concluded that the lack of enhancement resulted from competition for binding sites on the minerals between HA and the metals. For montmorillonite, metal sorption did not correlate with the concentration of sorbed HA (Liu 1999). The researchers also found no significant differences in the sorption of Pb or Cd when the montmorillonite was preequilibrated with the HA vs. simultaneous equilibration of metal-HA-montmorillonite. There was, however, a slight increase in Cu sorption in the coadsorption process. The authors proposed a bridging mechanism for the ternary complexes where the metal served to connect the HA to the montmorillonite surface. Huang and Yang (1995) examined the sorption of Cu on "synthetic" soils where kaolin was preequilibrated with FA or HA. They found that the HA-kaolin complex had a higher affinity for Cu. Evangelou et al. (1999) studied sorption of Ca, Cd, and Cu on illite-humic substances. The humic substances were fractionated, based on size. The relative stability of the metal–illite–humic complexes followed the order Cu > Cd > Ca for high affinity sites in the complex. They also found that the larger the humic fraction, the lower the metal-illite-humic stability constants for the more numerous (relative to the high affinity sites) low affinity sites. The authors stated that the low molecular weight fractions contained the most functional groups. Rocha et al. (1999) also observed Cu preferentially complexed by lower molecular weight aquatic humic substances, as did Han and Thompson (1999) for humic substances isolated from biosolids.

6.9 EFFECT OF HUMIC SUBSTANCES ON THE SOLID PHASE AND SOLUTION PHASE DISTRIBUTION OF METALS

The pH of a particular system (soil, surface water, groundwater, etc.) acts as a master variable, controlling the mobility, speciation, and binding of a particular metal (Sauvé et al. 2000). At lower pH values, proton competition and less negative surface potentials tend to inhibit binding of metals by solids (Yin et al. 2002). Thus as pH increases, an increase in the association of metals with solids generally occurs. However, in some instances, an increase in system pH has been observed to result in increased solution concentrations of the metal. An example of this effect is shown in Figure 6.1 where Cu (Figure 6.1b) and Pb (Figure 6.1d) solubility was found to

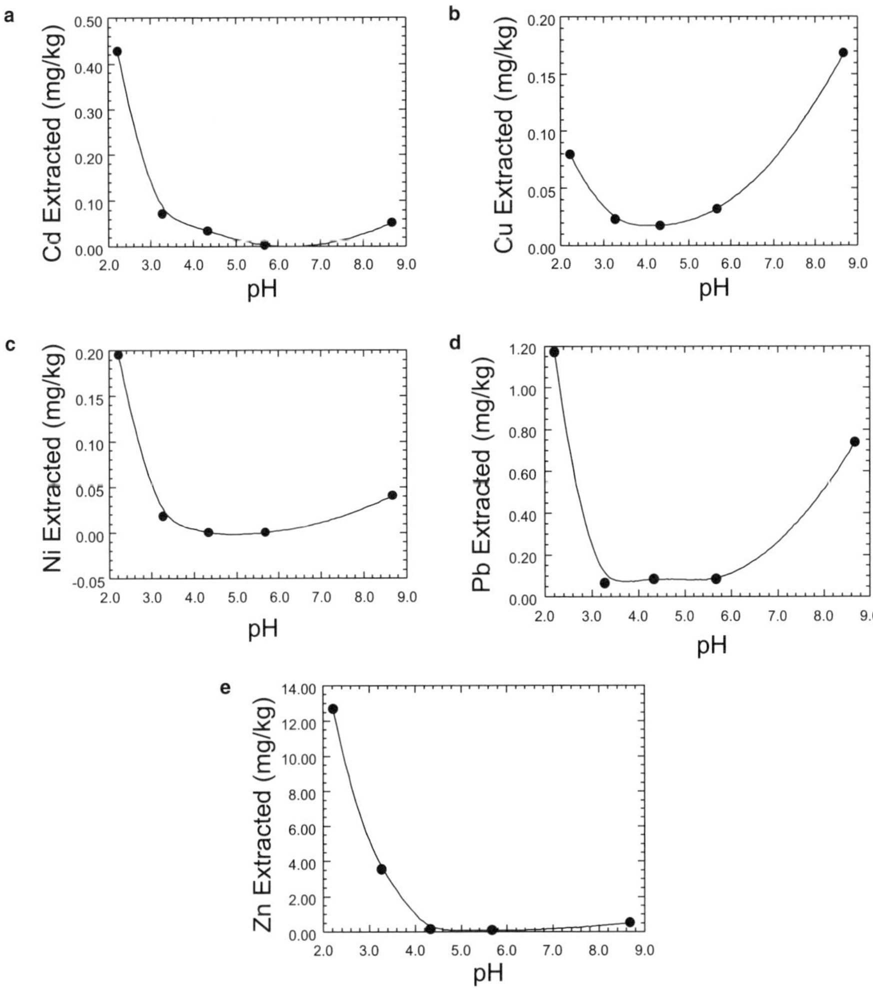

FIGURE 6.1 Metal extracted as a function of pH for a Dutch soil. (Data from Impellitteri CA, Lu Y, Saxe JK, Allen HE, Peijnenburg WJGM. 2002. Correlation of the partitioning of dissolved organic matter fractions with the desorption of Cd, Cu, Ni, Pb, and Zn from 18 Dutch soils. *Environ Int* 28:401–410. With permission.)

significantly increase with an increase in soil pH from pH 6 to pH 9. This seemingly contradictory behavior can be explained by the transfer of humic substances and associated metals from the solid to the solution phase. In natural soils, clay minerals tend to be coated by oxides and organic substances (Jenne 1968; Davis 1984), and oxides can also be coated by organic materials (Stumm 1992). Although there is a positive relationship between soil pH and sorption of metals on humic substances and other solid phase components, soil pH also affects sorption of humic substances onto clays and oxides, and affects the solubility of humic substances. A study on a series of Dutch soils with widely varying metal contamination backgrounds and

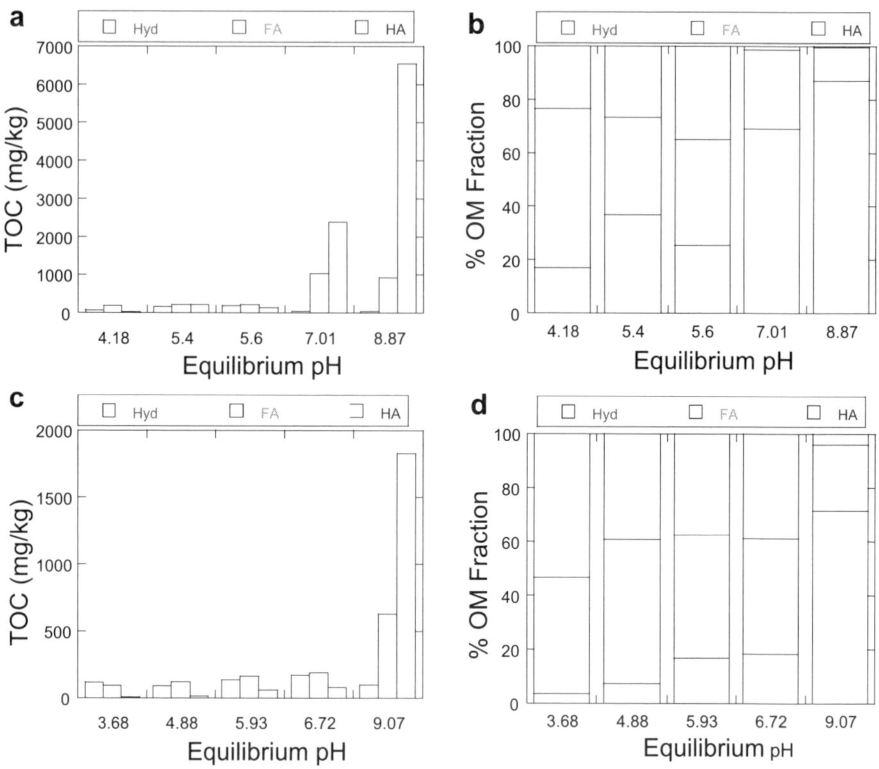

FIGURE 6.2 TOC in each fraction as a function of pH for 2 Dutch soils; (a) Budel soil, (2c) Valkenswaard soil); (2b) Budel soil and (2d) Valkenswaard soil show percent distribution of the 3 fractions as a function of 24-h equilibrium pH. (Data from Impellitteri CA, Lu Y, Saxe JK, Allen HE, Peijnenburg WJGM. 2002. Correlation of the partitioning of dissolved organic matter fractions with the desorption of Cd, Cu, Ni, Pb and Zn from 18 Dutch soils. *Environ Int* 28:401–410. With permission.)

concentrations serves as a clear example of the effects of pH on the interrelationships in the partitioning of metals and humic substances (Impellitteri et al. 2002). Data from this study are shown in Figure 6.2, where it can be seen that the pH greatly affects the solubility of operationally defined types of humic substances. This study found that the partitioning coefficients for Cu and Pb correlated well with partitioning coefficients for HA ($R^2 = 0.61$ and 0.45), respectively (Figure 6.3), suggesting that these metals were associated with HA and were brought into solution with the HA as the solubility of HA increased with increasing pH. There were no significant correlations between distributions of HA and Cd ($R^2 = 0.02$), HA and Ni ($R^2 = 0.13$), and HA and Zn ($R^2 = 0.02$). The correlation for Pb was observed to improve significantly when the data were normalised by total Ca concentrations (Figure 6.4). Thus, in addition to pH, it appears that Ca also plays an important role in metal–humic interactions. Monovalent cations disperse humic substances, whereas divalent cations flocculate them. Therefore, in systems where Ca is abundant (e.g., calcic soils), the concentration of Ca will impact the distribution of humic substances. In natural

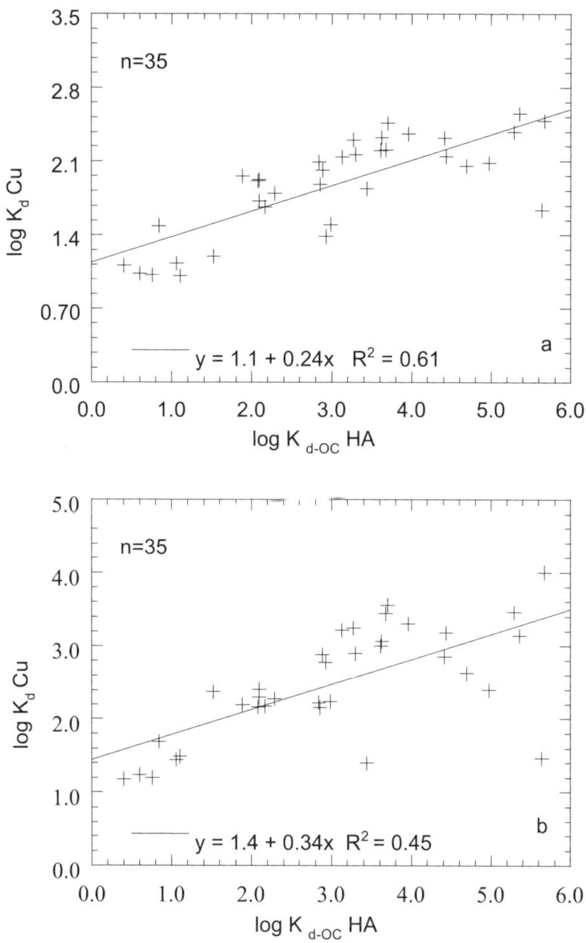

FIGURE 6.3 Partitioning of Cu (a) and Pb (b) vs. partitioning of humic acids. (Data from Impellitteri CA, Lu Y, Saxe JK, Allen HE, Peijnenburg WJGM. 2002. Correlation of the partitioning of dissolved organic matter fractions with the desorption of Cd, Cu, Ni, Pb and Zn from 18 Dutch soils. *Environ Int* 28:401–410. With permission.)

soils, organic matter was identified as the major source of Ca-preferring sites (Curtin et al. 1998). Calcium can outcompete cations such as Ni, causing the formation of very weak FA-Ni complexes that will potentially release Ni into solution (Mandal et al. 2000). Magnesium was also shown to decrease Ni binding on FA. Conversely, Romkens and Dolfing (1998) found that Cu remained bound to dissolved organic carbon that flocculated upon addition of Ca. Fifty percent of the dissolved organic carbon precipitated out of solution taking the Cu with it. The dissolved organic carbon in this study was fractionated into 2 sizes, low and high molecular weight. The low molecular weight carbon had a higher capacity for Cu binding (450 μmol Cu g^{-1} C vs. 250 μmol Cu g^{-1} C), but the high molecular weight component had higher affinities for Cu. The authors noted that after flocculation, 50% of the total dissolved

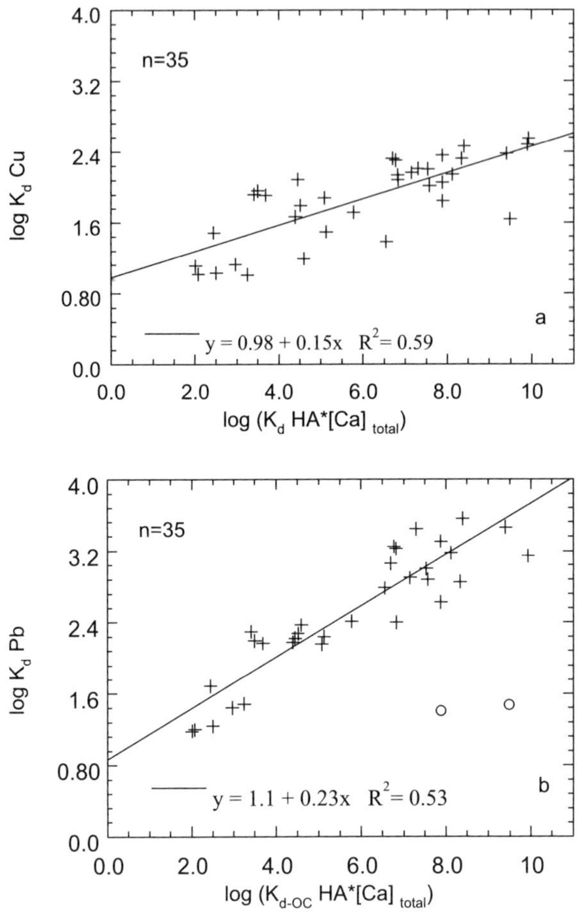

FIGURE 6.4 Partitioning of metals vs. the total Ca normalized partitioning of humic acids. Data points shown as circles in (4b) represent outliers. When these 2 outlier values were excluded, R^2 increased from 0.53 to 0.84. (Data from Impellitteri CA, Lu Y, Saxe JK, Allen HE, Peijnenburg WJGM. 2002. Correlation of the partitioning of dissolved organic matter fractions with the desorption of Cd, Cu, Ni, Pb and Zn from 18 Dutch soils. *Environ Int* 28:401–410. With permission.)

organic carbon that could potentially mobilize Cu remained. Oste et al. (2002) suggested that pH plays a relatively minor role compared to the solution concentration of Ca. Temminghoff et al. (1998) stated that coagulation of dissolved organic matter was affected primarily by the amounts of Al^{3+}, Ca^{2+}, and Cu^{2+} in solution as there was little effect of pH in the $4 < pH < 6$ range.

Thus, humic substances can transport metals in solution or, by flocculation, remove them from solution. For Cu these effects are well documented. Salam and Helmke (1998) showed that in sewage-sludge-amended soil, total dissolved Cu concentrations increased above pH 5.5. They attributed this to the chelation of Cu by organic substances with subsequent transfer to the solution phase because of the increasing solubility of the organic substances with increasing pH. Hsu and Lo (2000) found an increase in extractable Cu in swine manure composts. They attributed the increase in extractable Cu to the concurrent increase in dissolved organic matter with increasing pH. The increase in dissolved organic matter concentration at higher pH values (pH > 8) did not result in increased soluble concentrations of Zn and Mn. Temminghoff et al. (1997) observed increased Cu mobility at low and neutral pH values in a Cu contaminated sandy soil. They found that at pH 3.9 only 30% of Cu in solution was bound by dissolved organic carbon, whereas 99% of the Cu was bound by dissolved organic carbon at pH 6.6. Strobel et al. (2000) showed that Cu mobilization in a forest soil was related to increases in pH and dissolved organic carbon, whereas Cd mobilization was related solely to system pH. They also concluded that organic carbon inhibited Cu release at lower pH (3.8 < pH < 5). Cadmium mobility has been related to organic matter, where increasing equilibration time and temperature reduced the mobility of Cd caused by organic matter (Almas et al. 1999). Naidu and Harter (1998) found that sorption of Cd onto soils was decreased in the presence of organic ligands at low pH. Jordan et al. (1997) examined the increased mobility of Pb in the presence of natural organic matter in a sandy soil. They found that peat humic acids had a higher binding affinity for Pb than peat fulvic acids. They also illustrated the decreased binding of Pb to the sandy soil when dissolved organic matter was present in column and batch sorption studies.

It is important to note that after desorption from the solid phase, humic substances continue to play a role in the behavior of the metal. For example, Temminghoff et al. (1994) found that free Cu^{2+} ion in solution ranged between 2% (pH 5.7) and 9% (pH 4.4), with an additional 1% complexed by nitrate. They found that the remainder of the Cu in solution was bound to dissolved organic matter. Humic substances have been shown to ameliorate the toxic effects of Cu in aquatic systems (Ma et al. 1999). The bioavailability and toxicity of organically complexed metals to organisms remains a very important issue in need of further research.

6.10 HUMIC SUBSTANCES, METALS, AND MODELS

This section examines the inclusion or use of humic substances in predictive models describing the fate of trace metals in the environment. Models can play an important role in trace metal risk assessments by helping to predict the attenuation (or lack thereof) of trace metal chemical species in a particular setting. Though related to actual structure, no discussion on structural models (Schulten and Schnitzer 1995; Schulten and Schnitzer 1997; Kubicki and Apitz 1999) will be presented.

Partitioning coefficients (defined here as K_d), also known as distribution coefficients (Anderson and Christensen 1988), can be described as the ratio of the concentration of a metal associated with the solid phase to the concentration of that metal in the solution phase:

$$K_d = [M]_{solid}/[M]_{solution}$$

K_p has also been used to describe a partitioning coefficient (Janssen et al. 1997). Regardless of the terminology or how $[M]_{solid}$ and $[M]_{solution}$ are actually measured, the partitioning of metals has often been related to environmental parameters such as pH, organic matter, oxides, etc., in multiple regression analyses. Organic matter, usually measured by dry or wet combustion, is consistently identified in these models as an important factor affecting metal partitioning. Normalization of partitioning coefficients for Cd in soils by organic-matter content greatly improved the correlation between K_d and pH (from $R^2 = 0.799$ to $R^2 = 0.927$) (Lee et al. 1996). Janssen et al. (1997) included dissolved organic carbon in equations predicting the partitioning of Cr, Cu, and Ni. Dissolved organic carbon related negatively to the partitioning for all 3 metals (i.e., increased DOC decreased the partitioning coefficient, or more metal was associated with the solution phase). Organic matter was identified as an important factor for prediction of the K_d values for Cd, Co, and Ni, but not for Zn (Anderson and Christensen 1988). McBride et al. (1997) found that organic matter was often a significant parameter in equations that predicted soluble metals, especially Cu. They concluded that solid phase organic matter limited free Cu^{2+} activity, whereas dissolved organic matter promoted Cu solubility. Organic matter and pH accounted for 70% of the variability in Cu partitioning and 80% of the variability in plant-available copper in 40 soils from around the world (Impellitteri et al. 2003).

6.11 MODELS INCLUDING HUMIC SUBSTANCES

Surface complexation type models describing the interactions of metals and humic substances have helped to improve prediction of metal speciation and partitioning in the environment. It has been stated for chemical equilibrium models in general that the most critical variable explaining discrepancies from model to model are the fundamental thermodynamic data used for input into the model (Nordstrom et al. 1979). Differences in stability constants for metals and humic substances will introduce discrepancies in results generated by surface complexation models. Surface complexation models for humic substances as the sole sorbent include (but are not limited to) the Humic Ion Binding Model V/VI (Tipping and Hurley 1992; Tipping 1998), the nonideal competitive adsorption (NICA) model (Benedetti et al. 1995), and the NICA-Donnan model (Kinniburgh et al. 1996). Other models include humic substances as a sorbing component, such as the Windermere Humic Aqueous Model (WHAM) (Tipping 1994), MINTEQA2 (Allison et al. 1991), and the sediment water algorithm for metal partitioning (SWAMP) (Radovanovic and Koelmans 1998). We will include a brief description of WHAM, *NICA*, and *NICA*-Donnan here. For all of the models, a common requirement is an extreme simplification of functional groups and structure of the humic substance.

WHAM is a combination of Humic Ion Binding Model V (Tipping and Hurley 1992) or Humic Ion Binding Model VI (Tipping 1998), with other submodels describing metal speciation in solution, solid phases, sorption on FA and HA, and cation exchange on representative clays. The submodel for the humic component, Model V, was improved by Model VI which accounts for ionic strength effects, competition from major cations (e.g., Ca^{2+} and Mg^{2+}), and better describes the nonlinear binding of metals at low concentrations (Tipping 1998). The humic substance component in WHAM uses a discrete site and electrostatic binding approach to surface complexation of metals by humic substances. WHAM was created to model environmental systems where humic substances dominate chemical speciation (Tipping 1994). Christensen and Christensen (1999) compared numerous methods including WHAM to describe Cd, Ni, and Zn complexation by dissolved organic carbon in groundwater. They found that WHAM described Cd and Ni complexation well, but the authors needed to alter a default binding constant for Zn to improve results. Mandal et al. (2000) compared WHAM output favorably with actual data to illustrate the competitive effects of Ca and Mg on Ni sorption on FA. WHAM has also been used to model Al solubility in soil (de Wit et al. 2001) and the effects of aquatic organic ligands on Cu and Cd complexation in lake waters (Xue and Sigg 1999).

The NICA model assumes a continuous distribution of binding affinities for cations and protons. For modeling the binding of cations, the NICA model utilizes a bimodal distribution, presumed to coincide with carboxylic and phenolic functional groups, but can also be simplified to monomodal distributions (over a narrow pH range), or a 2-parameter Freundlich isotherm (Benedetti et al. 1995).

The NICA model describes specific binding of cations by surface functional groups. The Donnan model aims to explain nonspecific associations between cations and Coulombic residual negative charge (Kinniburgh et al. 1996). The Donnan model is based on ionic behavior in gel membrane media. Essentially, a cation exchange gel membrane is covered by a homogeneous, evenly distributed electronegative potential known as the Donnan potential. The membrane excludes transport of anions, but as long as the salt concentrations in the donor and acceptor solutions separated by the membrane are equal, equilibrium of free metal ions will occur. This is the Donnan equilibrium (Temminghoff et al. 2000). The Donnan model treats humic substances like a gel membrane to account for nonspecific interactions between cations and humic substances.

The NICA-Donnan model was first used to model H, Ca, Cd, Cu, and Pb binding by a purified peat humic acid (Kinniburgh et al. 1996). The model was also used to gauge the competitive effect of Ca on Cd and Cu binding at a variety of pH values. The researchers identified problems with the model which tended to underestimate the H^+/M^{2+} exchange ratio at high pH and underestimate the binding of Cu. These problems were addressed in later variations of the model (Kinniburgh et al. 1999).

WHAM Model V and the NICA-Donnan model were compared in acid titrations of FA from groundwater (Christensen et al. 1998). Proton binding properties for the FAs from polluted and uncontaminated groundwater were assessed and modeled by both procedures. The default proton binding parameters and the experimentally obtained binding parameters for the FAs were compared by simulating Cd complexation. They found that for WHAM Model V, the results for Cd complexation varied little using default or experimental values. Larger discrepancies were discovered for the NICA-Donnan model.

Further improvements, such as consideration of ionic strength and major ion effects in the WHAM Model VI upgrade, will continue for these models. Probably the largest drawback for the NICA-Donnan model at this point in time is its relative immaturity. As an input database grows, the model will only improve and serve to more accurately describe a wide range of possible scenarios.

6.12 CONCLUSION

Humic substances are an ill-defined composite of organic compounds whose structure is influenced by biological factors such as microbial degradation, and chemical factors such as pH and ionic strength. Their presence in soils can assist in the attenuation of metal mobility and bioavailability by sorption or chelation and sequestration. However, they can also transport metals into solution or bind them in solution, preventing the sorption of metal onto the solid phase, thus potentially enhancing their offsite transport and bioavailability. Hence, the effect of humic substances on attenuation of metals is important but transient and variable. Inclusion of humic substances in risk assessment of contaminant metals will hopefully become less problematic in the future as we gain more knowledge concerning the structure of these molecules and more exactly define their roles on trace elements in the environment.

ACKNOWLEDGMENTS

This manuscript has not been subjected to internal review by the U.S. Environmental Protection Agency. Therefore, the research results presented herein do not, necessarily, reflect Agency policy. Mention of trade names or commercial products does not constitute endorsement or recommendation for use.

REFERENCES

Aiken G, Leenheer J. 1993. Isolation and chemical characterization of dissolved and colloidal organic matter. Chem Ecol 8:135–151.

Aiken GR, Thurman EM, Malcolm RL. 1979. Comparison of XAD macroporous resins for the concentration of fulvic acid from aqueous solution. Anal Chem 51:1799–1803.

Allison JD, Brown DS, Novo-Gradac KJ. 1991. MINTEQA2/PRODEFA2. A geochemical assessment model for environmental systems: version 3.0 Users Manual EPA/600-3-91/021. Athens, GA: USEPA.

Almas A, Singh BR, Salbu B. 1999. Mobility of cadmium-109 and zinc-65 in soil influenced by equilibration time, temperature and organic matter. J Environ Qual 28:1742–1750.

Andersen DO, Alberts JJ, Takacs M. 2000. Nature of natural organic matter (NOM) in acidified and limed surface waters. Wat Res 34:266–278.

Anderson PR, Christensen TH. 1988. Distribution coefficients of Cd, Co, Ni, and Zn in soils. J Soil Sci 39:15–22.

Andersson S, Nilsson SI, Saetre P. 2000. Leaching of dissolved organic carbon (DOC) and dissolved organic nitrogen (DON) in mor humus as affected by temperature and pH. Soil Biol Biochem 32:1–10.

Ball DF. 1964. Loss-on-ignition as an estimate of organic matter and organic carbon in non-calcareous soils. J Soil Sci 15:84–92.

Benedetti MF, Milne CJ, Kinniburgh DG, Riemsdijk WHV, Koopal LK. 1995. Metal ion binding to humic substances: application of the non-ideal competitive adsorption model. Environ Sci Technol 29:446–457.

Bennett LE, Drikas M. 1993. The evaluation of colour in natural waters. Wat Res 27:1209–1218.

Bloom PR, Leenheer, JA. 1989. Vibrational, electronic, and high-energy spectroscopic methods for characterizing humic substances. In Hayes, MHB, MacCarthy, P, Malcolm, RL, Swift, RS editors. Humic substances: II. In search of structure. New York: J Wiley. p 410–446.

Bohn H. 1976. Estimate of organic carbon in world soils. Soil Sci Soc Am J 40:468–469.

Bolin B. 1977. Changes of land biota and their importance for the carbon cycle. Science 196:613–615.

Broadbent FE, Bradford GR. 1952. Cation exchange groupings in soil organic fraction. Soil Sci 74:447–457.

Buffle J, Stumm W. 1994. General chemistry of aquatic systems. In Buffle J, DeVitre RR, editors. Chemical and biological regulation of aquatic systems. Boca Raton, FL: CRC Pr. p 1–42.Brown GK, MacCarthy P, Leenheer JA. 1999. Simultaneous determination of Ca, Cu, Ni, Zn and Cd binding strengths with fulvic acid fractions by Schubert's method. Anal Chim Acta 402:169–181.

Bunzl K, Schmidt W, Sansoni B. 1976. Kinetics of ion exchange in soil organic matter. IV. Adsorption and desorption of Pb^{2+}, Cu^{2+}, Cd^{2+}, Zn^{2+}, and Ca^{2+} by peat. J Soil Sci 27:32–41.

Chancy RL, Ryan JA. 1993. Toxic metals and toxic organic pollutants in MSW-composts: research results on phyto-availability, bioavailability, fate, etc. In: Hoitink HA, Keener HM, editors. Science and engineering of composting: design, environmental, microbiological and utilization aspects. Worthington, OH: Renaissance. p 451–506.

Charberek S, Martell AE. 1959. Organic sequestering agents. New York: J Wiley.

Chefetz B, Hatcher PG, Hadar Y, Chen YN. 1998. Characterization of dissolved organic matter extracted from composted municipal solid waste. Soil Sci Soc Am J 62:326–332.

Chefetz B, Heemst JDHv, Chen Y, Romaine CP, Chorover J, Rosario R, Mingxin G, Hatcher PG. 2000. Organic matter transformations during the weathering process of spent mushroom substrate. J Environ Qual 29:592–602.

Chen Y, Schnitzer M. 1976. Scanning electron microscopy of a humic acid and of a fulvic acid and its metal and clay complexes. Soil Sci Soc Am J 40:682–686.

Chen Y, Senesi N, Schnitzer M. 1977. Information provided on humic substances by E_4/E_6 ratios. Soil Sci Soc Am J 41:352–358.

Chen Y, Inbar Y, Hadar Y, Malcolm RL. 1989. Chemical properties and solid-state CPMAS super(13)C-NMR of composted organic matter. Sci Total Environ 81–82:201–208.

Chien YY, Bleam WF. 1998. Two-dimensional NOESY nuclear magnetic resonance study of pH-dependent changes in humic acid conformation in aqueous solution. Environ Sci Technol 32:3653–3658.

Chiou CT, Kile DE, Rutherford DW, Sheng G, Boyd SA. 2000. Sorption of selected organic compounds from water to a peat soil and its humic-acid and humin fractions: potential sources of the sorption nonlinearity. Environ Sci Technol 34:1254–1258.

Christensen JB, Christensen TH. 1999. Complexation of Cd, Ni, and Zn by DOC in polluted groundwater: a comparison of approaches using resin exchange, aquifer material sorption, and computer speciation models (WHAM and MINTEQA2). Environ Sci Technol 33:3857–3863.

Christensen JB, Tipping E, Kinniburgh DG, Gron C, Christensen TH. 1998 Proton binding by groundwater fulvic acids of different age, origins, and structure modeled with the Model V and NICA-Donnan Model. Environ Sci Technol 32:3346–3355.

Christl I, Kretzschmar R. 2001. Interaction of copper and fulvic acid at the hematite-water interface. Geochim Cosmochim Acta 65:3435–3442.

Corey RB, King LD, Lue-Hing C, Fanning DS, Street JJ, Walker JM. 1987. Effects of sludge properties on accumulation of trace elements by crops. In: Page AL, Logan, TJ, Ryan, JA, editors. Land application of sludge. Food chain implications. Chelsea, MI: Lewis. p 25–51.

Curtin D, Selles F, Steppuhn H. 1998. Estimating calcium-magnesium selectivity in smectitic soils from organic matter and texture. Soil Sci Soc Am J 62:1280–1285.

Davis JA. 1984. Complexation of trace metals by adsorbed natural organic matter. Geochim Cosmochim Acta 48:679–691.

de Wit HA, Groseth T, Mulder J. 2001. Predicting aluminimum and soil organic matter solubility using the mechanistic equilibrium model WHAM. Soil Sci Soc Am J 65:1089–1100.

Du Q, Sun Z, Forsling W, Tang H. 1999. Complexation in illite-fulvic acid-Cu^{2+} systems. Wat Res 33:693–706.

Erich MS, Trusty GM. 1997. Chemical characterization of dissolved organic matter released by limed and unlimed forest soil horizons. Can J Soil Sci 77:405–413.

Evangelou VP, Marsi M, Vandiviere MM. 1999. Stability of Ca^{2+}, Cd^{2+}, Cu^{2+}-[illite-humic] complexes and pH influence. Plant Soil 213:63–74.

Gjessing ET. 1976. Physical and chemical characteristics of aquatic humus. Ann Arbor, MI: Ann Arbor Science Publishers.

Haiber S, Herzog H, Burba P, Gosciniak B, Lambert J. 2001.Two-dimensional NMR studies of size fractionated Suwanee river fulvic and humic acid reference. Environ Sci Technol 35:4289–4294.

Han N, Thompson ML. 1999. Copper-binding ability of dissolved organic matter derived from anaerobically digested biosolids. J Environ Qual 28:939–944.

Hatcher PG, Dria KJ, Kim S, Frazier SW. 2001. Modern analytical studies of humic substances. Soil Sci 166:770–794.

Hautala K, Peuravuori J, Pihlaja K. 1999. Measurement of aquatic humus content by spectroscopic analyses. Wat Res 34:246–258.

Hongve D, Akesson G. 1996. Spectrophotometric determination of water colour in Hazen units. Wat Res 30:2771–2775.

Hsu JH, Lo SL. 2000. Characterization and extractability of copper, manganese, and zinc in swine manure composts. J Environ Qual 29:447–453.

Huang C, Yang YL. 1995. Adsorption characteristics of Cu(II) on humus-kaolin complexes. Wat Res 29:2455–2460.

Huang WL, Young TM, Schlautman MA, Yu H, Weber WJ. 1997. A distributed reactivity model for sorption by soils and sediments. 9. General isotherm nonlinearity and applicability of the dual reactive domain model. Environ Sci Technol 31:1703–1710.

Impellitteri CA, Lu Y, Saxe JK, Allen HE, Peijnenburg WJGM. 2002. Correlation of the partitioning of dissolved organic matter fractions with the desorption of Cd, Cu, Ni, Pb and Zn from 18 Dutch soils. Environ Intern 28:401–410.

Impellitteri CA, Saxe JK, Cochran M, Janssen G, Allen HE. 2003. Predicting the bioavailability of copper and zinc in soils. Modeling the partitioning of potentially bioavailable copper and zinc from soil solid to soil solution. Environ Toxicol Chem 22:1380–1386.

Janssen RPT, Peijnenburg WJGM, Posthuma L, Hoop MAGTvd. 1997. Equilibrium partitioning of heavy metals in Dutch field soils. I. Relationship between metal partition coefficients and soil characteristics. Environ Toxicol Chem 16:2470–2478.

Jardine PM, Weber NL, McCarthy JF. 1989. Mechanisms of dissolved organic carbon adsorption on soil. Soil Sci Soc Am J 53:1378–1385.

Jenne EA. 1968. Controls on Mn, Fe, Co, Ni, Cu, and Zn concentrations in soils and water: The significant role of hydrous Mn and Fe oxides. In: Gould RF, editor. Trace inorganics in water. Advances in Chemistry Series Number 73. Washington, DC: American Chemical Society. p 337–387.

Jordan RN, Yonge DR, Hathhorn WE. 1997. Enhanced mobility of Pb in the presence of dissolved natural organic matter. J Contam Hydrol 29:59–80.

Kaiser K, Haumaier L, Zech W. 2000. The sorption of organic matter in soils as affected by the nature of soil carbon. Soil Sci 165:305–313.

Kalbitz K, Solinger S, Park, Michalzik B, Matzner E. 2000. Controls on the dynamics of dissolved organic matter in soils: A review. Soil Sci 165:277–304.

Karlik B. 1995. Liming effect on dissolved organic matter leaching. Water Air Soil Poll 85:949–954.

Kinniburgh DG, van Riemsdijk WH, Koopal LK, Borkovec M, Benedetti MF, Avena MJ. 1999 Ion binding to natural organic matter: competition, heterogeneity, stoichiometry and thermodynamic consistency. Coll Surf A-Physicochem Eng Asp 151:147–166.

Kinniburgh DG, Milne CJ, Benedetti MF, Pinheiro JP, Filius J, Koopal LK, Riemsdijk WHV. 1996. Metal ion binding by humic acid: application of the NICA-Donnan model. Environ Sci Technol 30:1687–1698.

Kononova MM. 1966. Soil organic matter. Oxford: Pergamon Pr.

Kubicki JD, Apitz SE. 1999. Models of natural organic matter and interactions with organic contaminants. Organic Geochem 30:911–927.

Kuwatsuka S, Watanabe A, Itoh K, Arai S. 1992. Comparison of two methods of preparation of humic and fulvic acids, IHSS method and NAGOYA method. Soil Sci Plant Nut 38:23–30.

Leboeuf EJ, Weber WJ. 1997. A distributed reactivity model for sorption by soils and sediments .9. Sorbent organic domains: Discovery of a humic acid glass transition and an argument for a polymer-based model. Environ Sci Technol 31:1697–1702.

Lee SZ, Allen HE, Huang CP, Sparks DL, Sanders PF, Peijnenburg WJGM. 1996. Predicting soil-water partitioning coefficients for cadmium. Environ Sci Technol 30:3418–3424.

Leenheer JA. 1981. Comprehensive approach to preparative isolation and fractionation of dissolved organic carbon from natural waters and wastewaters. Environ Sci Technol 15:578–587.

Leenheer JA, Huffman JEWD. 1976. Classification of organic solutes in water by using macroreticular resins. J Res U.S. Geol Surv 4:737–751.

Li Z, Ryan JA, Chen J-L, Al-Abed SR. 2001. Adsorption of cadmium on biosolids-amended soils. J Environ Qual 30:903–911.

Liu A. 1999. Adsorption/desorption in a system consisting of humic acid, heavy metals, and clay minerals. J Coll Interf Sci 218:225–232.

Liu A, Gonzalez RD. 2000. Modeling adsorption of copper(II), cadmium(II) and lead(II) on purified humic acid. Langmuir 16:3902–3909.

Lu Y, Allen HE. 2002. Characterization of copper complexation with natural dissolved organic matter (DOM)-link to acidic moieties of DOM and competition by Ca and Mg. Wat Res 36:5083–5101.

Ma H, Allen HE, Yin Y. 2001. Characterization of isolated fractions of dissolved organic matter from natural waters and a wastewater effluent. Wat Res 35:985–996.

Ma H, Kim SD, Cha DK, Allen HE. 1999. Effect of kinetics of complexation by humic acid on toxicity of copper to Ceriodaphnia dubia. Environ Toxicol Chem 18:828–837.

Malcolm RL, Aiken GR, Bowles EC, Malcolm JD. 1994. Isolation of fulvic and humic acids from the Suwanee River. In: Averett RC, Leenheer, JA, McKnight, DM, Thorn, KA, editors. Humic substances in the Suwanee River, Georgia: interactions, proper and proposed structures, Vol. Water Supply Paper 2373. Washington, DC: United States Geological Survey. p 13–19.

Mandal R, Salam MSA, Murimboh J, Hassan NM, Chakrabarti CL, Back MH, Gregoire DC. 2000. Competition of Ca(II) and Mg(II) with Ni(II) for binding by a well-characterized fulvic acid in model solutions. Environ Sci Technol 34:2201–2208.

McBride M, Sauve S, Hendershot W. 1997. Solubility control of Cu, Zn, Cd and Pb in contaminated soils. Eur J Soil Sci 48:337–346.

McBride MB. 1994. Environmental chemistry of soils. New York: Oxford University Pr, Inc.

McBride MB. 1995. Toxic metal accumulation from agricultural use of sludge: are USEPA regulations protective? J Environ Qual 24:5–18.

McBride MB. 1998. Growing food crops on sludge-amended soils: problems with the U.S. Environmental Protection Agency method of estimating toxic metal transfer. Environ Toxicol Chem 17:2274–2281.

McGrath SP, Zhao FJ, Dunham SJ, Crosland AR, Coleman K. 2000. Long-term changes in the extractability and bioavailability of zinc and cadmium after sludge application. J Environ Qual 29:875–883.

Mills GL, Mcfadden E, Quinn JG. 1987. Chromatographic studies of dissolved organic matter and copper organic complexes isolated from estuarine waters. Mar Chem 20:313–323.

Naes K, Axelman J, Naf C, Broman D. 1998. Role of soot carbon and other carbon matrices in the distribution of PAHs among particles, DOC, and the dissolved phase in the effluent and recipient waters of an aluminum reduction plant. Environ Sci Technol 32:1786–1792.

Naidu R, Harter RD. 1998. Effect of diferent organic ligands on cadmium sorption by and extractability from soils. Soil Sci Soc Am J 62:644–650.

Namjesnik-Dejanovic K, Maurice PA, Aiken GR, Cabaniss S, Chin Y-P, Pullin MJ. 2000. Adsorption and fractionation of a muck fulvic acid on kaolinite and goethite at pH 3.7, 6, and 8. Soil Sci 165:545–559.

Nordstrom DK, Plummer LN, Wigley TML, Wolery TJ, Ball JW, Jenne EA, Bassett RL, Crerar DA, Florence TM, Fritz B, and others. 1979. Comparison of computerized chemical models for equilibrium calculations in aqueous systems. In: Jenne EA, editor. Chemical modeling in aqueous systems: speciation, sorption, solubility and kinetics. Washington, DC: American Chemical Society.

Oste LA, Temminghoff EJM, Riemsdijk WHV. 2002. Solid-solution partitioning of organic matter in soils as influenced by an increase in pH or Ca concentration. Environ Sci Technol 36:208–214.

Pandey AK, Pandey SD, Misra V. 2000. Stability constants of metal-humic acid complexes and its role in environmental detoxification. Ecotoxicol Environ Safety 47:195–200.

Peuravuori J. 2000. Precision of isolation of aquatic humic matter by XAD-resin technology from NMR spectroscopy's point of view. Intern J Environ Anal Chem 76:179–198.

Peuravuori J, Pihlaja K. 1997. Molecular size distribution and spectroscopic properties of aquatic humic substances. Anal Chim Acta 337:133–149.

Radovanovic H, Koelmans AA. 1998. Prediction of in situ trace metal distribution coefficients for suspended solids in natural waters. Environ Sci Technol 32:753–759.

Robertson AP, Leckie JO. 1996. Goethite/humic interactions and their impact on copper partitioning as a function of pH, ionic strength and copper loading. 211th ACS National Meeting: Spring National Meeting of the American Chemical Society, New Orleans, LA. March 24–28, 1996.

Rocha JC, Sene JJD, Santos D, Toscano IAS, Zara LF. 1999. Aquatic humus from an unpolluted Brazilian dark-brown stream: General characterization and size fractionation of bound heavy metals. J Environ Monit 2:39–44.

Romkens PFAM, Dolfing J. 1998. Effect of Ca on the solubility and molecular size distribution of DOC and Cu binding in soil solution samples. Environ Sci Technol 32:363–369.

Salam AK, Helmke PA. 1998. The pH dependence of free ionic activities and total dissolved concentrations of copper and cadmium in soil solution. Geoderma 83:281–291.

Sauvé S, Hendershot W, Allen HE. 2000. Solid-solution partitioning of metals in contaminated soils: Dependence on pH, total metal burden, and organic matter. Environ Sci Technol 34:1125–1131.

Schnitzer M. 1978. Humic substances: chemistry and reactions. In: Schnitzer M, Khan SU, editors. Soil organic matter. Developments in soil science 8. New York: Elsevier Scientific Publishing. p 1–64.

Schnitzer M. 2000. A lifetime perspective on the chemistry of soil organic matter. Advan Agron 68:1–58.

Schnitzer M, Hansen EH. 1970. Organo-metallic interactions in soils: 8. An evaluation of methods for the determination of stability constants of metal-fulvic acid complexes. Soil Sci 109:333–340.

Schulten HR, Schnitzer M. 1995. Three-dimensional models for humic acids and soil organic matter. Naturwissenschaften 80:29–30.

Schulten HR, Schnitzer M. 1997. Chemical model structures for soil organic matter and soils. Soil Sci 162:115–130.

Shen YH. 1999. Sorption of natural dissolved organic matter on soil. Chemosphere 38:1505–1515.

Simpson AJ, Kingery WL, Spraul M, Humpfer E, Dvortsak P, Kerssebaum R. 2001. Separation of structural components in soil organic matter by diffusion ordered spectroscopy. Environ Sci Technol 35:4421–4425.

Spark KM, Wells JD, Johnson BB. 1997a. Sorption of heavy metals by mineral-humic acid substrates. Aust J Soil Res 35:113–122.

Spark KM, Wells JD, Johnson BB. 1997b. Characteristics of the sorption of humic acid by soil minerals. Aust J Soil Res 35:103–112.

Sparks DL. 1995. Environmental soil chemistry. San Diego, CA: Academic Pr.

Stevenson FJ. 1982. Humus chemistry genesis, composition, reactions. New York: J Wiley.

Stevenson FJ. 1994. Humus chemistry genesis, composition, reactions. 2nd ed. New York: J Wiley.

Strobel BW, Hansen HCB, Borggaard OK, Andersen MK, Raulund-Rasmussen K. 2000. Cadmium and copper release kinetics in relation to afforestation of cultivated soil. Geochim Cosmochim Acta 65:1233–1242.

Stumm W. 1992. Chemistry of the solid-water interface: Processes at the mineral-water and particle-water interface in natural systems. New York: J Wiley.

Subbalakshmi Y, Patti AF, Lee GSH, Hooper MA. 2000. Structural characterisation of macromolecular organic material in air particulate matter using Py-GC-MS and solid state super(13)C-NMR. J Environ Monit 2:561–565.

Swift RS. 1996. Organic matter characterization. In: Sparks DL, ed. Methods of soil analysis. Part 3. Chemical methods. Madison, WI: Soil Science Society of America. p 1018–1020.

Temminghoff EJM, Zee SEATMVD, Keizer MG. 1994. The influence of pH on the desorption and speciation of copper in a sandy soil. Soil Sci 158:398–408.

Temminghoff EJM, Zee SEATMVD, Haan FAMD. 1997. Copper mobility in a copper-contaminated sandy soil as affected by pH and solid and dissolved organic matter. Environ Sci Technol 31:1109–1115.

Temminghoff EJM, Zee SEATMVD, Haan FAMD. 1998. Effects of dissolved organic matter on the mobility of copper in a contaminated sandy soil. Eur J Soil Sci 49:617–628.

Temminghoff EJM, Plette ACC, Eck RV, Riemsdijk WHV. 2000. Determination of the chemical speciation of trace metals in aqueous systems by the Wageningen Donnan membrane technique. Anal Chim Acta 417:149–157.

Theng BKG, Churchman GJ, Newman RH. 1986. The occurrence of interlayer clay-organic complexes in two New Zealand soils. Soil Sci 142:262–266.

Thurman EM. 1985. Organic geochemistry of natural waters. Hingham, MA: Kluwer Academic Publishers.

Thurman EM, Malcolm RL. 1981. Preparative isolation of aquatic humic substances. Environ Sci Technol 15:463–466.

Tipping E. 1994. WHAM-A chemical equilibrium model and computer code for waters, sediments, and soils incorporating a discrete site/electrostatic model of ion-binding by humic substances. Comput Geosci 20:973–1023.

Tipping E. 1998. Humic ion-binding model VI: an improved description of the interactions of protons and metal ions with humic substances. Aquatic Geochem 4:3–48.

Tipping E, Hurley MA. 1992. A unifying model of cation binding by humic substances. Geochim Cosmochim Acta 56:3627–3641.

USEPA. 1994. A plain English guide to the EPA part 503 biosolids rule EPA/832/R-93/003. Washington, DC: Office of Wastewater Management.

Vermeer AWP, McCulloch JK, Riemsdijk WHV, Koopal LK. 1999. Metal ion adsorption to complexes of humic acid and metal oxides: deviations from the additivity rule. Environ Sci Technol 33:3892–3897.

Waksman SA. 1936. Baltimore, MD: Humus Williams and Wilkins.

Walkley A, Black IA. 1934. An examination of the Degtjareff method for determining soil organic matter and a proposed modification of the chromic acid titration method. Soil Sci 37:29–38.

Weber WJ, Huang WL. 1996. A distributed reactivity model for sorption by soils and sediments. 4. Intraparticle heterogeneity and phase-distribution relationships under nonequilibrium conditions. Environ Sci Technol 30:881–888.

Xue H, Sigg L. 1999. Comparison of the complexation of Cu and Cd by humic or fulvic acids and by ligands observed in lake waters. Aquatic Geochem 5:313–335.

Yin Y, Impellitteri CA, You SJ, Allen HE. 2002. The importance of organic matter distribution and extract soil:solution ratio on the desorption of heavy metals from soils. Sci Tot Environ 287:107–119.

Zhou X, Wangersky PJ. 1989. Study of copper-complexing organic ligands: Isolation by a Sep-Pak C18 column extraction technique and characterization by chromarod thin-layer chromatography. Mar Chem 26:21–40.

7 Attenuation of Metal Toxicity in Soils by Biological Processes

M.B. McBride

7.1 INTRODUCTION

Soils contain bio-organic materials that include dead and decaying plants and microbes, as well as live, actively-growing fungal, algal, and bacterial biofilms. These have a profound influence on physical-chemical properties of soils and the solubility of nutrients and trace elements. Nevertheless, modern soil chemistry still employs simplistic mineral solubility product and adsorption models in attempting to explain or predict toxic metal behavior in soils, despite accumulating evidence that complex biological processes can override physico-chemical equilibrium processes. Although semiempirical equations give crude estimates of toxic metal solubility based solely upon the soil physico-chemical properties pH and total metal concentration (McBride et al. 1997a; Sauve et al. 1997, 2000a; Gray et al. 1999), an important part of the spatial and temporal variability of metal solubility under environmentally realistic conditions remains unexplained.

The fact that specific mineralogical information on each soil is often not necessary to explain general trends in metal activity or solubility suggests that mineral surface reactions are often secondary to sorption on bio-organic materials. The most commonly used (but very crude) measure of bio-organic material in soils, total organic carbon, is frequently correlated strongly to metal solubility or activity, whereas "reactive" soil minerals such as Fe or Al oxides generally are not (Lee et al. 1996, 2001; Yin et al. 1996; McBride et al. 1997a). Nevertheless, the highly variable nature of organic matter implies a reactivity toward metals that must depend on origin, degree of decomposition (age), and elemental composition (see Chapter 6). For example, high N and S content in organic matter is likely to favor trace metal binding. Present models of metal binding in soils fail to capture this complex behavior of living and decomposed organic matter.

It is likely that in many soils, the bio-organic materials of soils dominate the chemistry of soil solution, rendering the underlying or intermingled mineral particles relatively inaccessible and inactive. Future progress in predicting long-term trends in metal toxicity and mobility will require a better understanding of the ability of biological processes and bio-organics to modify metal solubility or activity. At present, the dynamic response of soil biological systems to toxic metal inputs is not well understood. Evidence is presented in this paper which suggests that heavy metal solubility is "biologically mediated" in soils, with free metal ion activities maintained at levels lower than expected from metal sorption, desorption, or precipitation reactions. At the same time, however, biological processes may enhance total metal solubility and leachability as dissolved organic matter (DOM) is increased, and pH and redox potential are modified in the biologically active surface soil. Consequently, loss of particular heavy metals from contaminated soils into shallow groundwater by leaching or facilitated transport may be significant over a time period of decades.

The hypothesis is also presented that, besides organic carbon, organic S may be a critical soil component in limiting the solubility and toxicity of chalcophilic metals, with recent evidence proving that a substantial fraction of S in soil organic matter is in reduced (including thiol) form and able to form strong complexes with Zn, Hg, and other chalcophiles. The long-term implications of S oxidation in soils, and possible diminution of the biological protective effect against metal toxicity, are discussed.

7.2 THE BIOLOGICAL RESPONSE TO METAL STRESS

When soils are contaminated with heavy metals, numerous biological processes are provoked, some of which are defenses against metal toxicity. These can have the effect of increasing metal mobility or, conversely, increasing sequestration and retention in soils. They may also be involved in the mitigating effect of "aging" on metal toxicity, observed when soluble forms of metals (i.e., metal salts) are added to soils. Consequently, there are both positive and negative "feedback" reactions to metal pollution. Some of the biological responses known to modify metal mobility and toxicity are now discussed.

7.2.1 Bioconcentration by Soil Biota

Accumulation of bioavailable metals by soil invertebrates (e.g., Cd in earthworms) and plants can potentially mobilize metals selectively into the ecosystem or food chain of animals. This can occur even after the metals have been rendered unavailable to plants by liming (Osté et al. 2001). The overall effect results in transfer of metals from soil-bound forms (possibly mineral) to biomass-bound forms, and finally into humus. Some transfer from the site of contamination is possible via mobility of the organisms.

7.2.2 Soil Organic Matter Accretion

Toxic metal contamination of soils reduces the ability of soil organisms to decompose complex organics (e.g., lignins, cellulose), causing a buildup of soil organic matter, especially fibrous material (Cotrufo et al. 1995; Sauvé et al. 1997; Aoyama 1998; Balabane et al. 1999), and potentially immobilizing metals. Although the living soil biomass is actually lowered by metal stress, at the same time the stress induces an increase in metabolic activity per unit weight of biomass (Brookes and McGrath

1984). The heavy metal stress leads to an overall inhibition of microbial activity and reduced release of metals into soil solution (Chander and Brookes 1991; Hattori 1992; Kuperman and Carreiro 1997).

7.2.3 Generation of Metal-Chelating Compounds

Metal stress provokes the generation of metallothioneins and phytochelatins (cysteine-rich low-molecular weight proteins) in addition to other metal- or metalloid-complexing organics (e.g., citric acid, phytin, polymers) by microbes and plants, potentially mobilizing metals by active efflux, or sequestering them into detoxified forms within cells (Silver and Phung 1996; Rauser 1999). Graminaceous plants under Fe deficiency stress release more phytosiderophores, Fe-chelating molecules, from roots, which in turn induces greater heavy metal uptake by these plants in contaminated soils (Romheld and Awad 2000). Bacterial extracellular polymers, which can mobilise soil-bound Cd and Pb (Chen et al. 1995; Czajka et al. 1997), may provide a feasible means of "cleansing" contaminated soils. For example, Moreno et al. (1999) showed that the incubation of low-Cd and high-Cd sewage sludge compost in soils resulted in higher DOC with the high-Cd compost, attributed to the inhibition of mineralization of this soluble fraction by Cd. It would be interesting to further study the DOC fractions released from metal-contaminated soils to determine if there is a high concentration of cysteine-rich proteins.

7.2.4 Metal Release in Volatile Form

Biological activity can convert metals and metalloids to volatile forms and release them to the atmosphere. This occurs in soils by the microbial methylation of Hg, Se, and As, and probably tin (Sn) as well, a process which might be viewed as a biological mechanism of self-defense. In many cases, however, the metal is converted to a more toxic form, as is the case under anaerobic conditions for the conversion of inorganic Sn to butyltin or methyltin (Donard et al. 1993; Hamasaki et al. 1995), and inorganic Hg to methylmercury (Porvari and Verta 1995). The enhanced release of metallic mercury, Hg^0, to the atmosphere following the application of organic wastes such as sewage sludge to soils has also been found, but this process appears to be driven by physical-chemical conditions at the soil surface (Carpi and Lindberg 1997).

7.2.5 Metal Binding on Cell Walls and Biogenic Minerals

Accumulation of metals on bacterial and fungal cell walls or biogenic Fe and Mn oxides (Douglas and Beveridge 1998), can potentially immobilize metals. For several important heavy metals, such as Ag, Cd, Cr, Cu, Ni, Pb, and Zn, bacterial cell walls and envelopes have a higher binding capacity than clays (Walker et al. 1989; Beveridge et al. 1995). Formation of metal sulfides (e.g., ZnS), carbonates, and phosphates upon accumulation of metals at bacterial surfaces (Schultze-Lam et al. 1996; Labrenz et al. 2000) is also believed to have a role in immobilizing metals.

7.3 EXPERIMENTAL EVIDENCE FOR BIOLOGICAL CONTROL OF METAL SOLUBILITY

There is a considerable body of evidence pointing to biologically mediated control, rather than direct chemical control, of heavy metal solubility. In this section, I will emphasize several key points about metal behavior in soils that support the hypothesis of biological control.

7.3.1 IMPORTANCE OF DOM IN METAL SOLUBILITY AND FACILITATED TRANSPORT

There is often a close correspondence of metal concentration to DOM in groundwater, with a large fraction of the dissolved metals (particularly Cd, Cu, and Pb) in organically complexed form (McBride et al. 1997b, 1999, 2000; Romkens and Dolfing 1998; Temminghoff et al. 1998). At long-term sewage sludge application sites, DOM in shallow groundwater may remain elevated for decades, presumably due to continued microbial activity, resulting in higher dissolved concentrations of a wide range of metals. This is illustrated in Figure 7.1a and Figure 7.1b for several trace elements at an old sludge application site, indicating that even metals considered to have low intrinsic solubility and mobility can migrate to shallow groundwater, attributable to facilitated transport by organic molecules and colloids. The DOM in shallow groundwater at this site remained elevated compared to a nearby control site for some 17 years after a heavy sludge application.

Despite the fact that older metal speciation models have predicted a low tendency for Cd^{2+} and Zn^{2+} to complex with fulvic acids, experimental measurements in Cd-contaminated soils have shown a large fraction of Cd and Zn in many cases to be complexed (Sauve et al. 2000a; Almas et al. 2000). The addition of organic matter such as manure to soils can therefore increase Cd and Zn solubility by increasing the DOM in soil water (Almas et al. 2000). In naturally high-Zn, high-Cd peat soils, we have found a large fraction of the dissolved Zn (but less of the Cd) to be in a nonlabile (strongly complexed) form (Martinez et al. 2002).

As soil pH values are increased above 6.5, the solubility of a number of metals, including Cd, Cu, and Pb, may increase even as the free metal cation activity decreases (McBride 1989). This results from the higher pH increasing DOM, complexing the metals to a higher degree, and inducing metal release from solid phases (see Chapter 6). Biological processes in the soil which generate DOM are therefore likely to increase dissolved concentrations of some heavy metals. DOM in soil solution can solubilize or prevent precipitation of the most "insoluble" phases of metals such as Pb phosphates (Sauve et al. 1998a, 1998b) and Hg sulfide (Ravichandran et al. 1999). Soluble thiol ligands such as thiosulfate can act as carriers to increase Ag^{+} uptake by biota even as the ligand reduces the free Ag^{+} activity (Fortin and Campbell 2001). This indicates that the free ion activity model (FIAM) does not always give the best prediction of biological availability or toxicity.

For metals in the form of oxyanions, such as arsenate, DOM tends to enhance solubility by competing for anion sorption sites on soil minerals (Xu et al. 1991). Thus, microbial activity which generates DOM can indirectly influence oxyanion solubility.

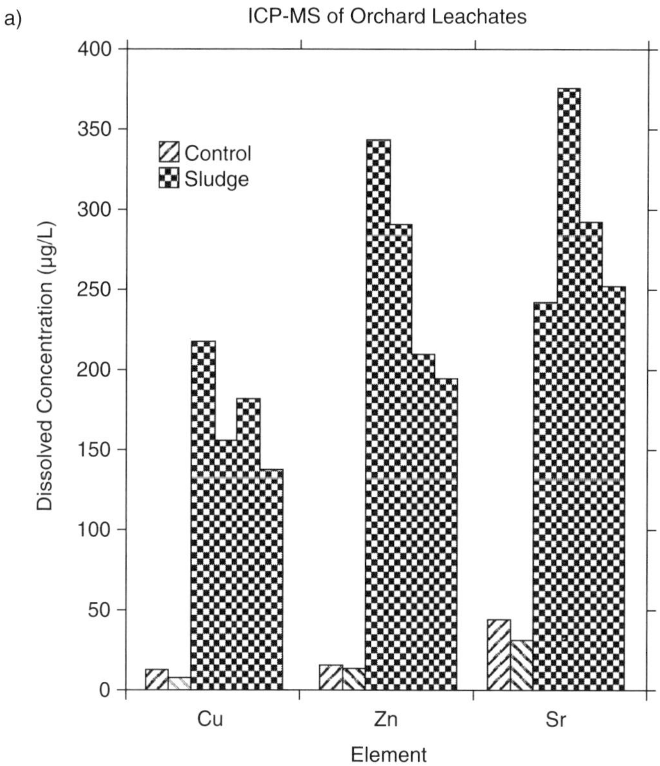

FIGURE 7.1 Dissolved concentrations (μg/ l) of the trace metals Cu, Zn, and Sr (a) and Ag, Hg, and Sn (b) in the shallow groundwater collected from lysimeters at an old sewage-sludge application site and nearby control site. All analyses by ICP-MS spectrometry, with site description given in McBride et al. (Data from McBride MB, Richards BK, Steenhuis T, Spiers G. 1999. Long-term leaching of trace elements in a heavily sludge-amended silty clay loam soil. *Soil Sci* 164:613–623.) A number of the control groundwater samples were below detection, denoted "nd." *Continued.*

Studies of metal "aging" in model mineral systems and sterile mineral soils may be overly simplistic in attempting to predict the long-term fate of toxic metals in soil environments because of the complicating effects of DOM. Metal release from strongly bound and precipitated forms is possible when biological activity is high.

7.3.2 Temperature-Induced Metal Release with Aging

If metals in soils were retained predominantly by high-surface area and porous mineral surfaces, aging would be predicted to allow a greater fraction of the metals to diffuse into inaccessible nanopores or to slowly form more stable bonds (McBride 2000). This is the usual explanation for the well-known "aging" effect, typified by a marked decrease in free metal activity and toxicity in soils following the incorporation of the metal in soluble form. It follows from the traditional explanation of aging, that higher temperatures would accelerate these physical-chemical processes,

b)

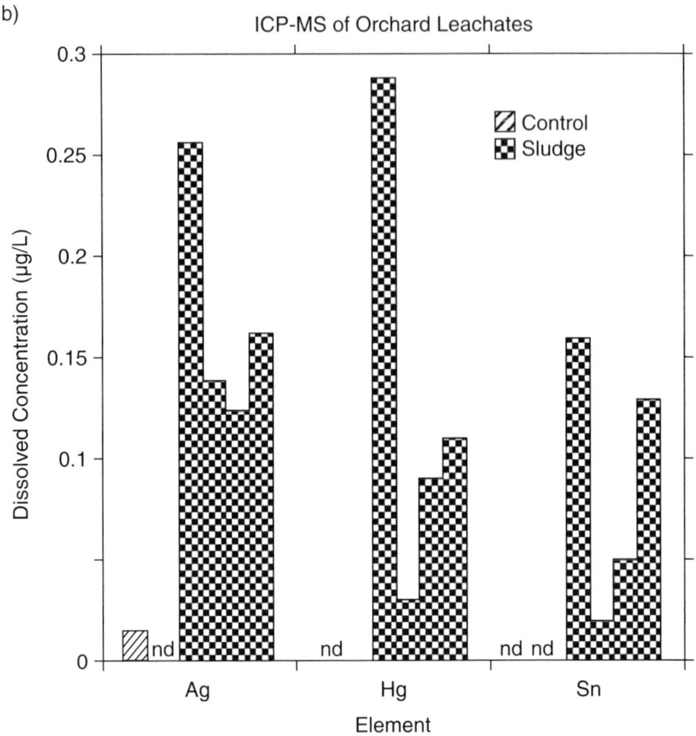

FIGURE 7.1 (Continued.)

thereby reducing metal solubility. In fact, the reverse happens in soils with significant organic matter content (Martinez et al. 1999, 2001), presumably because of enhanced microbial activity at somewhat elevated temperature, and heat-induced release of metals from lysed cells, hydrolyzing organics, and oxidized carbon, nitrogen, and S at more extreme temperatures.

Strictly physico-chemical processes of sorption and diffusion are not highly sensitive to temperature changes of 20 to 40 °C, unlike biological processes. Recent experiments (Qureshi et al. 2001) have demonstrated that the release of trace and heavy metals from organic (peat) soils and sewage sludge into solution is decreased by both low (4 °C) and high (37 °C) temperatures, and by the addition of the biocide, silver (Ag^+). Maximum dissolution and leaching of heavy metals occurred in the 16 to 28 °C range, correlating with the extent of acidification and mineralization of S. These results are consistent with biological, not chemical, control on metal dissolution.

Incubation of soils amended with organic residues such as manures and sewage sludges in closed systems (in the absence of plants) increases metal solubility (Hooda and Alloway 1993, 1994a, 1994b, 1994c; Sadovnikova et al. 1996; Romkens et al. 1999). Incubation of soils amended with sewage sludges can cause the plant-available fraction of Cd and Zn to increase over several months (Stacey et al. 2001). According to Chanmugathas and Bollag (1987), mobilization of strongly bound or fixed Cd is a microbially mediated process which is, however, affected by abiotic soil environmental factors including moisture and aeration. Cadmium mobilization is more pronounced under anaerobic conditions, presumably due to the buildup of organic acids able to complex and solubilize metals, but a marked release of Cd also occurs following soil drying (Chanmugathas and Bollag 1987).

A slow and delayed increase in leachability and bioavailability of some elements (e.g., Mo) applied to soils in sewage sludges (Richards et al. 2000) may reflect the oxidation of organic C and S, with release of S-bound forms.

When these observations on aged soils are considered, it is apparent that aging does not necessarily result in decreased metal solubility, toxicity, or leaching. In any event, the biological influence must be considered as a contributor to the "aging" effect, and in fact may be responsible for the mitigation of the activity and toxicity of recently added toxic metals. As soil biota acclimatize to the introduced metal stress, they may accumulate metal-binding polymers and lower the activity of the toxic metal cation.

7.3.3 High Affinity of Most Metals for Organic Matter

There is ample circumstantial evidence for a close association of several trace and heavy metals with soil organic matter. For example, we have found a correlation of toxic metal concentrations in field-aged metal-contaminated soils to elements in organic matter, particularly carbon and S (McBride et al. 2000). In fact, the partitioning coefficients (K_d) of soils for metals such as Cd^{2+} and Hg^{2+} are to a large degree determined by the organic carbon content of soils (Lee et al. 1996, 2001; Yin et al. 1996), not the clay or oxide content. Using a multisite sorption model, Weng et al. (2001) found that organic matter was the most important metal adsorbent and explained the activities of Cd^{2+}, Cu^{2+}, Ni^{2+}, and Zn^{2+} in the soil solutions of field-contaminated soils, whereas sorption to iron oxide appeared not to be important. Only for Pb^{2+} was strong sorption to Fe oxide in addition to organic matter predicted.

Why then is there a tendency to assume that minerals such as Fe and Mn oxides ("free oxides"), often at low levels in soils, act as the primary reservoirs for heavy metals and control solubility? This seems in part to be due to the assumption that soil oxides have the same adsorptive power as pristine laboratory-synthesized oxides, the latter often having very high surface areas (low crystallinity). In reality, oxides prevalent in highly weathered soils frequently have low surface areas and are not very reactive. In younger soils with oxides of high surface area, such as ferrihydrite, the oxide surface is occupied by bound silicate, organic matter, phosphate, and other "impurities." Trace metals may adsorb on these surfaces, but not with the selectivity expected, based on laboratory studies with synthetic oxides, and metal sorption may actually be in association with adsorbed organic matter (Davis 1984; Murphy and Zachara 1995). For example, the well-known high affinity of laboratory-synthesized pure iron oxide for Pb^{2+} cations is not demonstrated by natural iron oxides collected from spodosols (Sauvé et al. 2000b). Adsorbed silica on iron oxides alters the surface properties, reducing the affinity of the surface for arsenic and DOM (Davis et al. 2001), and probably for other metals as well. Much of the total Pb in the surface soils of industrialized countries is now known to be from anthropogenic sources, and this "*de novo*" Pb is strongly associated with organic matter (Semlali et al. 2001).

Conversely, "native" Pb originating from the soil's parent material is not associated strongly with organic matter, presumably because it remains largely retained within unweathered primary minerals (Semlali et al. 2001). It is reasonable to expect that other metal contaminants introduced into soils could strongly associate with organic matter, whereas lower concentrations of native metals remain largely sequestered in the mineral fraction.

A second source of belief in the importance of oxide minerals in the strong retention of heavy metals is the uncritical interpretation of the results of sequential extraction procedures for metals in soils. These often seem to show that a large fraction of most heavy metals are either sorbed to, or entrained within, the operationally-defined "oxide" fraction (Chapter 1). However, despite the intended purpose of such extractions, most are unable to differentiate among the various bound forms of metals in soils. This is due in part to the inherently aggressive nature of the extracting reagents, which are not very selective in dissolving metals from various minerals and organic substances, and by metal repartitioning during the multistep extraction procedure (Martin et al. 1987; Nirel and Morel 1990; Kim and Fergusson 1991; Bermond 1992; Xiao-Quan and Bin 1993). A recent study showed that there is no relationship between any of the sequentially extracted Cd fractions and a separate direct measure of Cd lability based on isotope exchange (Ahnstrom and Parker 2001), further undermining the physico-chemical meaning of sequential extraction and whether it should be assumed to have any bearing on the bioavailability of metals in soils.

There is empirical evidence that under natural conditions, retention of toxic metals against losses by leaching over the long term is highly dependent on soil organic matter content and pH. For example, Boekhold (1992) demonstrated, in the smelter-contaminated Kempen region of the Netherlands, that a major proportion of the variability of Cd content in the topsoil was explained by soil pH and organic matter content. Based upon a statistical analysis of data reported by Holmgren et al. (1993), statewide mean concentrations of Cd, Zn, Cu, Ni, and Pb in agricultural mineral soils in the U.S. were found to be significantly correlated to soil organic matter content ($p < 0.05$). Other important soil properties, CEC and pH, were less consistently significant, but higher pH and CEC correlated with higher metal concentrations in soils. The strongest association with organic carbon (OC) was obtained for Cd:

$$\text{Soil Cd (mg/kg total)} = 0.102 + 0.094 \times \text{OC (\%)} \quad r = 0.51 \quad (p < 0.01),$$

which is interesting given the common belief that Cd binds only weakly to soil organic matter compared to other heavy metals. Soil texture (clay content) was nevertheless quite significant to soil metal content, as ranking the soils by texture revealed that fine-textured soils were generally higher in all metals analyzed (Holmgren et al. 1993). It is possible that the generally higher Cd in high-clay soils is due to retention on clay exchange or specific-sorption sites, but the ability of clays to protect organic matter from decomposition (Six et al. 2002) could provide an indirect basis for explaining the clay–Cd correlation.

7.3.4 The Important Role of Sulfur in Strong Metal Bonding

There is now clear evidence for the preferential binding of chalcophilic heavy metals to reduced forms of S, including thiols and secondary sulfide minerals, in the organic matter of both aerobic and flooded or wet soils (Morra et al. 1997; Xia et al. 1997, 1998, 1999; Zhou et al. 1999; Skyllberg et al. 2000), as well as in sediments and oxic fresh water (Rozan et al. 1999, 2000). Research has provided direct spectroscopic evidence (EXAFS) for preferential chalcophilic metal (e.g., Hg, Zn) bonding to thiol ligands in soil organic matter. Evidence for heavy metal retention by reduced S (sulfides, thiol groups) has been found in naturally metalliferous peat soils that have been drained for agricultural use for decades. In these soils, the log K_d values are at least 0.5 units higher for Cd than for Zn (Martinez et al. 2002), a result that can only reasonably be explained by the preferential bonding of Cd by reduced S. Generally, Zn would be expected to bond more strongly to minerals (Fe and Al oxides, silicate clays) and organic matter than Cd at any particular soil pH, so the selective retention of Cd suggests the involvement of S. The importance of reduced S species in heavy metal bonding is least expected for aerobic soil environments and may be an indication that reduced organic S is somehow chemically or physically protected from oxidation.

7.3.5 Behavior of Metals in Model Mineral-Organic Systems

Experiments with model mineral-organic systems and with soils indicate that, at least for some metals, the free metal cation activity (and predicted toxicity) of the metal is reduced by strong complexation to both insoluble (humics, living biomass) and soluble (fulvic acids, simple organic acids) forms of soil organic matter. Commonly, biological activity lowers free metal cation (e.g., Cd^{2+}, Cu^{2+}, and Pb^{2+}) activity, yet raises the total dissolved metal solubility relative to that which would be predicted, based on mineral adsorption or precipitation processes. For example, when the relative ability of a high-surface Fe oxide was compared to organic matter, the organic matter was able to retain a given amount of Cu^{2+} at lower free Cu^{2+} activity than the oxide (McBride et al. 1998).

In multiphase model mineral-organic systems, organic matter has been demonstrated to have an intrinsically stronger affinity for metals such as Cd, Cu, and Co than the common silicate and oxide minerals (McLaren et al. 1983; Zachara et al. 1992, 1994). Martinez and McBride (1998, 1999) measured the ability of noncrystalline iron hydroxide to control Cd^{2+}, Cu^{2+}, Pb^{2+}, and Zn^{2+} solubility by sorption or coprecipitation in the absence and presence of natural organic matter. It was found that Fe hydroxide coprecipitation, while very effective in reducing Pb solubility, failed to lower the solubility of free Cu^{2+}, Cd^{2+}, or Zn^{2+} cations below phytotoxic levels. The presence of organic matter lowered both the labile (potentially toxic) and dissolved concentrations of these 3 metals. However, over longer periods of equilibration with organic matter, both Cu and Pb solubility increased concomitantly with a spontaneous increase in DOM, although the lability of dissolved Cu and Pb remained low throughout the aging period of about 200 d.

TABLE 7.1
Properties of filtered water extracts after amending soil from an old sewage-sludge application site

Soil extract property	Untreated soil	Soil + peat (10%)	Soil + Fe oxide (10%)
pH	6.7	6.6	6.7
Light absorbance (650 nm)	0.069	0.080	0.022
pCu	8.2	8.8	8.1
Soluble Cu (mg/L)	0.30	0.36	0.04
Soluble Zn (mg/L)	0.18	0.29	0.13
Soluble Cd (μg/L)	9	7	16
Soluble Cr (μg/L)	67	63	23
Soluble Ni (mg/L)	0.22	0.19	0.08
Soluble Mo (μg/l)	63	61	2
Soluble As (mg/L)	0.25	0.34	0.02
Soluble P (mg/L)	0.86	1.26	0.17

In a soil "remediation" experiment, when synthetic Fe oxide was added to a long-contaminated sludge-amended soil, there was a relatively beneficial effect in terms of reducing the solubility of As, Cu, and Mo; however, the solubility of Cd and Zn were not lowered (Table 7.1). As there was evidence of effective removal of DOM from the soil solution by the Fe oxide (based on the loss of color and reduced light absorbance), those dissolved metals largely in the form of DOM complexes (e.g., Cu, Ni) may have been fairly effectively sorbed, whereas those with lower affinity for DOM (e.g., Cd, Zn) were not. In addition, the oxide showed affinity for metals likely to occur as anions in soil solution (As, Mo). The activity of free Cu^{2+} was not reduced by the oxide amendments but was reduced by peat (organic matter) addition, as would be expected for a metal with a known high affinity for complexation sites on solid and DOM. However, in general, the peat amendment did not effectively lower heavy metal solubility, probably attributable to the fact that this amendment actually raised the level of DOM and may have enhanced the formation of soluble metal-organic complexes.

The addition of Al and Fe hydroxide amendments to a severely Cu-contaminated soil were unable to lower Cu^{2+} activity or phytotoxicity (McBride and Martinez 2000). It is possible that the surfaces of these pure oxide minerals, once added to soils, became coated by DOM, phosphate, and silica, or become saturated by metals, and were quickly rendered ineffective in lowering metal toxicity. Surface saturation would be particularly likely for soils with high levels of metal contamination, and therefore large reserves of potentially labile metals could be released from the soil.

Numerous alkaline materials have been used to "remediate" contaminated soils by immobilizing heavy metals. There have been various claims made for the sequestration or fixation of the metals, but recent work suggests that the reduction in metal solubility and phytoavailability can be explained by the pH change alone (liming effect) (Osté et al. 2001).

The conclusion from these studies is that metal lability and toxicity, but not necessarily metal solubility or mobility, is lowered by organic matter. Biological activity in these aged model systems may be the cause of increasing DOM and soluble metals.

The advantage of soil organic matter over minerals for metal binding is greatest under acidic conditions and tends to diminish at higher pH, in part because of the alkaline dissolution of organic matter which brings metals into solution. Although oxides of Al and Fe added to soils show promise in lowering the solubility of As, Cu, Mo, and DOM in the short-term, they are relatively ineffective in lowering the solubility of Cd or Zn. A reduction in metal solubility, however, does not necessarily translate into a reduction of free metal activity, toxicity, or bioavailability.

7.3.6 Rhizosphere Effects on Metal Solubility

Enhanced solubility of metals such as Cd and Zn in the rhizosphere has been demonstrated (Hamon et al. 1995; Whiting et al. 2001). Higher Cd solubility in the soil water as a result of plant root growth is presumably a response to increase in DOM which occurs when biological activity in the soil is higher. The higher DOM was actually found to reduce free Cd^{2+} ion activity, the probable consequence of Cd^{2+} complexation by fulvic acids.

Romkens et al. (1999) found a large effect of plant growth on soluble Cu species in a Cu-contaminated soil, but dissolved Cu was reduced in this case, and free Cu^{2+} activity was lowered even more dramatically. This was at least partly due to an increase in soil pH caused by plant growth, which led to higher DOM. Root exudates, especially low molecular weight organic acids (e.g., lactic, oxalic, acetic, etc.), have an essential role in making sparingly soluble trace metals available to growing plants. The amendment of soils with organic-rich materials such as sewage sludges enhances this effect (Romheld and Awad 2000; Koo et al. 2001).

Biological activity in the vicinity of plant roots (the rhizosphere effect) is known to modify pH locally. The uptake of nitrate tends to raise pH, whereas ammonium uptake lowers it. In poorly buffered soils, biological activity of plant roots and microbes can lower pH by 0.5 units when environmental conditions are favorable. While pH can be considered a master variable in controlling soil chemical processes, it is often under the influence of biological activity in soils. Temporal changes in soil pH in response to the intensity of biological activity can have marked effects on trace and heavy metal activity and solubility.

7.3.7 Biological Control of Bioavailability

Bioassay tests of metal toxicity in soils are sensitive to the organism chosen for the test. It is well known, for example, that plant uptake of metals from a particular soil is sensitive to plant species, and even to cultivars within species. This suggests that plants modify conditions in the rhizosphere to alter metal solubility; that is, plants exert some control over solubility and uptake (as discussed with the rhizosphere effect described previously). Thus, for a particular soil contamination level of a toxic metal, uptake potential by different plant species can be dramatically different. Biota which absorb metals by mechanisms different from those of plants may show an entirely different relationship of metal bioavailability to soil physico-chemical factors. For example, Beyer et al. (1982) found that, in soils contaminated by Cd from sludge application, plant uptake of Cd was very much reduced by liming the soil to raise pH, yet bioaccumulation of Cd by earthworms was only slightly affected by liming. In agreement with this result, Ost́e et al. (2001) found that, while the addition of alkaline materials to soils reduced plant availability of Cd and Zn, uptake by earthworms was barely affected, indicating that the soil solution concentration of a toxic metal does not predict exposure of these soil animals to the metal. Other studies have also found that for various soil organisms exposed to metal-contaminated soils, including snails (*Helix aspersa*) and Collembolans (Crommentuijn et al. 1997; Pedersen et al. 1997; Scheifler et al. 2003), there is not a strong relationship between soil solution concentration of the metal and the degree of metal bioaccumulation in the organism. Snails were shown to accumulate Cd to some degree from the nonlabile soil Cd (as defined by isotopic exchange), contrary to the "pore-water hypothesis" of metal bioavailability (Scheifler et al. 2003).

These results reveal that the definition of metal "bioavailability" is problematic, and may have to include more than the soluble or labile fraction of metals as determined by chemical extractions. Bioavailability must take into account the species-specific physiological processes that modify the conditions of absorption by the organism.

7.3.8 Sensitivity of Metal Solubility to Oxidation Status

Chalcophilic metals such as Cd, Cu, Pb, and Zn are retained in sediments and in marine-influenced, submerged soils (acid sulfate soils) in sulfidic and organo-sulfur forms. Upon aeration or drying of these materials, there is a release of these metals, particularly Cd, into more soluble and leachable forms associated with the oxidation of S and a concomitant pH decrease (Gambrell et al. 1991; Saeki et al. 1993; Tack et al. 1996). A similar release of metals, although less extreme, can be expected in aerated soils that contain organic matter with reduced S (thiol) groups. The S oxidation reaction is actually harnessed under extreme conditions of pH and metal toxicity in "bioleaching," a process where metals such as Ni, Cd, Cu, and Zn from low-grade sulfide ores or sewage sludges are scavenged by *Thiobacillus ferrooxidans* bacteria, converting insoluble metal sulfides into soluble metal sulfates (Bosecker 1997; Kitada et al. 2001). In drained peat bogs high in S and Zn, Zn release into soil solution leads to phytotoxicity as bacteria oxidize reduced S to sulfate (Martinez et al. 2002).

For some metals, release into solution under anoxic conditions might be expected due to reduction of Fe or Mn oxides, potential scavengers of trace metals. However, waterlogging of mineral soils appears to decrease the most soluble and bioavailable forms of Cu and Zn even though increases in dissolved Fe and Mn indicate reductive dissolution of Fe and Mn oxides (Lu et al. 1981; Dutta et al. 1989). Soil pH rises under anaerobic conditions; hence, it is possible that sorption or precipitation reactions of carbonates or hydroxides are involved in limiting metal solubility, but metal sulfide formation appears likely to be involved (see Chapter 8).

For elements with several possible oxidation states in soil, such as As, Cr, Hg, Mn, and Se, a redox-sensitive solubility could also indicate microbial or chemical transformation to a more or less soluble species. Thus, for example, As tends to be more soluble and bioavailable in flooded soils, either because it is released from Fe oxide sorption sites as Fe^{3+} is reduced to Fe^{2+} or because arsenate (As(V)) is reduced to arsenite (As(III)), a more soluble and mobile form of As (Xu et al. 1991). Conversely, Cd generally becomes less soluble when soils are strongly reduced under flooding because of the formation of CdS, a very insoluble precipitate. It is not clear whether these conversions of metal form and solubility are chemical or biological, but it is apparent that biological activity modifies the redox conditions that cause such changes.

7.4 IMPLICATIONS OF BIOLOGICAL CONTROL: EXPLAINING METAL LOSSES FROM SOILS

Any attenuation of toxic metal bioavailability that might be observed over time is usually attributed to physical-chemical processes such as slow diffusion into nanopores or into the solids themselves. Attenuation by actual metal migration out of the topsoil has generally been assumed to be negligible. However, trace and heavy metal losses from soils contaminated by sewage sludges by leaching, facilitated transport, or biological activity may be significant over a time period of decades, as has been shown by numerous field studies that have failed to account for all or most of the metals applied on these sites (e.g., Baveye et al. 1999; McBride et al. 1999, 2000, and references therein). This conclusion of long-term metal losses in the field is controversial because there are numerous errors that enter into attempts to calculate metal mass balances on old sewage sludge application sites. These include uncertainties in metal application rates in the field, variability in metal concentration of the applied sludges, sampling errors arising from imperfect mixing of the sludge with the topsoil, uncertainties about soil bulk densities required for converting from soil metal concentrations (mg/kg, weight basis) to area- or volume-based (kg/ha) metal loading, and potential losses from field plots by tillage or erosion. Thus, attempts at metal mass balances in the field have often found such high variability that clear confirmation of losses has not been possible.

In 1 experiment, Berti (1992) reported mass balances ranging from 45 to 150% for several heavy metals applied in sewage sludges at 3 field sites and concluded that losses had not occurred by leaching (Berti and Jacobs 1996). Although it is true that the large variability prevented any clear conclusions, it should be noted that out of 36 estimates of metal mass balance (6 metals at 3 sites for 2 years), Berti (1992) recorded 19 cases of metal recoveries substantially below 100% (< 80%) and only 2 cases of recoveries substantially greater than 100% (>120%). Thus, we must conclude that the evidence for metal loss is stronger than the evidence for complete retention in the soil profile.

Many of the errors inherent to mass balance calculations can be eliminated by estimating metal loss from the topsoil relative to loss of 1 particular "index" metal, preferably a metal expected to have very low leachability (McBride et al. 1997b, 1999). I have reanalyzed Berti's data using Pb as the "index" metal and found metal deficits in the topsoils at all 3 sites and for all 5 metals analyzed, as shown in Figure 7.2. This result strongly suggests that there has been preferential loss of Cd,

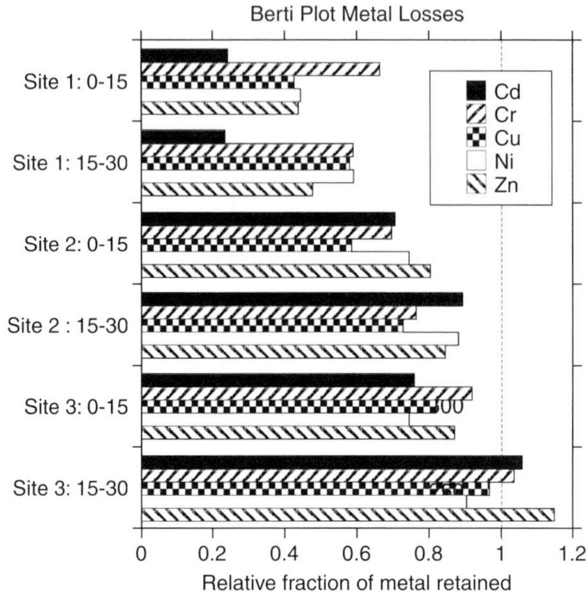

FIGURE 7.2 Calculated fractions of total Cd, Cr, Cu, Ni, and Zn retained relative to Pb in the 0 to 15 and 15 to 30 cm soil layers at 3 long-term sewage sludge application sites described in Berti. (Data from Berti WR. 1992. Soil trace elements loaded in high amounts from sewage sludge applications: chemical fractionation, movement, and bioavailability. Ph.D. thesis. Michigan State University. East Lansing, MI.)

Cr, Cu, Ni, and Zn relative to Pb, particularly in the topsoil (0 to 15 cm), and to a lesser degree in the subsoil (15 to 30 cm). The deficits based on this "index" metal ratio method cannot be explained by tillage, erosion, sampling bias, errors in metal loading estimates, or variations of bulk density, all potentially important sources of error in standard mass balance calculations. It is perhaps interesting that the greatest relative deficits of the metals are found at site 1, which had received the smallest heavy metal loading (with the exception of Cd). This result raises the possibility that metal release from solid phases and preferential loss is biological, as biological activity at the severely heavy metal-contaminated sites 2 and 3 may have been limited by toxicity. Conversely, had chemical processes been dominant in controlling metal release and leaching, greater relative losses from the most highly contaminated sites might have been expected. However, because biological activity in the soils at these sites was not measured, it is not possible to test the biological hypothesis of metal mobilization.

Gomez et al. (1992) described another long-term (1974 to 1989) field experiment attempting a mass balance on the trace and heavy metals applied with long-term sewage sludge application. They found large deficits in the top 1.0 meter of the soil profiles for several metals, based on a careful accounting of all excess metals in the excavated profiles. Their method avoided the problem of uncertain bulk densities by determining the total metals in the excavated soil on a volume rather than weight basis. As shown in Figure 7.3, some of the metal mass balances for the 2 sludges

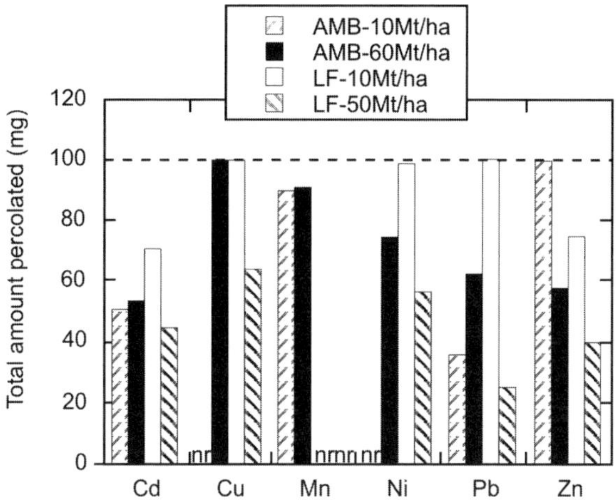

FIGURE 7.3 Percentage of total metals applied at a long-term sewage sludge application site recovered in the 0 to 100 cm soil depth. Data for 2 application rates (10 and 50 Mt/ha) of 2 sludges (AMB and LF), taken are shown. (Data from Gomez A, Solda P, Lambrot C, Wilbert J, Juste C. 1992. Bilan des elements traces metalliques transferes dans un sol sableux apres 16 annees d'apports continus et connus de boues de station d'epuration et de fumier de ferme en monoculture irriguee de mais. Ministere de l'Environnement. Convention de Recherche No. 89–256.) Those metal recoveries which could not be estimated because of the low loadings relative to the background levels of metals in the soils are indicated by "nr."

applied (AMB and FA) at 2 rates (averaging 10 and 50 metric tonnes (Mt)/ hectare/year) were near 100% (indicating no losses by leaching or other processes) but most were not. Cadmium deficits averaged close to 50% across all treatments. The highest metal recoveries tended to be for the lower sludge (and metal) loadings, as might be expected if the sorptive capacity of the light-textured soil was being overwhelmed at the higher application rates. Some of the highest metal recoveries were found for the LF sludge at the low application rate. The LF sludge was severely contaminated with Cd and Ni, and may have been toxic to soil biota. The authors noted that, whereas the AMB 10 Mt/ha sludge application over 16 years (160 Mt/ha total) resulted in no increase in soil organic matter, the LF 10 Mt/ha produced an increase in organic matter even though the cumulative application (50 Mt/ha) was much less. This could mean that metal toxicity suppressed microbiological activity in the LF plots, limiting the extent of leaching from biological mobilization. However, it is also possible that the nature of organic matter in the 2 different sewage sludges, LF and AMB, was different, explaining the difference in decomposition rate.

Although Gomez et al. (1992) concluded that the mass balances "clearly evidenced an incomplete metal recovery," they argued that the deficits were "due neither to the leaching of trace elements downward ... nor to their removal by plant offtake" and that the most plausible explanation of the metal deficits was soil displacement during cultivation practices or water runoff. However, the fact that some metals, such as Cd, were lost to much greater degree than others (there was a near-complete recovery of certain of the applied metals, especially at the lower application rate)

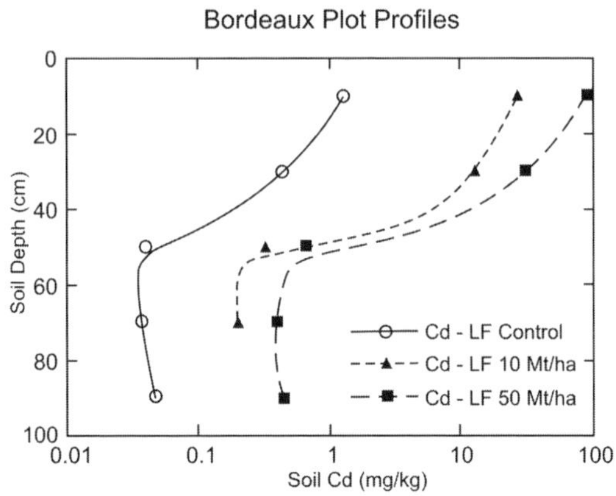

FIGURE 7.4 Soil profile of Cd concentration at the long-term LF sewage sludge application site described in Figure 7.3. Data for 2 application rates (10 and 50 Mt/ha) of LF sludge are shown.

argues against a physical process such as tillage or erosion accounting for the losses. Such processes would not be selective in removing particular metals from the amended topsoils. The researchers' dismissal of leaching as a possible explanation, despite the very coarse texture of soils at the site, was based on the appearance of the metal profiles with depth on the sludge plots, which indicated that leaching would have to have been very deep (> 1 m) to explain the metal losses. An example is given in Figure 7.4 for the Cd profile in the top 1.0 m of the LF sludge-amended and control plots. Very high Cd concentrations in the top 30 to 40 cm of the sludge-amended plots decrease to much lower concentration below 50 cm. It is common to assume from a steep profile such as this that virtually no metal migration or leaching below the cultivated layer has occurred. However, it is also clear from Figure 7.4 that the deep subsoil, to 1.0 m depth at least, is contaminated with Cd (by comparison to the control plot). How can we be sure that a large fraction of the total Cd has not migrated through the subsoil, either in soluble or colloidal form?

There are several factors that would tend to prevent Cd (and other metals), once mobilized, from accumulating substantially in the subsoil, and could explain the metal profile in Figure 7.4 even if substantial downward migration of Cd had occurred

1) Facilitated transport: Migrating metals may be in the form of colloidal or dissolved metal-organic complexes that readsorb less readily in the subsoil than the free metal cations,
2) Preferential flow: Much of the subsoil matrix may be bypassed by metal-contaminated water moving downward,
3) Lack of sorptive materials: The subsoils are much lower in organic matter than the topsoil and have less capacity to chemisorb metals, and
4) Subsoil pH: At this particular site, soil pH becomes acidic at depth, further decreasing the chance of metal readsorption, particularly on minerals.

All of these factors would suggest that, once mobilized, metals would tend to migrate deeply, not readsorb in the shallow subsoil as predicted by the convective-dispersive equation. This behavior and profile of heavy metals has been found in a fine-textured soil (Richards et al. 1998), even though the subsoil had a higher potential to adsorb metals (near-neutral pH, clayey texture) than the sandy soil described by Gomez et al. (1992). The lack of accumulation of heavy metals in the subsoil can be explained largely by preferential flow processes.

Even where tillage and erosion were not significant factors in the migration of heavy metals in long-term sludge-amended plots, such as the no-till continuous-sod plots at the University of Guelph, marked decreases of metal concentrations in the organic-rich surface layer were observed 17 years after the last surface application of sludge (see Figure 7.5). This was the case for all metals investigated (Cd, Cu, Ni, and Zn) for 3 different sludge types applied (data for only 1 sludge type, the high Ca-sludge, is shown in Figure 7.5). The organic matter content of the 0 to 5 cm soil surface layer decreased over the 17 years for the 2 most organic-rich sludge amendments (data not shown) but not for the Ca-sludge, consistent with partial decomposition of the applied material. It is possible that organic matter mineralization or the action of invertebrates had a role in the mobilization of metals from the surface.

Metal losses can be related qualitatively to the K_d values of individual metals (McBride et al. 1999; Steenhuis et al. 1999), as metals with the largest K_d values are least mobile. For metals which are largely in the complexed form (especially Cu), most of the mobilized metal is in the form of metal-organic complexes (McBride et al. 1997b), unlikely to be a form in reversible equilibrium with soil solids. This fact suggests that the convective-dispersive equation would be unlikely to predict adequately the mobility of a number of trace and heavy metals. The assumption of local equilibrium in the convective-dispersive model, even if preferential flow were not a serious

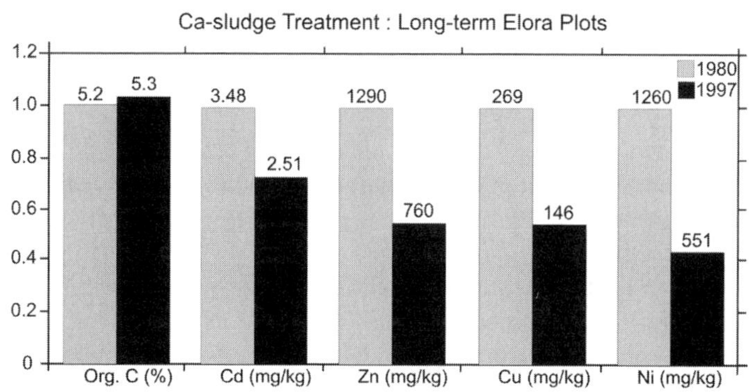

FIGURE 7.5 Relative concentrations of organic carbon, Cd, Cu, Ni, and Zn in surface soil (0 to 5 cm) of a plot amended with high-Ca sewage sludge from 1973 to 1980, measured in 1980 and again in 1997. Numbers labeling the bars are the absolute soil concentrations of the metals (mg/kg) and organic C (%). (Data from McBride MB, Martínez CE, Topp E, Evans L. 2000. Trace metal solubility and speciation in a calcareous soil 18 years after no-till sludge application. *Soil Sci* 165:646–656.)

complication, renders this model untenable for real situations where biological processes, colloidal particles, and DOM strongly influence metal mobilization.

At numerous sites where metal losses have been observed over a period of several years to decades, estimated losses have frequently fallen in the 40 to 60% range for several heavy metals (Baveye et al. 1999; McBride et al. 1999, 2000). The magnitude of these losses cannot, however, be explained by the K_d values of the metals at the time these losses were detected. They therefore suggest a biological process of metal mobilization that was most active shortly after application of metal-contaminated waste. Steenhuis et al. (1999) derived an exponential function to describe metal leaching loss from contaminated topsoil; this model predicts most of the leaching losses to occur relatively rapidly, with metal solubility diminishing to low levels after several years. Thus, the K_d values measured for metals on old sewage sludge application sites are not indicative of the K_d values operative during and shortly after application. Nevertheless, the K_d values measured for a number of metals and other elements at an old sludge application site were correlated to the estimated extent of loss from the topsoil. That is, elements with a strong tendency to bind to soil solids (high K_d) such as Ag had small relative tendency to be lost, whereas other elements with weaker soil binding (e.g., Mo, S, Sr) were lost to a greater degree (McBride et al. 1999).

7.5 SUMMARY

Natural microbial processes in aerobic soils oxidize organic carbon, nitrogen, and S, generate acidity, and reduce the metal-binding capacity of soils, which could be expected to have the long-term consequence of increasing toxicity of several heavy metals. When soils are contaminated with heavy metals, biological defenses against metal toxicity are provoked. These can have the effect of increasing metal mobility or, conversely, increasing metal sequestration and retention in soils. Consequently, there are both positive and negative biological "feedback" reactions to metal pollution. Biological responses to metal stress that are known to modify metal mobility and toxicity are

1) reduced decomposition of complex organic molecules and buildup of soil organic matter,
2) generation of metallothioneins and other metal complexing agents,
3) conversion of metals and metalloids to volatile forms, and
4) accumulation and immobilization of metals on bacterial and fungal cell walls by sorption or by formation of insoluble biogenic minerals.

It is difficult to ascertain what the cumulative effect of biological processes on solubility and bioavailability would be, as some of the defences against toxicity mobilize metals into the groundwater or atmosphere. It seems, however, that the role of organic matter and living biomass in converting toxic metals into less toxic (or more mobile) forms has been underappreciated and is an important part of the mitigation of toxicity observed with aging. Misconceptions have tended to persist about the strength of heavy metal bonding to organic molecules, due in part to early work that measured metal bonding to organic matter at unrealistically high metal loadings (near saturation), saturating and overwhelming thiol groups and other highly selective ligands. Even now, databases used to estimate metal binding to soil organic matter predict low affinity of Cd for organic ligands in soils, despite considerable experimental evidence from soils to the contrary. Sequential extraction methods, used with the intent of determining the forms of trace elements in the soil solids, are nonselective and unreliable and cannot determine the extent of metal binding to different soil constituents. Experiments with model mineral-organic systems and with soils indicate that, at least for some metals, the toxicity of the metal is reduced by strong complexation to both insoluble (humics, living biomass) and soluble (fulvic acids, simple organic acids) forms of soil organic matter. Commonly, biological activity lowers free metal cation (e.g., Cd^{2+}, Cu^{2+}, and Pb^{2+},) activity, yet raises the total dissolved metal solubility, relative to that which would be predicted based on mineral adsorption or precipitation processes. Consequently, heavy metal losses from contaminated soils into shallow groundwater by leaching or facilitated transport may be significant over a time period of decades.

If organic S ligands prove to have a critical function in limiting chalcophilic metal toxicity in soils, as hypothesized in this chapter, this raises the question of whether this protective effect can be expected to persist indefinitely. Although heavy metals appear to stabilize organic matter against decomposition, oxidation of the organic C, N, and S is inevitable in the long run, and metal toxicity could eventually be expected to increase as the strongly metal-complexing ligands are lost. Thus, the maximization of soil organic matter levels in soils (e.g., by reduced tillage, choice of cropping system, and maintenance of a high water table) may be useful in suppressing the toxicity and mobility of several heavy metals. Conversely, bringing undisturbed metal-contaminated soils into cultivation might be expected to promote organic matter (and S) oxidation and increase metal bioavailability, as Zhao et al. (1997) noted for Cd on a long-term sludge application site.

REFERENCES

Ahnstrom ZAS, Parker DR. 2001. Cadmium reactivity in metal-contaminated soils using a coupled stable isotope dilution-sequential extraction procedure. Environ Sci Technol 35:121–126.

Almas AR, McBride MB, Singh BR. 2000. Solubility and lability of cadmium and zinc in two soils treated with organic matter. Soil Sci 165:250–259.

Aoyama M. 1998. Effects of heavy metal accumulation in apple orchard soils on the mineralization of humified plant residues. Soil Sci Plant Nutr 44:209–215.

Balabane M, Faivre D, van Oort F, Dahmani-Muller H. 1999. Mutual effects of soil organic matter dynamics and heavy metal fate in a metallophyte grassland. Environ Pollut 105:45–54.

Baveye P, McBride MB, Bouldin D, Hinesly TD, Dahdoh MSA, Abdel-Sabour MF. 1999. Mass balance and distribution of sludge-borne trace elements in a silt-loam soil following long-term applications of sewage sludge. Sci Tot Environ 227:13–28.

Bermond AP. 1992. Thermodynamics applied to the study of the limits of sequential extraction procedures used for the speciation of trace elements in sediments and soils. Environ Technol 13:1175–1179.

Berti WR. 1992. Soil trace elements loaded in high amounts from sewage sludge applications: chemical fractionation, movement, and bioavailability [PhD Thesis]. Michigan State University, East Lansing, MI.

Berti WR, Jacobs LW. 1996. Chemistry and phytotoxicity of soil trace elements from repeated sewage sludge applications. J Environ Qual 25:1025–1032.

Beveridge TJ, Schultze-Lam S, Thompson JB. 1995. Detection of anionic sites on bacterial walls, their ability to bind toxic heavy metals and form sedimentable flocs and their contribution to mineralization in natural freshwater environments. In: Allen HE, Huang CP, Bailey GW, Bowers AR, editors. Metal speciation and contamination of soil. Boca Raton, FL: Lewis Publishers. p 184–205.

Beyer WN, Chaney RL, Mulhern BM. 1982. Heavy metal concentrations in earthworms from soil amended with sewage sludge. J Environ Qual 11:381–385.

Boekhold AE. 1992. Field scale behavior of cadmium in soil [PhD Thesis]. Wageningen Agricultural University, the Netherlands.

Bosecker K. 1997. Bioleaching: metal solubilization by microorganisms. FEMS Microbiol Rev 20:591–604.

Brookes PC, McGrath SP. 1984. Effects of metal toxicity on the size of the soil microbial biomass. J Soil Sci 35:341–346.

Carpi A, Lindberg SE. 1997. Sunlight-mediated emission of elemental mercury from soil amended with municipal sewage sludge. Environ Sci Technol 31:2085–2091.

Chander K, Brookes PC. 1991. Effects of heavy-metals from past applications of sewage-sludge on microbial biomass and organic-matter accumulation in a sandy loam and silty loam UK soil. Soil Biol Biochem 23:927–932.

Chanmugathas P, Bollag J-M. 1987. Microbial mobilization of cadmium in soil under aerobic and anaerobic conditions. J Environ Qual 16:161–167.

Chen JH, Lion LW, Ghiorse WC, Shuler ML. 1995. Mobilization of adsorbed cadmium and lead in aquifer material by bacterial extracellular polymers. Wat Res 29:421–430.

Cotrufo MF, DeSanto AV, Alfani A, Bartoli G, DeCristofaro A. 1995. Effects of urban heavy metal pollution on organic matter decomposition in Quercus ilex L. woods. Environ Pollut 90:81–87.

Crommentuijn T, Doornekamp A, VanGestel CAM. 1997. Bioavailability and ecological effects of cadmium on *Folsomia candida* (Willem) in an artificial soil substrate as influenced by pH and organic matter. Appl Soil Ecol 5:261–271.

Czajka DR, Lion LW, Shuler ML, Ghiorse WC. 1997. Evaluation of the utility of bacterial extracellular polymers for treatment of metal-contaminated soils: polymer persistence, mobility, and the influence of lead. Wat Res 31:2827–2839.

Davis CC, Knocke WR, Edwards M. 2001. Implications of aqueous silica sorption to iron hydroxide: mobilization of iron colloids and interference with sorption of arsenate and humic substances. Environ Sci Technol 35:3158–3162.

Davis JA. 1984. Complexation of trace metals by adsorbed natural organic matter. Geochim Cosmochim Acta 48:679–691.

Donard OFX, Quevauviller P, Bruchet A. 1993. Tin and organotin speciation during wastewater and sludge treatment processes. Wat Res 27:1085–1089.

Douglas S, Beveridge TJ. 1998. Mineral formation by bacteria in natural microbial communities. FEMS Microbiol Ecol 26:79–88.

Dutta D, Mandal B, Mandal LN. 1989. Decrease in availability of zinc and copper in acidic to near neutral soils on submergence. Soil Sci 147:187–195.

Fortin C, Campbell PGC. 2001. Thiosulfate enhances silver uptake by a green alga: role of anion transporters in metal uptake. Environ Sci Technol 35:2214–2218.

Gambrell RP, Wiesepape JB, Patrick Jr. WH, Duff MC. 1991. The effects of pH, redox, and salinity on metal release from a contaminated sediment. Water Air Soil Pollut 57–58:359–367.

Gomez A, Solda P, Lambrot C, Wilbert J, Juste C. 1992. Bilan des elements traces metalliques transferes dans un sol sableux apres 16 annees d'apports continus et connus de boues de station d'epuration et de fumier de ferme en monoculture irriguee de mais. Ministere de l' Environnement. Convention de Recherche No. 89–256.

Gray CW, McLaren RG, Roberts AHC, Condron LM. 1999. Solubility, sorption and desorption of native and added cadmium in relation to properties of soils in New Zealand. Eur J Soil Sci 50:127–137.

Hamasaki T, Nagase H, Yoshioka Y, Sato T. 1995. Formation, distribution, and ecotoxicity of methylmetals of tin, mercury, and arsenic in the environment. Crit Rev Environ Sci Technol 25:45–91.

Hamon RE, Lorenz SE, Holm PE, Christensen TH, McGrath SP. 1995. Changes in trace-metal species and other components of the rhizosphere during growth of radish. Plant Cell Environ 18:749–756.

Hattori H. 1992. Influence of heavy metals on soil microbial activities. Soil Sci Plant Nutr 38:93–100.

Holmgren GGS, Meyer MW, Chaney RL, Daniels RB. 1993. Cadmium, lead, zinc, copper and nickel in agricultural soils of the United States of America. J Environ Qual 22:335–348.

Hooda PS, Alloway BJ. 1993. Effects of time and temperature on the bioavailability of Cd and Pb from sludge-amended soils. J Soil Sci 44:97–110.

Hooda PS, Alloway BJ. 1994a. The plant availability and DTPA extractability of trace metals in sludge-amended soils. Sci Tot Environ 149:39–51.

Hooda PS, Alloway BJ. 1994b. Changes in operational fractions of trace metals in two soils during two-years of reaction time following sewage sludge treatment. Intern J Environ Anal Chem 57:289–311.

Hooda PS, Alloway BJ. 1994c. Sorption of Cd and Pb by selected temperate and semi-arid soils-effects of sludge application and aging of sludged soils. Water Air Soil Pollut 74:235–250.

Iu KL, Pulford ID, Duncan HJ. 1981. Influence of waterlogging and lime or organic matter additions on the distribution of trace metals in an acid soil. Plant Soil 59:327–333.

Kim ND, Fergusson JE. 1991. Effectiveness of a commonly used sequential extraction technique in determining the speciation of cadmium in soils. Sci Tot Environ 105:191–210.

Kitada K, Ito A, Yamada K, Aizawa J, Umita T. 2001. Biological leaching of heavy metals from anaerobically digested sewage sludge using indigenous sulfur-oxidizing bacteria and sulfur waste in a closed system. Water Sci Technol 43:59–65.

Koo B-J, Chang AC, Page AL, Crowley DE. 2001. Organic acids in root exudates and their effects on plant uptake of biosolids-borne metals. In: Evans L, editor. Proceedings of the 6th International Conference on the Biogeochemistry of Trace Elements (6th ICOBTE); July 2001; University of Guelph, Ontario, Canada. p 561.

Kuperman RG, Carreiro MM. 1997. Soil heavy metal concentrations, microbial biomass and enzyme activities in a contaminated grassland ecosystem. Soil Biol Biochem 29:179–190.

Labrenz M, Druschel GK, Thomsen-Ebert T, Gilbert B, Welch SA, Kemner KM, Logan GA, Summons RE, De Stasio G, Bond PL, Lai B, and others. 2000. Formation of sphalerite (ZnS) deposits in natural biofilms of sulfate-reducing bacteria. Science 290:1744–1747.

Lee S-Z, Allen HE, Huang CP, Sparks DL, Sanders PF, Peijnenburg WJGM. 1996. Predicting soil-water partition coefficients for cadmium. Environ Sci Technol 30:3418–3424.

Lee S-Z, Chang L, Chen C-M, Tsai YI, Liu M-C. 2001. Predicting soil-water partition coefficients for Hg(II) from soil properties. Water Sci Technol 43:187–196.

Martin JM, Nirel P, Thomas A. 1987. Sequential extraction techniques: promises and problems. Mar Chem 22:313–341.

Martínez CE, Jacobson A, McBride MB. 2001. Thermally induced changes in metal solubility of contaminated soils is linked to mineral recrystallization and organic matter transformations. Environ Sci Technol 35:908–916.

Martínez CE, McBride MB. 1999. Dissolved and labile concentrations of Cd, Cu, Pb and Zn in aged ferrihydrite-organic matter systems. Environ Sci Technol 33:745–750.

Martínez CE, McBride MB 1998. Solubility of Cd, Cu, Pb and Zn in aged coprecipitates with amorphous iron hydroxides. Environ Sci Technol 32:743–748.

Martínez CE, McBride MB, Kandianis MT, Duxbury JM, Yoon S-J, Bleam WF. 2002. Zinc-sulfur and cadmium-sulfur association in metalliferous peats: Evidence from spectroscopy, distribution coefficients and phytoavailability. Environ Sci Technol 36:3683–3689.

Martínez CE, Sauvé S, Jacobson A, McBride MB. 1999. Thermally induced release of adsorbed Pb upon aging ferrihydrite and soil oxides. Environ Sci Technol 33:2016–2020.

McBride MB. 2000. Chemisorption and precipitation reactions. pp B-265–B-302. In: Sumner ME, editor. Handbook of soil science. Boca Raton, FL: CRC Pr.

McBride MB. 1989. Reactions controlling heavy metal solubility in soils. Advan Soil Sci 10, 1–56.

McBride MB, Martínez CE. 2000. Copper phytotoxicity in a contaminated soil: remediation tests with adsorptive materials. Environ Sci Technol 34:4386–4391.

McBride MB, Martínez CE, Sauvé S. 1998. Copper (II) activity in aged suspensions of goethite and organic matter. Soil Sci Soc Am J 62:1542–1548.

McBride MB, Martínez CE, Topp E, Evans L. 2000. Trace metal solubility and speciation in a calcareous soil 18 years after no-till sludge application. Soil Sci 165:646–656.

McBride MB, Richards BK, Steenhuis T, Russo JJ, Sauvé S. 1997b. Mobility and solubility of toxic metals and nutrients in soil fifteen years after sludge application. Soil Sci 162:487–500.

McBride MB, Richards BK, Steenhuis T, Spiers G. 1999. Long-term leaching of trace elements in a heavily sludge-amended silty clay loam soil. Soil Sci 164:613–623.

McBride M, Sauvé S, Hendershot W. 1997a. Solubility control of Cu, Zn, Cd and Pb in contaminated soils. Eur J Soil Sci 48:337–346.

McLaren RG, Williams JG, Swift RS. 1983. Some observations on the desorption and distribution behaviour of copper with soil components. J Soil Sci 34:325–331.

Moreno JL, Hernandez T, Garcia C. 1999. Effects of a cadmium-contaminated sewage sludge compost on dynamics of organic matter and microbial activity in an arid soil. Biol Fert Soils 28:230–237.

Morra MJ, Fendorf SE, Brown PD. 1997. Speciation of sulfur in humic and fulvic acids using x-ray absorption near-edge structure (XANES) spectroscopy. Geochim Cosmochim Acta 61:683–688.

Murphy EM, Zachara JM. 1995. The role of sorbed humic substances on the distribution of organic and inorganic contaminants in groundwater. Geoderma 67:103–124.

Nirel PVM, Morel FMM. 1990. Pitfalls of sequential extractions. Wat Res 24:1055–1056.

Osté LA, Dolfing J, Ma W, Lexmond TM. 2001. The effect of beringite on Cd and Zn uptake by plants and earthworms: More than a liming effect? Environ Toxicol Chem 20:1339–1345.

Pedersen MB, Temminghoff EJM, Marinussen MPJC, Elmegaard N, vanGestel CAM. 1997. Copper accumulation and fitness of Folsomia candida Willem in a copper contaminated sandy soil as affected by pH and soil moisture. Appl Soil Ecol 6:135–146.

Porvari P, Verta M. 1995. Methylmercury production in flooded soils: a laboratory study. Water Air Soil Pollut 80:765–773.

Qureshi, Richards SB, Akhtar MS, Steenhuis T, McBride M. 2001. Release and leaching of heavy metals from muck soil as affected by temperature and microbial activity. In: Evans L, editor. Proceedings of the 6th International Conference on the Biogeochemistry of Trace Elements (6th ICOBTE); July 2001; University of Guelph, Ontario, Canada. p 351.

Ravichandran M, Aiken GR, Ryan JN, Reddy MM. 1999. Inhibition of precipitation and aggregation of metacinnabar (mercuric sulfide) by dissolved organic matter isolated from the Florida Everglades. Environ Sci Technol 33:1418–1423.

Richards BK, Steenhuis TS, Peverly JH, McBride MB. 1998. Metal mobility at an old, heavily-loaded sludge application site. Environ Pollut 99:365–377.

Romheld V, Awad F. 2000. Significance of root exudates in acquisition of heavy metals from a contaminated calcareous soil by graminaceous species. J Plant Nutr 23:1857–1866.

Romkens PFAM, Bouwman LA, Boon GT. 1999. Effect of plant growth on copper solubility and speciation in soil solution samples. Environ Pollut 106:315–321.

Romkens PFAM, Dolfing J. 1998. Effect of Ca on the solubility and molecular size distribution of DOC and Cu binding in soil solution samples. Environ Sci Technol 32:363–369.

Rozan TF, Benoit G, Luther GW. 1999. Measuring metal sulfide complexes in oxic river waters with square wave voltammetry. Environ Sci Technol 33:3021–3026.

Rozan TF, Lassman ME, Ridge DP, Luther GW. 2000. Evidence for iron, copper and zinc complexation as multinuclear sulphide clusters in oxic rivers. Nature 406:879–882.

Sadovnikova L, Otabbong E, Iakimenko O, Nilsson I, Persson J, Orlov D. 1996. Dynamic transformation of sewage sludge and farmyard manure components. 2. Copper, lead, and cadmium forms in incubated soils. Agric Ecosys Environ 58:127–132.

Saeki K, Okazaki M, Matsumoto S. 1993. The chemical phase change in heavy metals with drying and oxidation of the lake sediments. Wat Res 27:1243–1251.

Sauvé S, Martinez CE, McBride M, Hendershot W. 2000b. Adsorption of free lead (Pb2+) by pedogenic oxides, ferrihydrite and leaf compost. Soil Sci Soc Am J 64:595–599.

Sauvé S, McBride M, Hendershot W. 1998a. Soil solution speciation of lead(II): effects of organic matter and pH. Soil Sci Soc Am J 62:618–621.

Sauvé S, McBride M, Hendershot W. 1998b. Lead phosphate solubility in water and soil suspensions. Environ Sci Technol 32:388–393.

Sauvé S, McBride MB, Norvell WA, Hendershot W. 1997. Copper solubility and speciation of in situ contaminated soils: effects of copper level, pH and organic matter. Water Air Soil Pollut 100:133–149.

Sauvé S, Norvell WA, McBride M, Hendershot W. 2000a. Speciation and complexation of cadmium in extracted soil solutions. Environ Sci Technol 34:291–296.

Scheifler R, Schwartz C, Echevarria G, de Vaufleury A, Badot P-M, Morel J-L. 2003. "Non-available" soil cadmium is bioavailable to snails: evidence from isotopic dilution experiments. Environ Sci Technol 37:81–86.

Schulze-Lam S, Fortin D, Davis BS, Beveridge TJ. 1996. Mineralization of bacterial surfaces. Chem Geol 132:171–181.

Semlali RM, van Oort F, Denaix L, Loubet M. 2001. Estimating distributions of endogenous and exogenous Pb in soils by using Pb isotopic ratios. Environ Sci Technol 35:4180–4188.

Silver S, Phung LT. 1996. Bacterial heavy metal resistance: new surprises. Ann Rev Microbiol 50:753–789.

Six J, Conant RT, Paul EA, Paustian K. 2002. Stabilization mechanisms of soil organic matter: implications for C-saturation of soils. Plant Soil 241:155–176.

Skyllberg UL, Xia K, Bloom PR, Nater EA, Bleam WF. 2000. Binding of Hg (II) to reduced sulfur in soil organic matter along upland-peat soil transects. J Environ Qual 29:855–865.

Stacey S, Merrington G, McLaughlin MJ. 2001. The effect of aging biosolids on the availability of cadmium and zinc in soil. Eur J Soil Sci 52:313–321.

Steenhuis TS, McBride MB, Richards BK, Harrison E. 1999. Trace metal retention in the incorporation zone of land-applied sludge. Environ Sci Technol 33:1171–1174.

Tack FM, Callewaert OWJJ, Verloo MG. 1996. Metal solubility as a function of pH in a contaminated, dredged sediment affected by oxidation. Environ Pollut 91:199–208.

Temminghoff EJM, Van der Zee SEATM, De Haan FAM. 1998. Effects of dissolved organic matter on the mobility of copper in a contaminated sandy soil. Eur J Soil Sci 49:617–628.

Walker SG, Flemming CA, Ferris FG, Beveridge TJ, Bailey GW. 1989. Physicochemical interaction of *Escherichia-coli* cell envelopes and *Bacillus-subtilis* cell-walls with 2 clays and ability of the composite to immobilize heavy-metals from solution. Appl Environ Microbiol 55:2976–2984.

Weng L, Temminghoff EJM, van Riemsdijk WH. 2001. Contribution of individual sorbents to the control of heavy metal activity in sandy soil. Environ Sci Technol 35:4436–4443.

Whiting SN, de Souza MP, Terry N. 2001. Rhizosphere bacteria mobilize Zn for hyperaccumulation by Thlaspi caerulescens. Environ Sci Technol 35:3144–3150.

Xia K, Bleam W, Helmke PA. 1997. Studies of the nature of binding sites of first row transition elements bound to aquatic and soil humic substances using x-ray absorption spectroscopy. Geochim Cosmochim Acta 61:2223–2235.

Xia K, Skyllberg UL, Bleam WF, Bloom PR, Nater EA, Helmke PA. 1999. X-ray absorption spectroscopic evidence for the complexation of Hg(II) by thiol and disulfane ligands in soil humic substances. Environ Sci Technol 33:257–261.

Xia K, Weesner F, Bleam WF, Bloom PR, Skyllberg UL, Helmke PA. 1998. XANES studies of oxidation states of sulfur in aquatic and soil humic substances. Soil Sci Soc Am J 62:1240–1246.

Xiao-Quan S, Bin C. 1993. Evaluation of sequential extraction for speciation of trace metals in model soil containing natural minerals and humic acid. Anal Chem 65:802–807.

Xu H, Allard B, Grimvall A. 1991. Effects of acidification and natural organic materials on the mobility of arsenic in the environment. Water Air Soil Pollut 57–58:269–278.

Yin Y, Allen HE, Li Y, Huang CP, Sanders PF. 1996. Adsorption of mercury (II) by soils: effects of pH, chloride and organic matter. J Environ Qual 25:37–844.

Zachara JM, Resch CT, Smith SC. 1994. Influence of humic substances on Co^{2+} sorption by a subsurface mineral separate and its mineralogical components. Geochim Cosmochim Acta 58:553–566.

8 Redox Processes and Attenuation of Metal Availability in Soils

Neal Menzies

8.1 INTRODUCTION

The redox (reduction-oxidation) potential is one of the primary factors controlling chemical reactions in soils. Well-drained soils are typically considered well aerated and hence oxidizing, whereas waterlogging is considered to produce reducing conditions. However, this generalization ignores the dynamic and heterogeneous nature of soil systems. Oxygen depletion can occur wherever oxygen consumption by plant roots and microorganisms exceeds its rate of supply. This condition is encountered in a range of situations within well-drained soils, for example, inside aggregates, in compacted subsoils, when organic wastes are applied to land, and within plant rhizospheres. In view of the ubiquitous nature of changing redox conditions in soil, the potential for redox reactions to alter trace metal chemistry must be considered.

A number of trace elements (e.g., As, Cr, Hg, Mo, and Se) may change their oxidation state within the range of redox conditions commonly encountered in soils, and this can directly influence their mobility and bioavailability. For example, As may be reduced from arsenate (AsO_4^{3-}) to the more toxic form arsenite (AsO_3^{3-}) and can be converted by microorganisms to volatile forms and lost from the soil system, whereas Se may be reduced from the relatively mobile form selenate (SeO_4^{2-}) to strongly adsorbed selenite (SeO_3^{2-}). However, the effects of redox on trace element availability are not limited to these redox-active elements. Changed redox conditions can indirectly influence the availability of a wide range of trace elements by altering other soil characteristics, as itemized here

1) Under reducing conditions, soil pH typically trends toward neutrality, decreasing the availability of metals in acid soils.
2) In strongly reduced soils, metal availability may be reduced by precipitation as low solubility sulfide minerals or, in less strongly reducing environments, as carbonates. Conversely, the oxidation of sediments and mine wastes containing metal sulfide minerals will result in the release of these metals, a problem often accentuated through the acidification caused by pyrite oxidation.

3) Reductive dissolution of Fe(III) and Mn(IV) oxy-hydroxides may release adsorbed and coprecipitated trace elements, increasing solution concentrations. On reoxidation, Fe and Mn are again oxidized and precipitate from solution as oxy-hydroxides, affording the opportunity for trace element adsorption or coprecipitation.
4) The increased soil solution Fe^{2+} and Mn^{2+} concentrations resulting from the reducing conditions in item 3, may displace trace elements from exchange sites on the soil, increasing soil-solution concentrations. However, uptake competition from Fe^{2+} and Mn^{2+} may also limit the bioavailability of trace elements.
5) Volatile organic acids (e.g., acetic, formic, propionic, and butyric), produced during the anaerobic decomposition of carbohydrates, may form metal complexes, increasing the total metal concentration in soil solution.

The effects of redox reactions on trace metal availability are largely reversible. This characteristic is successfully exploited in the use of anaerobic conditions, generated in wetlands, to treat acidity and capture sulfate and metals released by oxidation reactions in acid mine drainage. However, some permanent (or semipermanent) attenuation of metal toxicity may result from the fluctuation of redox conditions that occurs in soils. In the short-to-medium term, attenuation may result from slow reaction kinetics (e.g., slow dissolution and oxidation of some metal sulfides), whereas coprecipitation of metals within Mn nodules may provide a long-term attenuation of metal availability.

8.2 REDOX CONDITIONS IN SOILS

The redox state of soil is dictated by the availability of oxygen for chemical and microbiological oxidative processes. In aerobic soils (oxidized), oxygen supply is comparable to its rate of consumption, whereas in anaerobic (reduced) soils, chemical and microbial demand for oxygen greatly exceeds its rate of supply, and anaerobiosis may develop. A primary factor controlling oxygen supply is the extent to which the soil's pore space is occupied by water. Water-filled pore space contains about 30 times less O_2 than the equivalent air-filled pore space at the same O_2 partial pressure, and O_2 diffusion in water is about 10,000 times lower than in air (Greenwood 1961).

Upon flooding, the access of oxygen to the soil is limited, and the oxygen content of the soil solution is lowered by root and microbial respiration, and chemical oxidation reactions. Soil-solution oxygen is depleted within several hours to a few days, depending on its rate of consumption (Ponnamperuma 1972). Once oxygen is depleted, soil microorganisms will use another substance, such as nitrate (NO_3) or manganese dioxide (MnO_2), as an electron acceptor. Thus, development of reducing conditions depends both on a reduction in the oxygen supply and on biological activity; the severity or intensity of the reducing conditions achieved is determined by the balance of these 2 factors. The intensity of oxidizing or reducing conditions is expressed as a redox potential (Eh) and typically measured using a platinum

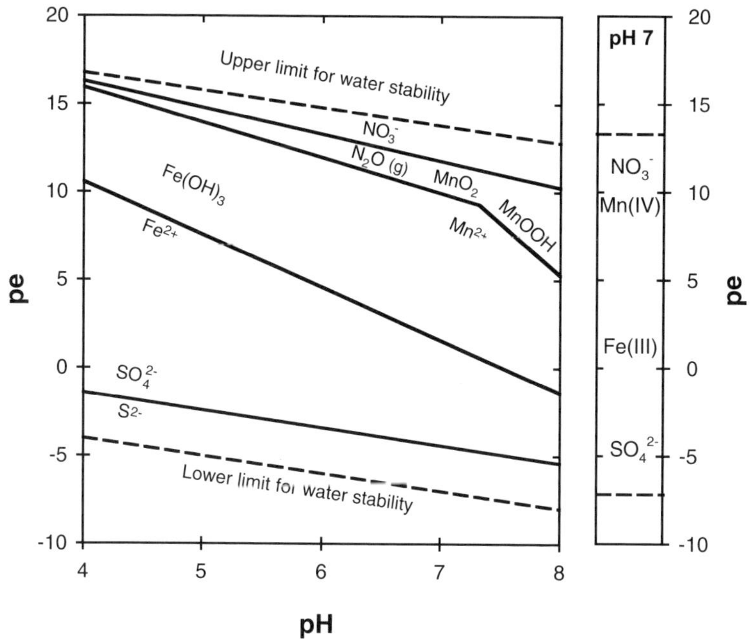

FIGURE 8.1 Equilibrium pe–pH relationships for important reactions in water. The upper broken line denotes the pe at which water is oxidized to O_2, whereas the lower broken line denotes the pe at which it is reduced to H_2. The side bar indicates the pe positions of the major redox couples at pH 7 (pH 7 is used as the pH of waterlogged soils tends toward neutrality). This side bar is used on subsequent figures to indicate the approximate position of these important redox couples.

electrode (Ponnamperuma 1972). In an aqueous system, the redox limits are imposed by the dissociation of water into $O_{2(g)}$ at the most oxidized condition, and dissociation into $H_{2(g)}$ at the most reduced condition (Baas Becking et al. 1960) (Figure 8.1).

Anaerobic conditions develop in permanently or seasonally flooded soils, such as swamp and marsh soils, and in agronomic systems such as rice production, in which flooding is used to enhance productivity (Ponnamperuma 1972) (Figure 8.2). Temporary waterlogging of soils with poor internal drainage can occur following rain or irrigation, leading to short periods of oxygen deficit (Meek and Stolzy 1978) (Figure 8.2). Redox conditions in soils are typically highly heterogeneous. In waterlogged soil, where the overlying surface water is shallow, a gradient of reducing conditions develops within the profile. Oxygen diffusing across the surface water–soil interface produces a thin, oxidized surface horizon (Ponnamperuma 1972; Patrick et al. 1985). Well-drained soils are generally considered aerobic; macropore space in the soil is air-filled, providing a diffusion path for oxygen from the atmosphere. However, even in unsaturated soils, reduction of air-filled porosity and limited continuity of air-filled pores can result in oxygen depletion within some zones (Greenwood 1961). This effect has been demonstrated for the interior of aggregates (Zausig et al. 1993) and in compacted soils (Taylor and Burnett 1963;

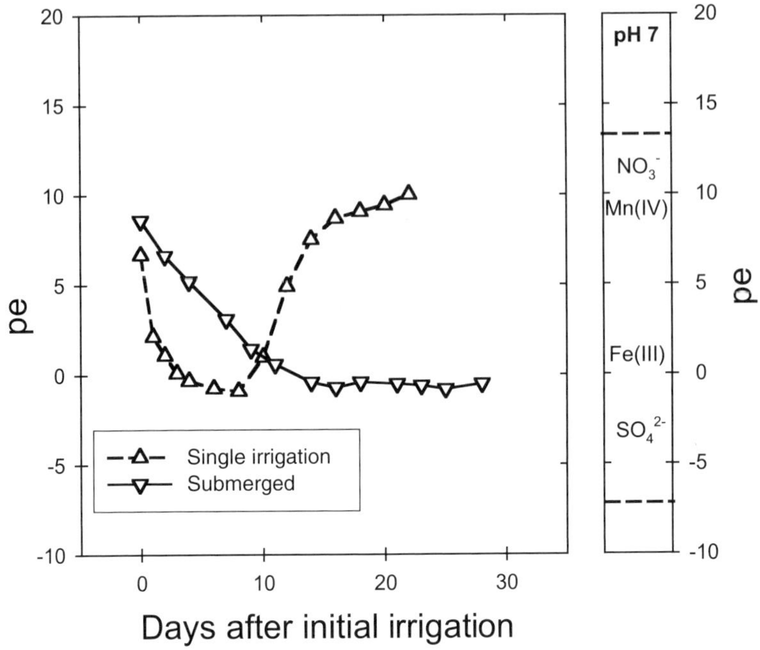

FIGURE 8.2 The redox state of 2 soils following waterlogging. Irrigation with effluent (—) resulted in a rapid onset of strongly reducing conditions, a consequence of the materials' high BOD. The pe increased as the oxygen supply was restored, once the soil drained in 8 d. In the marsh soil, oxygen consumption was lower and, hence, the rate of pe decrease was lower. In both systems, the pe appeared to be poised at the Fe(III)/Fe(II) couple.

Mathers et al. 1971). Variability of redox conditions within the soil is further accentuated by the differences in the rate of oxygen consumption (or electron supply) resulting from heterogeneous distribution of available organic carbon and active microbial populations (Greenland 1962).

An additional degree of redox heterogeneity is added to soil through the action of plant roots. In aerated soils, plant roots directly consume O_2 in respiration and indirectly result in an increased rate of microbial O_2 consumption by supplying labile organic matter into the rhizosphere (Drew and Lynch 1980). Conversely, in wetland systems, adapted plants may limit the development of reducing conditions. The presence of aerenchyma in the roots and stems of adapted plants enhances O_2 movement into the roots and its radial loss into the rhizosphere (Jackson and Armstrong 1999). The input of O_2 into the rhizosphere is sufficient to substantially alter the soil's redox condition (Jackson and Armstrong 1999), and hence alter metal bioavailability. Cacador et al. (1996) observed that the concentration of Cu, Pb, and Zn in rhizosphere soil of estuary salt marshes was greater than in nonvegetated areas. However, the availability of these metals, as assessed by acid and DTPA extraction, was lower in the rhizosphere soil than in soil from nonvegetated areas. Accumulation of As within rhizosphere soil has also been reported (Doyle and Otte 1997). In contrast, the distribution of Cr and Ni does not appear to be influenced by vegetation (Cacador et al. 1996).

This review is intended to provide an overview of the processes affecting metal bioavailability, which occur as soils move between oxidized and reduced conditions. These possible processes are separated out and considered individually, but it is important to remember that they will potentially occur simultaneously. Thus, an observed change in metal availability may be the result of any number of processes, even the net result of several processes acting simultaneously.

8.3 REDOX-ACTIVE TRACE ELEMENTS IN SOILS

A number of trace elements (e.g., As, Cr, Hg, Mo, Se, and V) may change their oxidation state within the range of redox conditions commonly encountered in soils. Aspects of the behavior of As, Cr, and Se are used here to illustrate how these changes may influence the bioavailability of the elements.

8.3.1 ARSENIC

Arsenic is present in oxidized environments as arsentate, As(V), with $Ca_3(AsO_4)_2$ and $Mn_3(AsO_4)_2$ as stable solid phases likely to limit the soil-solution concentration (Sadiq et al. 1983). Under moderately reduced conditions (pe + pH < 8), arsenite, As(III), becomes the more stable phase (Sadiq et al. 1983) (Figure 8.3); laboratory

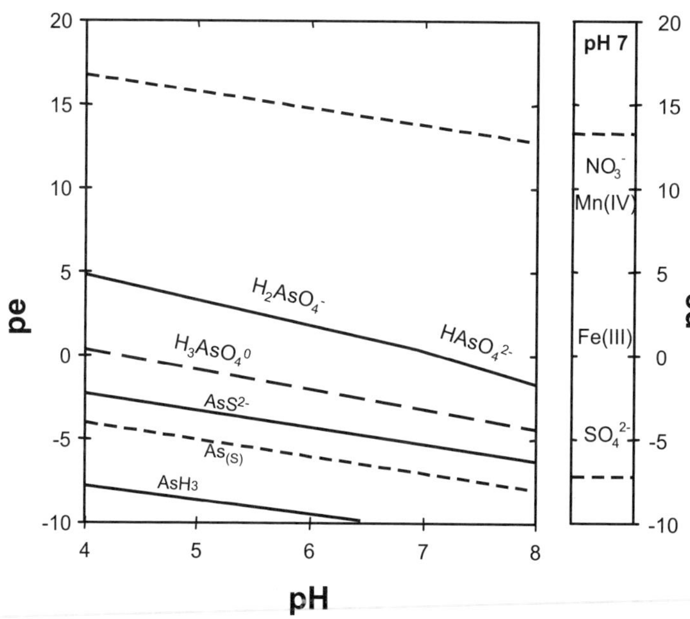

FIGURE 8.3 A pe/pH diagram for arsenic species. Note that formation of AsS^2 occurs if sulfate is present, to be reduced to H_2S.

and glasshouse studies of reduced soils have reported As(III) as the predominant species in soil solution (Masscheleyn et al. 1991a, b; Marin et al. 1993). Nevertheless, these studies have reported that a proportion of As remained as As(V) under reduced conditions (Masscheleyn et al. 1991b; Marin et al. 1993). It has also been suggested that this As(V) may be maintained by competition with Fe(III) as a terminal electron acceptor in microbial respiration (Masscheleyn et al. 1991b) or by Mn(IV) oxidation of As(III) (Oscarson et al. 1981).

Under strongly reducing conditions, metallic As is thermodynamically stable (Ferguson and Gavis 1972). However, in many situations, sufficient sulfur will be present for the solution concentration of As to be limited by the formation of low-solubility sulfide minerals (Sadiq et al. 1983; Sadiq 1997) (Figure 8.3) rather than As metal. Arsine (AsH_3) is stable only under conditions more strongly reducing than the redox limit imposed by the dissociation of water (Ferguson and Gavis 1972).

Soil microorganisms are recognized as contributing to both oxidation and reduction transformations between As(V) and As(III) (Cullen and Reimer 1989). Of particular interest here is the capacity of some microorganisms to reduce As to volatile forms that can be lost from the soil. Generation of dimethylarsine and trimethylarsine from As-contaminated soil has been demonstrated in laboratory studies (Woolson 1977). In the field, Woolson and Isensee (1981) found that the accumulation of As in soil from the use of arsenical pesticides was not as great as predicted, and suggested that 14 to 15% was lost each year as a result of microbially mediated volatilization.

Soil-solution As concentrations are typically reported to rise under reducing conditions (Deuel and Swoboda 1972; Masscheleyn et al. 1991a,b), an effect generally attributed to the reductive dissolution of Fe(III) and Mn(IV) oxides (see Subsection 8.4.3) (Deuel and Swoboda 1972; Masscheleyn et al. 1991b). Under moderately reducing conditions, the toxicity risk of As is increased both by its increased solubility and by the higher Toxicity of the As(III) form. Marin et al. (1992) reported that for rice (*Oryza sativa*), toxicity decreased in the order arsenite > monomethyl arsenic acid > arsenate > dimethyl arsenic acid. Similarly, Yan-Chu (1994) considered reduced forms of As most toxic; AsH_3 > As(III) > As(V). To humans, As(III) is about 60 times more toxic than As(V), an effect attributed to the capacity of As(III) to react with sulfhydryl (–SH) groups (Webb 1966).

8.3.2 Selenium

Selenium can exist in 4 oxidation states: in the Se(VI) oxidation state as selenate (SeO_4^{2-}), in the Se(IV) oxidation state as selenite (SeO_3^{2-}), in the zero oxidation state as elemental Se, and in the Se(II) oxidation state as selenide (Se^{2-}). Chemical thermodynamics predict that a reduction sequence, NO_3–SeO_4–MnO_2, should occur in soils at pH > 5, and this sequence has been demonstrated in soil (Sposito et al. 1991). Although the reduction of selenate to selenite is relatively rapid, the oxidation of selenite to selenate is a slow process (Jayaweera and Biggar 1996).

The solubility of selenate and selenite minerals in oxidized systems is sufficiently high that they will not limit Se availability. Elrashidi (1987) showed that elemental Se is more soluble than a range of selenide minerals under strongly reducing conditions, and suggested that selenide minerals will act as a major sink of Se in the environment. Selenide minerals are extremely insoluble under strongly reducing conditions, and Se presents a minimal hazard under these conditions (Elrashidi et al. 1987; Masscheleyn et al. 1991a).

Selenate is only weakly adsorbed to soil surfaces at neutral pH, and its adsorption is strongly suppressed by competition from SO_4^{2-} (White and Dubrovsky 1994). However, under moderately reducing conditions, Se availability may be limited by adsorption, as selenite is relatively strongly adsorbed. Furthermore, selenite adsorption is not subject to competition from SO_4^{2-} (Balistrieri and Chao 1987). Thus, under moderately reduced conditions, the availability and mobility of Se will be low as a result of selenite adsorption (Alemi et al. 1991; Neal and Sposito 1991). Plants show a strong uptake preference for selenate over selenite (Asher et al. 1977), but despite this, selenite shows a higher phytotoxicity (Smith and Watkinson 1984).

8.3.3 Chromium

Chromium may exist in many different oxidation states with Cr (III) and Cr (VI) the most stable under environmental conditions (Shupack 1991). In contrast to As and Se, where a reduced form is most toxic, Cr(VI) is toxic to many plants, animals, and bacteria, and is a confirmed human carcinogen, whereas Cr(III) is generally benign and even a known micronutrient in organic form (Kimbrough et al. 1999). Naturally occurring Cr is typically present as Cr(III), with Cr(VI) rare and only found in highly oxidizing environments. Thus, the presence of Cr(VI) in soils and sediments is almost always the result of human activity (Kimbrough et al. 1999).

Under the redox and pH conditions encountered in most soils, Cr(III) is the most stable form and is present in solution as Cr^{3+}, $Cr(OH)^{2+}$, $Cr(OH)_3^0$, and $Cr(OH)_4^-$ (Bartlett and Kimble 1976; Rai et al. 1987). At slightly acidic to alkaline pH, Cr(III) precipitates as amorphous $Cr(OH)_3$ (Rai et al. 1987) or as a solid solution ((Fe, Cr)$(OH)_3$) if Fe^{3+} is present (Francoise and Bourg 1991). In contrast, the Cr(VI) ions chromate (CrO_4^{2-}) and dichromate ($Cr_2O_7^{2-}$) are soluble at all pH values. As Cr(VI) is both more toxic and more mobile than Cr(III), the reduction and oxidation reactions controlling the relative abundance of these 2 oxidation states dictate the risk presented by Cr contamination.

Abiotic oxidation of Cr(III) to Cr(VI) by Mn oxides has been demonstrated (Bartlett and James 1979; Fendorf and Zasoski 1992). Relative to this process, the oxidation of Cr(III) by dissolved oxygen has been found to be insignificant (Rai et al. 1989). Biological oxidation of Cr(III) to Cr(VI) has not been reported. Bartlett and James (1979) demonstrated in laboratory experiments that a proportion, typically $\approx 15\%$, of Cr(III) added to soils was oxidized to Cr(VI). The extent of oxidation was related to the readily reducible Mn content of the soil and to the form of the added Cr. Aged precipitates of $Cr(OH)_3$ were less prone to oxidation than soluble or freshly precipitated $Cr(OH)_3$; thus, the risk of Cr oxidation will decrease with time, following contamination.

Chromium (VI) can be reduced to Cr(III) by microorganisms through direct microbial enzymatic reduction (Chen and Hao 1998) or, indirectly, by the depletion of oxygen and reduction of Fe and S (Eary and Rai 1988; Pettine et al. 1994). Thus,

Cr reduction is strongly dependent on the presence of organic matter to act as a microbial energy source, and remediation strategies for Cr(VI)-contaminated soils are frequently based on the addition of organic amendments (Losi et al. 1994; Cifuentes et al. 1996).

8.4 INDIRECT EFFECTS ON TRACE ELEMENT AVAILABILITY

Many trace elements are not subject to redox state changes within the range of redox conditions encountered in soils. Nevertheless, the bioavailability of such trace elements can be altered by the redox state of the soil. Mikkelsen and Brandon (1975) reported that Zn deficiency was more severe in submerged soils, whereas Jugsujinda and Patrick (1977) reported lower Zn uptake from anaerobic soils than from aerobic soils. Similar effects have also been reported for Cd uptake by rice (Bingham et al. 1976; Reddy and Patrick 1977). Such changes in availability may be the result of a number of interacting processes, and it is generally not possible either to discriminate between the effects of these processes or to predict their net result. The aim here is to describe the various processes and the underlying chemistry.

8.4.1 EFFECTS OF pH CHANGE

Baas Becking et al. (1960) observed that many reactions in natural environments involve both protons and electrons, and that a relationship exists between these parameters in most reactions. They described this relationship under standard state conditions as follows:

$$Eh = E_0 - 59(a / n)pH \tag{8.1}$$

where

Eh = redox potential,
E_0 = standard potential at equal activities of reduced and oxidized species,
a = number of protons transferred, and
n = number of electrons involved in the reaction.

In soil systems, in which there are many redox couples with widely varying concentration ranges, there is little basis for expecting pH/redox potential slopes to coincide with theoretical predictions (Gambrell and Patrick 1978). Furthermore, soil pH is buffered by dissolution/precipitation of silicate, carbonate, and oxide minerals, which are insensitive to changes in redox potential (Bohn 1971). Nevertheless, the relationship between pe and pH is sufficiently strong that, despite the limitations, it has been used with reasonable success in making comparisons between different media (Bohn 1971).

The pH of both acid and alkaline soils tend toward neutrality on waterlogging (Ponnamperuma 1972; Patrick et al. 1985) (hence the use of the pH 7 in the sidebars of Figure 8.1 to Figure 8.3). Ponamperuma (1972) attributed the pH increase observed in acid soils to the reduction of NO_3, Mn, and Fe, and identified organic matter and reactive Fe content as the major determinants of the rate of pH change and its magnitude. In sodic and calcareous soils, the observed pH decrease is attributed to the increased partial pressure of CO_2 (Ponnamperuma 1972).

The effect of pH on trace metals has been extensively studied in aerobic soils; solution concentrations and bioavailability decreasing through precipitation and adsorption (McBride 1989). This behavior is also reflected in studies of anaerobic systems. Dutta et al. (1989) reported that, for a range of rice soils, flooding increased pH and decreased the DTPA-extractable Cu and Zn, with the change in Zn well correlated with pH. Welch and Lund (1987) noted less leaching of Ni in saturated soil columns than in unsaturated columns for 5 of the 6 sludge-amended soils they studied. In these 5 soils, the pH was increased by the saturated conditions, whereas in the sixth soil, in which no change in Ni mobility was observed, the pH was similar under saturated and unsaturated conditions (Welch and Lund 1987). Reduced metal bioavailability due to increasing pH has been reported for Cd and Pb in rice (Reddy and Patrick 1977).

On reoxidation of soil previously inundated with fresh water, the pH of soils tends back to its initial condition (Ponnamperuma 1972). In contrast, when coastal and estuarine sediments, or seasonally inundated coastal soils are drained, strongly acid conditions can result through the oxidation of pyrite. During pyrite formation, sulfate from seawater is "split" into mobile alkalinity (HCO_3) and immobile potential acidity (FeS) that is left behind in the sediment. When metal-contaminated dredged sediments are land-disposed, acid generated by pyrite oxidation can result in greatly increased mobility of the metal contaminants (Gambrell et al. 1991). The extent of metal mobilization is primarily dependent on the pH produced by the sulfide oxidation (Tack et al. 1996; Astrom 1998).

8.4.2 Precipitation of Carbonates and Sulfides

At the onset of waterlogging, gaseous exchange with the atmosphere is restricted and the concentration of carbonate/bicarbonate in solution increases. In metal-contaminated soils, the combination of increased CO_2 and the pH increase observed on waterlogging acid soils results in precipitation of metals as carbonates. This process has been reported to lower solution activities of Cd (Street et al. 1978), Pb (Elliot et al. 1986), and Zn (Haldar and Mandal 1979; Ma and Lindsay 1993).

A range of experimental strategies is employed to achieve reducing conditions, and the effects of these strategies on soil chemical processes need to be considered when interpreting the results produced. For example, the study of Chuan et al. (1996) indicated that the solubility of heavy metals in contaminated rice soils was slightly higher under reducing conditions than under oxidizing conditions. However, the potential for immobilization by carbonate precipitation was reduced by the experimental approach; reducing conditions were achieved by bubbling N_2 gas through soil suspensions, thus maintaining a low CO_2 partial pressure and precluding metal carbonate precipitation. Studies using this approach typically show increasing concentrations of Cu, Ni, and Pb with decreasing pe, until sulfate is reduced to sulfide and the metal concentrations drop dramatically as metal sulfides precipitate (Carbonell et al. 1998).

Under strongly reducing conditions, sulfate is reduced to sulfide. Under such conditions, the availability of metals such as Cd, Cu, Hg, Ni, Pb, and Zn will be lowered by precipitation of low-solubility sulfides (Allen et al. 1993; Morse and Arakaki 1993; Brennan and Lindsay 1996). Arsenic also has a strong affinity for sulfide (Ferguson and Gavis 1972), with As solubility limited by the formation of As_2S_3 (orpiment) and/or FeAsS (arsenopyrite) under strongly reducing conditions. As the solubility of sulfide minerals is lower than that of carbonates, the carbonate minerals precipitated early in the reduction sequence will dissolve, supplying the metal to sulfide precipitation. Brennan and Lindsay (1996) demonstrated this sequence of carbonate then sulfide precipitation for mountain bog soil contaminated with Cd, Pb, and Zn. The soil was allowed a period of oxidation, then reduced using either organic matter or a reducing gas mixture as the source of electrons. Where reducing conditions were induced by organic matter decomposition, the Cd^{2+}, Pb^{2+}, and Zn^{2+} activity in solution approached the activity supported by the dissolution of the carbonate minerals octavite ($CdCO_3$), cerrusite ($PbCO_3$), and smithsonite ($ZnCO_3$), respectively (Brennan and Lindsay 1996). When lower pe + pH conditions were achieved using the reducing gas mixture, solution metal activities approached those supported by sulfide minerals. The predicted sequence of sulfide precipitation as pe + pH decreases is CdS (greenockite, 4.85), ZnS (sphalerite, 4.70), PbS (galena, 4.40), FeS_2 (pyrite, 4.35), FeS_2 (marcasite, 4.30), FeS (troilite, 3.97), FeS (mackinawite, 3.94), FeS (amorphous, 3.85), and Fe_3S_4 (greigite, 3.72) (Brennan and Lindsay 1996). An important implication of this sequence is that the trace metal sulfides would be expected to precipitate before pe + pH is poised by pyrite formation. However, this sequence is not normally observed under natural conditions in which heavy metals are found coprecipitated with the more abundant iron sulfides (Huerta-Diaz and Morse 1992).

Evidence for the formation of heavy metal sulfide minerals under reducing conditions has also been provided by electron microscopy. Van den Berg et al. (1998), in a study of periodically inundated wetlands, found Zn and S associations with molar ratios of almost 1, indicating precipitation of ZnS. Traces of As, Cd, Pb, and Cu were also coprecipitated. The sulfide precipitates formed where roots were decaying, the substitution of sulfide precipitate for the cell wall material preserving the root structure (formation of pseudomorphs) (Van den Berg et al. 1998).

Increase in metal availability resulting from the reverse of this process, oxidation of metal sulfide minerals, has been widely reported from mine sites and acid sulfate soils. The increase in metal availability results from greater solubility of sulfate compounds, and is greatly accentuated in many situations by the acidification that results from pyrite (FeS) oxidation and precipitation of Fe (III).

8.4.3 Reductive Dissolution of Mn and Fe Oxides

Oxides of Fe(III) and Mn(IV) represent the major electron-accepting pools in an anoxic soil, and descriptions of reducing conditions in soils typically provide half reactions describing the reduction of these oxides (Ponnamperuma 1972; Bartlett and James 1993).

$$MnO_2 \text{ (pyrolusite)} + 4H^+ + e^- \rightleftharpoons Mn^{2+} + 2H_2O \quad (8.2)$$

$$Fe(OH)_3 \text{ (amorph.)} + 3H^+ + e^- \rightleftharpoons Fe^{2+} + 3H_2O \quad (8.3)$$

As these reduced forms, Fe(II) and Mn(II), are more soluble than the oxidized forms, the solution concentration of Fe and Mn is typically observed to increase under waterlogged conditions. The solution concentration of other trace elements is often observed to increase in parallel with Fe and Mn, and dissolution of the oxides has been suggested as an explanation for the trace element release (Chuan et al. 1996). The basis of this contention is that many trace elements are specifically adsorbed to, or coprecipitated with, Fe and Mn oxides, and that the dissolution of these oxides would result in the release of the associated trace elements.

A consideration of the reduction reactions (Equation 8.2 and Equation 8.3) alone provides a poor description of the processes occurring in a reduced soil. Soil-solution concentrations of Mn^{2+} and Fe^{2+} do not continue to rise indefinitely, but reach a maximum concentration dictated by the solubility of their respective oxy-hydroxide and carbonate minerals (Lindsay 1979). The reduction of Mn(IV) and Fe(III) oxides can be described as (Lindsay 1979)

$$MnO_2 + H^+ + e^- \rightleftharpoons MnOOH \text{ (manganite)} \quad (8.4)$$

$$3MnO_2 + 4H^+ + 4e^- \rightleftharpoons Mn_3O_4 + 2H_2O \text{ (hausmannite)} \quad (8.5)$$

$$MnO_2 + 2H^+ + 2e^- + CO_2(g) \rightleftharpoons MnCO_3 \text{ (rhodochrosite)} \quad (8.6)$$

$$3Fe(OH)_3 + H^+ + e^- \rightleftharpoons Fe_3O_4 \text{ (magnetite)} + 5H_2O \quad (8.7)$$

$$Fe(OH)_3 + H^+ + e^- + CO_2(g) \rightleftharpoons FeCO_3 \text{ (siderite)} + 2H_2O \quad (8.8)$$

Thus, the reduction of Fe(III) and Mn(IV) results in a conversion of minerals from one form to another, rather than a dissolution reaction. The increase in soil-solution Fe concentrations typically reported in reduced soils ($\approx$100 mg Fe/l) (Ponnamperuma 1972), represents the dissolution of $\approx$100 mg oxide/kg of soil (0.01% of soil mass). As most soils have free Fe concentrations of over 10 g/kg, the dissolution of $\approx$1% of this is unlikely to markedly alter trace element retention.

Evidence for the effects of Fe reduction is contradictory. Uptake of Zn (Jugsujinda and Patrick 1977; Mandal et al. 1988) and Cd (Reddy and Patrick 1977) has been reported to increase under waterlogged conditions, whereas other studies report decreased Cd (Ito and Iimura 1975) and Zn availability (Iu et al. 1981; Ghanem and Mikkelsen 1987). Francis and Dodge (1990) have demonstrated that reductive-dissolution of goethite results in the release of copreciptated Cd, Ni, Zn, and, to a lesser extent, Cr and Pb, confirming this release mechanism. However, the wide solid/solution ratio used (50 mg goethite:40 ml culture media) resulted in a large proportion of the reduced Fe remaining in solution (60% of added Fe), rather than precipitating as magnetite or siderite. Release of coprecipitated metals was proportionally less than Fe, implying either nonstoichiometric dissolution, or a high affinity of the trace metals for the Fe(II) minerals formed. Under the conditions encountered in waterlogged soils, only a small proportion of the reduced Fe and Mn would remain in solution, the remainder being converted to new solid phases. The production of new oxide surfaces will present fresh sites for metal adsorption. Release of metal from crystalline oxides and recapture by amorphous oxides under waterlogged conditions is well demonstrated by Mandal et al. (1988).

Although the total Mn oxide content of soils is typically lower than the Fe content, the solution concentration of Mn observed in reduced soils is also lower ($\approx$ 10 mg/l), thus the proportion of Mn oxides solubilized is small. Reductive dissolution of Mn(IV) oxides is likely to have a marked effect on Co, an element closely associated with Mn in soil (McKenzie 1975), and increased Co availability has been observed under reduced conditions (Adams and Honeysett 1964; Harvey et al. 1997; Phillips 1999; Quantin et al. 2001). Following drainage of waterlogged soil, reoxidation of Mn(II) may be slow, and increased plant availability of Co and Mn may persist. Iu et al. (1982) demonstrated elevated Co and Mn uptake by French beans (*Phaseolus vulgaris*) and maize (*Zea mays*) during a 45-d growth period in soil that had been allowed to drain and dry for 30 d prior to planting.

An important example of oxidation and precipitation of Fe and Mn in waterlogged soils occurs in the oxidized rhizosphere of plants with aerenchyma to supply O_2 to the root system. Iron(II) and Mn(II) transported toward the root by mass flow and diffusion are oxidized and precipitated, forming a plaque around the root (Figure 8.4). Field-collected plaque samples have been found to contain elevated concentrations of Cd, Cu, Pb, and Zn relative to the surrounding soil (Sundby et al. 1998), and adsorption or coprecipitation of metals within the plaque has been suggested as a means of reducing plant metal uptake (Taylor and Crowder 1983). In a study of *Phalaris arundinacea* roots, Hansel et al. (2001) demonstrated that metal sequestration at the soil–root interface was the combined result of the presence of the anoxic-oxic boundary (Fe-oxide precipitation), plant/bacterial respiration (Mn- and Zn-carbonate nucleation), and microbial biofilm formation (Pb complexation). Solution culture experiments have demonstrated that the presence of Fe plaque reduced plant Cu uptake by *Phragmites australis* (Batty et al. 2000) and Cu, Zn, and Ni uptake by rice (Greipsson and Crowder 1992; Greipsson 1994, 1995). However, the metals did not accumulate in the plaque, suggesting that the reduction in metal uptake was not a result of metal capture by the plaque.

8.4.4 Altered Soil-Solution Composition

The establishment of reducing conditions in soils results in a marked increase in ionic strength of the soil solution. This increase is, in part, the result of increased soil-solution concentrations of Mn^{2+}, Fe^{2+}, and HCO_3^-, as described in earlier sections. In addition, soil-solution concentrations of Ca and Mg increase as a result of their displacement from exchange sites by Mn^{2+} and Fe^{2+} (Ponnamperuma 1972). Displacement from exchange sites may also partly account for the increase in the soil-solution concentration of cationic trace elements. This release mechanism is

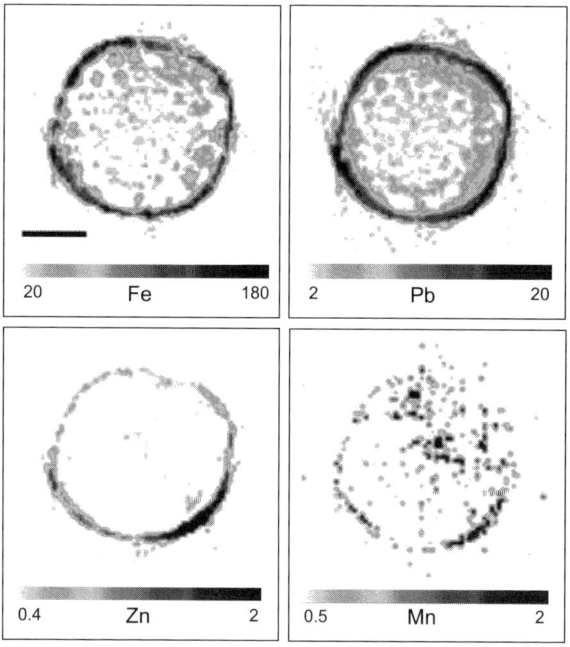

FIGURE 8.4 Metal distributions on and within the roots of *Phalaris arundinacea* as depicted by fluorescence microtomography. The spatial distribution of Fe, Pb, Zn, and Mn within a cross-sectional slice of the grass root is illustrated. Scale bar represents 150 μm, and the gray scale units are fg/μm^3. (Reprinted with permission from *Environ Sci Technol* 2001, **35**: 3863–3868. ©2001 American Chemical Society.)

especially applicable to ions such as Cd^{2+} and Zn^{2+}, which are not strongly adsorbed by soil colloids.

Although increased soil-solution Fe^{2+} may result in increased trace element concentrations through exchange, metal bioavailability may not be increased. Plant uptake of the released trace elements may be inhibited by competition with Fe^{2+}.

8.4.5 Organic Matter

Reducing conditions result in an increase in the concentration of organic ligands present in soil solution, and in the alteration of solid-phase organic matter. Both these changes have the potential to alter trace element availability. Anaerobic degradation of organic matter results in an increase in the concentration of fulvic acid and a number of volatile organic acids (Sinha 1972), the most important being acetic, formic, propionic, and butyric acids (Ponnamperuma 1972). These organic acids can complex trace metals, and elevated concentrations of organically complexed Cd, Pb, and Zn have been reported under reduced conditions (Charlatchka et al. 1997; Charlatchka and Cambier 2000). Dudley et al. (1986) reported that, in sewage sludge-amended soil, during the first week following sludge application, anaerobic conditions were induced in the soil, and the mobility of Cu and, to a lesser extent,

Ni was increased, despite an increase in soil pH. Soil-solution organic carbon concentrations also increased during this initial period, and the Cu and Ni mobilization was attributed to complexation. In a similar experiment, Lamy et al. (1993) reported elevated concentrations of organic matter and Cd in leachate, but no measurements were made of the soil's redox state.

On the reestablishment of oxidizing conditions, the volatile organic acids will rapidly be metabolized by soil microorganisms, lowering the solution metal concentration. The complexation of metals by dissolved organic matter is strongly pH dependent. Acidification of the system, for example, through the oxidation of pyrite, results in a release of metal from organic ligands (Reddy et al. 1995). Although the mobility of trace metals is increased by their complexation to soluble organic matter, these complexed metals may be less bioavailable.

Alteration of solid-phase organic matter and, hence, its ability to hold trace elements, has also been suggested. This suggestion has not been extensively investigated, and little consideration given to the mechanism by which the organic matter is altered. Furthermore, the effect of reoxidation of the soil is unknown. Evidence for altered retention of trace metals by solid-phase organic matter comes from sequential extraction studies, but there is little agreement between studies. Using this approach, Sims and Patrick (1978) found that the total concentration of metal retained by organic sites, as assessed by extraction with pyrophosphate, increased, but that this was primarily from increased retention of Fe; the organic pool of Zn was scarcely altered, though Cu increased. Similar results have been reported by other workers (Bjerre and Schierup 1985; Mandal et al. 1988). In contrast, Iu et al. (1981) and Ghanem and Mikkelsen (1987) reported from similar experiments reduced organic matter retention of Zn.

8.5 ATTENUATION OF METAL AVAILABILITY BY REDOX REACTIONS

Most of the changes that occur when oxidized soils are placed under reducing conditions are largely reversible on oxidation. Similarly, the effects of oxidation on reduced systems are readily reversed when reducing conditions are reestablished, an effect exploited in the use of wetlands to capture sulfate and metals released during the oxidation of pyritic mine wastes (Machemer and Wildeman 1992; Groudeva et al. 2001). There may be short-term effects, but the system is likely to return to a condition comparable to that which existed prior to the onset of waterlogging.

There is little information to suggest that alternation of oxidizing and reducing conditions attenuates trace element contamination. One indication that this does occur is the observation that oxidation/reduction cycles can accentuate trace element deficiency. Until now, the mechanism for such effects has not been clearly explained.

One clear piece of evidence for the attenuation of trace elements through oxidation/reduction cycling is the accumulation of these elements in Fe/Mn nodules. These nodules form in soils that are regularly inundated; Fe and Mn are reduced and solubilized and then preferentially oxidized and precipitated on the developing nodule (Figure 8.5). Palumbo et al. (2001) showed that in nodules from Sicilian

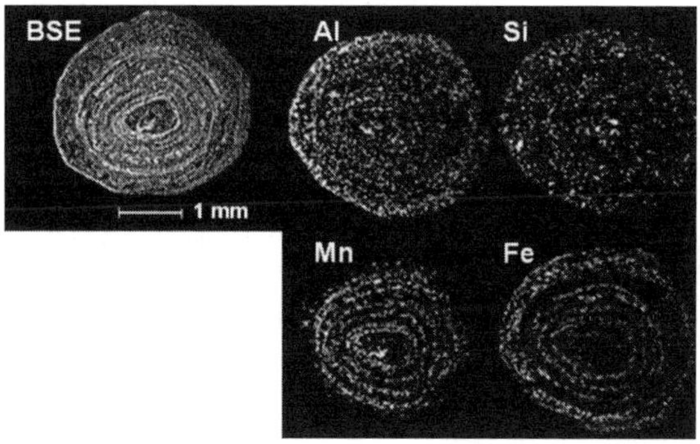

FIGURE 8.5 The banding in a Fe-Mn nodule cross-section revealed by back-scatter electron imaging (BSE) and the elemental distribution in this nodule as assessed by energy dispersive x-ray spectrometer analysis. (From Palumbo et al. 2001. *Chem Geol* 173:257–269. With permission.)

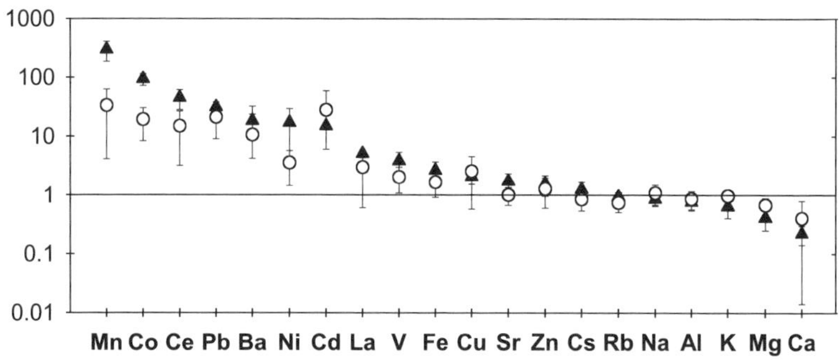

FIGURE 8.6 Average enrichment factor for elements in nodules relative to soil. (From Palumbo et al. 2001. *Chem Geol* 173:257–269. With permission.)

soils, Ba, Cd, Ce, Co, Cd, Ni, and Pb are markedly enriched relative to the host soil (Figure 8.6). In nodules showing strong Fe-Mn banding, Ba, Cd, Cu, Ni, and Sr were accumulated in the Mn bands, whereas Pb and V were accumulated in the Fe-enriched bands. This effect will not be limited to nodules but will occur wherever redox cycling leads to an accumulation of Fe or Mn within 1 part of the soil matrix, for example, within aggregates in which there is differential mobilization of Mn in response to seasonal waterlogging (Bundt et al. 1997). Clearly, attenuation of trace metal availability by this mechanism will be a relatively slow process.

REFERENCES

Adams SN, Honeysett SL. 1964. Some effects of soil waterlogging on the cobalt and copper status of pasture plants grown in pots. Aust J Agric Res 7:29–42.

Alemi MH, Goldhamer DA, Nielsen DR. 1991. Modeling selenium transport in steady-state, unsaturated soil columns. J Environ Qual 20:89–95.

Allen HE JFu G, Deng B. 1993. Analysis of acid-volatile sulfide (AVS) and simultaneously extracted metals (SEM) for the estimation of potential toxicity in aquatic sediments. Environ Toxicol Chem 12:1441–1453.

Asher CJ, Butler GW, Peterson PJ. 1977. Selenium transport in root systems of tomato. J Expt Bot 28:279–291.

Astrom M. 1998 Mobility of Al, Co, Cr, Cu, Fe, Mn, Ni and V in sulphide-bearing fine-grained sediments exposed to atmospheric O-2: an experimental study. Environ Geol 36:219–226.

Baas Becking LGM, Kaplan IR, Moore D. 1960. Limits of the natural environment in terms of pH and oxidation-reduction potentials. J Geol 68:243–284.

Balistrieri LS, Chao TT. 1987. Selenium adsorption by goethite. J Environ Qual 51:1145–1151.

Bartlett RJ, James BR. 1979. Behaviour of chromium in soils: III. Oxidation. J Environ Qual 8:31–35.

Bartlett RJ, James BR. 1993. Redox chemistry of soils. Adv Agron 50:151–208.

Bartlett RJ, Kimble JM. 1976. Behaviour of chromium in soils: I. Trivalent forms. J Environ Qual 5:379–383.

Batty LC, Baker AJM, Wheeler BD, Curtis CD. 2000. The effect of pH and plaque on the uptake of Cu and Mn in *Phragmites australis* (Cav.) Trin ex. Steudel. Annal Bot 86:647–653.

Bingham FT, Page AL, Mahler RJ, Ganje TJ. 1976. Cadmium availability to rice in sludge-ammended soil under "flood" and "nonflood" culture. Soil Sci Soc Am J 40:715–719.

Bjerre GK, Schierup HH. 1985. Influence of waterlogging on availability and uptake of heavy metals by oat grown in different soils. Plant Soil 88:45–56.

Bohn HL. 1971. Redox potentials Soil Sci 112:39–46.

Brennan EW, Lindsay WL. 1996. The role of pyrite in controlling metal ion activities in highly reduced soils. Geochim Cosmoch Acta 60:3609–3618.

Bundt M, Kretzschmar S, Zech W, Wilcke W. 1997. Seasonal redistribution of manganese in soil aggregates of a Costa Rican coffee field. Soil Sci 162:641–647.

Cacador I, Vale C, Catarino F. 1996. Accumulation of Zn, Pb, Cu, Cr and Ni in sediments between roots of the Tagus estuary salt marshes, Portugal. Estuar Coast Shelf Sci 42:393–403.

Carbonell AA, DeLaune RD, Patrick WH. 1998. Effect of phosphogypsum and barite amendments on heavy metals and trace elements chemistry in Mississippi river alluvial sediment. J Environ Sci Health A-Toxic/Hazard Subst Environ Eng 33:1–21.

Charlatchka R, Cambier P. 2000. Influence of reducing conditions on solubility of trace metals in contaminated soils. Water Air Soil Pollut 118:143–167.

Charlatchka R, Cambier P, Bourgeois S. 1997. Mobilization of trace metals in contaminated soils under anaerobic conditions. In: Prost R, editor. Third International Conference on the Biogeochemistry of Trace Elements (3rd ICOBTE), Paris. INRA Editions. p 159–174.

Chen JM, Hao OJ. 1998. Microbial chromium (VI) reduction. Crit Rev Environ Sci Technol 28:219–251.

Cifuentes FR, Lindermann WC, Barton LL. 1996. Chromium sorption and reduction in soil with implications to bioremediation. Soil Sci 161:233–241.Chuan MC, Shu GY, Liu JC. 1996. Solubility of heavy metals in a contaminated soil: effects of redox potential and pH. Water Air Soil Pollut 90:543–556.

Cullen WR, Reimer KJ. 1989. Arsenic speciation in the environment. Chem Rev 89:713–764.

Deuel LE, Swoboda AR. 1972. Arsenic solubility in a reduced environment. Soil Sci Soc Am Proc 36:276–278.

Doyle MO, Otte ML. 1997. Organism-induced accumulation of iron, zinc and arsenic in wetland soils. Environ Pollut 96:1–11.

Drew MC, Lynch JM. 1980. Soil anaerobiosis, microorganisms and root function. Ann Rev Phytopathol 18:37–66.

Dudley LM, McNeal BL, Baham JE. 1986. Time-dependent changes in soluble organics, copper, nickel, and zinc from sewage sludge amended soils. J Environl Qual 15:188–192.

Dutta D, Mandal B, Mandal LN. 1989. Decrease in availability of zinc and copper in acidic to near neutral soils on submergence. Soil Sci 147:187–195.

Eary LE, Rai D. 1988. Chromate removal from aqueous wastes by reduction with ferrous ion. Environ Sci Technol 22:972–977.

Elliot HA, Liberati MR, Huang CP. 1986. Competitive adsorption of heavy metals by soils. J Environ Qual 15:214–219.

Elrashidi MA, Adriano DC, Workman SM, Lindsay WL. 1987. Chemical equilibria of selenium in soils: A theoretical development. Soil Sci 144:141–152.

Fendorf SE, Zasoski RJ. 1992. Chromium (III) oxidation by δ-MnO_2. 1. Characterization. Environ Sci Technol 26:79–85.

Ferguson JF, Gavis J. 1972. A review of the arsenic cycle in natural waters. Wat Res 6:1259–1272.

Francis JA, Dodge CJ. 1990. Anaerobic microbial remobilization of toxic metals coprecipitated with iron oxide. Environ Sci Technol 24:373–378

Francoise CR, Bourg ACM. 1991. Aqueous geochemistry of chromium: a review. Wat Res 25:807–816.

Gambrell RP, Patrick WH. 1978. Chemical and microbiological properties of anaerobic soils and sediments. In: Hook DD, Crawford RMM, editors. Plant life in anaerobic environments. Ann Arbor, MI: Ann Arbor Science Publishers. p 375–423.

Gambrell RP, Wiesepape JB, Patrick WH, Duff MC. 1991. The effects of pH, redox, and salinity on metal release from a contaminated sediment. Water Air Soil Pollut 57–58:359–367.

Ghanem SA, Mikkelsen DS. 1987. Effect of organic matter on changes in soil zinc fractions found in wetland soils. Comm Soil Sci Plant Anal 18:1217–1234.

Greenland DJ. 1962. Denitrification in some tropical soils. J Agric Sci 58:227.

Greenwood DJ. 1961. The effect of oxygen concentration on the decomposition of organic materials in soils. Plant Soil 14, 360–369.

Greipsson S. 1994. Effects of iron plaque on roots of rice on growth and metal concentration of seeds and plant-tissues when cultivated in excess copper. Comm Soil Sci Plant Anal 25:2761–2769.

Greipsson S. 1995. Effect of iron plaque on roots of rice on growth of plants in excess zinc and accumulation of phosphorus in plants in excess copper or nickel. J Plant Nutr 18:1659–1665.

Greipsson S, Crowder AA. 1992. Amelioration of copper and nickel toxicity by iron plaque on roots of rice (Oryza sativa). Can J Bot 70:824–830.

Groudeva VI, Groudev SN, Doycheva AS. 2001. Bioremediation of waters contaminated with crude oil and toxic heavy metals. Int J Mineral Proc 62:293–299.

Haldar M, Mandal B. 1979. Influence of soil moisture regimes and organic matter application on the extractable Zn and Cu content of rice soils. Plant Soil 53:203–213.

Harvey NW, Shaw G, Bell NJB. 1997. Influence of plant roots upon the mobility of radionuclides in soil, with respect to location of contamination below the surface. J Radioanal Nucl Chem 226:159–173.

Hansel CM, Fendorf S, Sutton S, Newville M. 2001. Characterization of Fe plaque and associated metals on the roots of mine-waste impacted aquatic plants. Environ Sci Technol 35:3863–3868.

Huerta-Diaz MA, Morse JW. 1992. Pyritization of trace-metals in anoxic marine sediments. Geochim Cosmochim Acta 56:2681–2702.

Ito H, Iimura K. 1975. Absorption of cadmium by rice plants in response to change of oxidation-reduction conditions of soils. Nippon Rozyohiryogaku Zasshi 46:82–88.

Iu KL, Pulford ID, Duncan HJ. 1981. Influence of waterlogging and lime or organic matter additions on the distribution of trace metals in an acid soil. Plant Soil 59:327–33.

Iu KL, Pulford ID, Duncan HJ. 1982. Influence of soil waterlogging on subsequent plant growth and trace metal content. Plant Soil 66:423–427.

Jackson MB, Armstrong W. 1999. Formation of aerenchyma and the processes of plant ventilation in relation to soil flooding and submergence. Plant Biol 1:274–287.

Jayaweera GR, Biggar JW. 1996. Role of redox potential in chemical transformations of selenium in soils. Soil Sci Soc Am J 60:1056–1063.

Jugsujinda A, Patrick WH. 1977. Growth and nutrient uptake by rice in a flooded soil under controlled aerobic-anaerobic and pH conditions. Agron J 69:705–710.

Kimbrough DE, Cohen Y, Winer AM, Creelman L, Mabuni C. 1999. A critical assessment of chromium in the environment. Crit Rev Environ Sci Technol 29:1–46.

Lamy I, Bourgeois S, Bermond A. 1993. Soil cadmium mobility as a consequence of sewage sludge disposal. J Environ Qual 22:731–737.

Lindsay WL. 1979. Chemical equilibria in soils. New York: J Wiley.

Losi ME, Amrhein C, Frankenberger WT. 1994. Factors affecting chemical and biological reduction of Cr(VI) in soil. Environ Toxicol Chem 13:1727–1735.

Ma QY, Lindsay WL. 1993. Measurements of free Zn^{2+} activity in uncontaminated and contaminated soils using chelation. Soil Sci Soc Am J 57:963–967.

Machemer SD, Wildeman TR. 1992. Adsorption compared with sulfide precipitation as metal removal processes from acid-mine drainage in a constructed wetland. J Contam Hydrol 9:115–131.

Mandal B, Hazra GC, Pal AK. 1988. Transformations of zinc in soils under submerged conditions and its relation with zinc nutrition of rice. Plant Soil 106:121–126.

Marin AR. 1992. The influence of chemical form and concentration of arsenic on rice growth and tissue arsenic concentration. Plant and Soil 139:175–183.

Marin AR, Masscheleyn PH, Patrick WH. 1993 Soil redox-pH stability of arsenic species and its influence on arsenic uptake by rice. Plant Soil 152:245–253.

Masscheleyn PH, Delaune RD, Patrick WH. 1991a. Arsenic and selenium chemistry as affected by sediment redox potential and pH. J Environ Qual 20:522–527.

Masscheleyn PH, Delaune RD, Patrick WH. 1991b. Effect of redox potential on arsenic speciation and solubility in a contaminated soil. Environ Sci Technol 25:1414–1419.

Mathers AC, Wilson GC, Schneider AD, Scott P. 1971. Sugarbeet response to deep tillage, nitrogen, and phosphorus on Pullman clay loam. Agron J 63:474–477.

McBride MB. 1989. Reactions controlling heavy metal solubility in soils. Advan Soil Sci 10:1–56.

McKenzie RM. 1975. Mineralogy and chemistry of soil cobalt. In: Nicholas DJD, Eagan AR editors. Trace elements in soil-plant-animal systems. New York: Academic Pr. p 83–93.

Meek BD, Stolzy LH. 1978. Short-term flooding. In: Hook DD, Crawford RMM, editors. Plant life in anaerobic environments. Ann Arbor, MI: Ann Arbor Science Publishers. p 351–373.

Mikkelsen DS, Brandon DM. 1975. Zinc deficiency in California rice. California Agric 29:8–9.

Morse JW, Arakaki T. 1993. Adsorption and coprecipitation of divalent metals with mackinawite (FeS). Geochim Cosmochim Acta 57:3635–3640.

Neal RH, Sposito G. 1991. Selenium mobility in irrigated soil columns as affected by organic carbon amendment. J Environ Qual 20:808–814.

Oscarson DW, Huang PM, Defosse C, Herbillion A. 1981. The oxidative power of Mn(IV) and Fe(III) oxides with respect to As(III) in terrestrial and aquatic environments. Nature 291:50–51.

Palumbo B, Bellanca A, Neri R, Roe MJ. 2001. Trace metal partitioning in Fe-Mn nodules from Sicilian soils, Italy. Chem Geol 173:257–269.

Patrick WH, Mikkelsen DS, Wells BR. 1985. Plant nutrient behaviour in flooded soil. In: Fertilizer technology and use. Madison, WI: Soil Science Society of America. p 197–228.

Pettine M, Millero FJ, Passino R. 1994. Reduction of chromium(VI) with hydrogen-sulfide in NaCl media. Mar Chem 46:335–344.

Phillips IR. 1999. Copper, lead, cadmium, and zinc sorption by waterlogged and air-dry soil. J Soil Contam 8:343–364.

Ponnamperuma FN. 1972. The chemistry of submerged soils. Advan Agron 24:29–96.

Quantin C, Becquer T, Rouiller JH, Berthelin J. 2001. Oxide weathering and trace metal release by bacterial reduction in a New Caledonia Ferralsol. Biogeochem 53:323–340.

Rai D, Eary LE, Zachara JM. 1989. Environmental chemistry of chromium. Sci Tot Environ 86:15–23.

Rai D, Sass BM, Moore DA. 1987. Chromium (III) hydrolysis constants and solubility of chromium (III) hydroxide. Inorg Chem 26:345–349.

Reddy CN, Patrick WH. 1977. Effect of redox potential and pH on the uptake of Cd and Pb by rice plants. J Environ Qual 6:259–262.

Reddy KL, Wang L, Gloss SP. 1995. Solubility and mobility of copper, zinc and lead in acidic environments. Plant Soil 171:53–58.

Sadiq M. 1997. Arsenic chemistry in soils: An overview of thermodynamic predictions and field observations. Water Air Soil Pollut 93:117–136.

Sadiq M, Zaidi TH, Mian AA. 1983. Environmental behaviour of arsenic in soils: theoretical. Water Air Soil Pollut 20:369–377.

Shupack SI. 1991. The chemistry of chromium and some resulting analytical problems. Environ Health Persp 92:7–11.

Sims JL, Patrick WH. 1978. The distribution of micronutrient cations in soil under conditions of varying redox potential and pH. Soil Sci Soc Am J 42:258–262.

Sinha MK. 1972. Organic matter transformation in soils I. Humification of C^{14}-tagged oat roots. Plant Soil 36:283–293.

Smith GS, Watkinson JH. 1984. Selenium toxicity in perennial ryegrass and white clover. New Phytol 97:557–564.

Sposito G, Yang A, Neal RH, Mackzum A. 1991 Selenate reduction in an alluvial soil. Soil Soil Sci Soc Am J 55:1597–1602.

Street JJ, Sabey BR, Lindsay WL. 1978. Influence of pH, phosphorus, cadmium, sewage sludge, and incubation time on the solubility and plant uptake of cadmium. J Environ Qual 7:286–290.

Sundby B, Vale C, Cacador I, Catarino F, Madureira MJ, Caetano M. 1998. Metal-rich concretions on the roots of salt marsh plants: Mechanism and rate of formation. Limnol Oceanogr 43:245–252.

Tack FM, Callewaert O, Verloo MG. 1996. Metal solubility as a function of pH in a contaminated, dredged sediment affected by oxidation. Environ Pollut 91:199–208.

Taylor GJ, Crowder AA. 1983. Use of the DCB technique for extraction of hydrous iron oxides from roots of wetland plants. Am J Bot 70:1254–1257.

Taylor HM, Burnett E. 1963. Some effects of compacted soil pans on plant growth in the southern Great Plains. J Soil Water Cons 18:235–236.

Van den Berg GA, Loch JPG, Winkels HJ. 1998. Effect of fluctuating hydrological conditions on the mobility of heavy metals in soils of a freshwater estuary in the Netherlands. Water Air Soil Pollut 102:377–388.

Webb JL. 1966. Arsenicals. In: Enzyme and metabolic inhibitors. Vol. III, Iodoacetate, Maleate, N-Ethylmaleimide, Atloxan, Quinones, Arsenicals, Chapter 6. Webb JL, editor. New York: Academic Pr. p 595–819.

Welch JE, Lund LJ. 1987. Soil properties, irrigation water quality, and soil moisture level influences on the movement of nickel in sewage sludge-treated soils. J Environ Qual 16:403–410.

White AF, Dubrovsky NM. 1994. Chemical oxidation-reduction controls on selenium mobility in groundwater systems. In: Frankenberger WT, Benson S editors. Selenium in the evironment. New York: Marcel Dekker. p 185–221.

Woolson EA. 1977. Generation of alkylarsines from soil. Weed Sci 25:412–416.

Woolson EA, Isensee AR. 1981. Soil residue accumulation for three applied arsenic sources. Weed Sci 29:17–21.

Yan-Chu H. 1994. Arsenic distribution in soils. In: Nriagu JO, editor. Arsenic in the environment I. Cycling and characterization. New York: J Wiley p 17–50.

Zausig J, Stepniewski W, Horn R. 1993. Oxygen concentration and redox potential gradients in unsaturated model soil aggregates. Soil Sci Soc Am J 57:908–916.

9 Fixation of Cadmium and Zinc in Soils: Implications for Risk Assessment

Erik Smolders and Fien Degryse

9.1 INTRODUCTION

This chapter collates the evidence for fixation of cadmium (Cd) and zinc (Zn) and discusses the relevance of fixation reactions in the risk assessment of these elements in soil.

After a trace element is added to soil, several reactions may occur, which change its concentration in the soil-pore water and the fraction available to organisms. Immobilization reactions in soil have a characteristic initial fast reaction (minutes to hours), which is then followed by a slow reaction (days to years). The slow immobilization reaction removes the trace element from the pool that is in dynamic equilibrium with the soil solution to a pool from which desorption may be very slow, if detectable at all within a realistic time frame (Figure 9.1). This slow immobilization reaction gradually reduces the availability of trace elements with increased aging time in soil and is hereafter termed "fixation." Fixation of potentially toxic trace elements is one of the factors causing natural attenuation of the risk associated with trace elements in soil.

One of the most striking examples of natural attenuation of trace element availability was identified for radiocaesium (^{137}Cs) in the aftermath of the 1986 Chernobyl accident. This accident released ^{137}Cs, which was deposited on a large area in western Europe. Food and crop surveys have identified a reduction in ^{137}Cs activity concentrations at a faster pace than the physical decay of this radionuclide (half-life of 30.2 years). As an example, milk powder ^{137}Cs concentrations in Austria steadily decreased, with a half-life of about 2 years, in the period between 500 days and 6 years after the 1986 Chernobyl accident (Mück 1995). Other studies of contaminated agricultural products have indicated that the initial rate of decline in ^{137}Cs activity may be described by an effective ecological half-life of around 1 year although, concurrently, long-term reduction rates may be modelled using a 10-year half-life (Absalom et al. 1999). These trends may be attributed to fixation and not to leaching, because migration rates of ^{137}Cs are typically less than 1 cm per year in soil. Fixation of ^{137}Cs in soils has been demonstrated with desorption studies, and

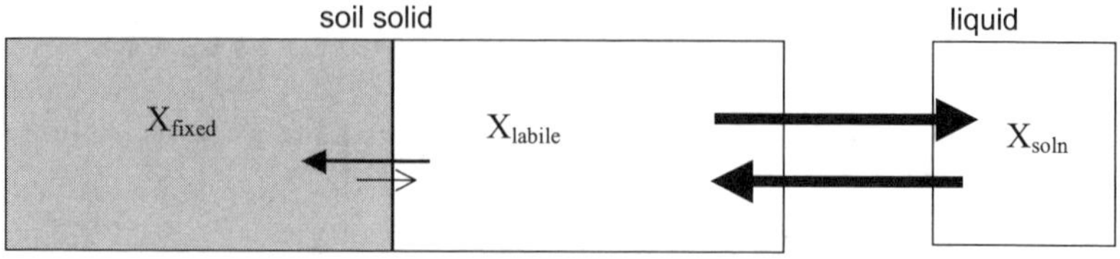

FIGURE 9.1 Labile and fixed pools of trace element X in soil. The trace element in solution (X_{soln}) is in dynamic contact with the labile pool. Fixation reactions remove the element from the labile pool to a pool from which desorption is very slow, if detectable at all within a realistic time frame.

increased aging time increases the fraction of ^{137}Cs that is fixed in soil (Wauters et al. 1994). Fixation is thought to occur as diffusion into the collapsed interlayer of illitic and micaceous minerals.

To our knowledge, there are no long-term data collected directly from the field that provide direct evidence of fixation of heavy metals such as Cd and Zn. Fixation of these metals occurs, but may be undetectable in field data. As an example, annual variations in grain Cd concentrations can be up to 5-fold and dominate the variance in long-term trends in crop Cd concentrations (Andersson and Bingefors 1985). Fixation of trace elements in soil can even be difficult to detect in long-term observations in the laboratory. As an example, prolonged Cd adsorption studies on 2 soils were inconclusive, because even minor changes of pH could explain apparent fixation after 35 weeks of contact (Christensen 1984a). It should be recalled that fixation is defined as the reaction that removes the element from a labile pool that is in equilibrium with the solution phase to a fixed pool from which desorption is slow. Any drift in soil properties affecting the partitioning of that element between the labile pool and the soil solution (e.g., soil pH) may obscure effects of fixation on trace element mobility (Figure 9.1).

To overcome the difficulties in detecting fixation of trace elements in long-term data, methods have been developed to detect fixed and labile fractions in soil. These methods rarely yield direct evidence for the fixation process, because a fixed fraction of an element may also be related to an insoluble source of that element in soil. The isotope dilution technique has been widely used in the last decade to detect fixed fractions of metals in soil. This method is based on the addition of small quantities of an isotope to the indigenous trace element in soil. The ratio of the isotope to indigenous trace element in soil extracts or in plants gives information on the lability of that element in soil. Details of this technique can be found in Chapter 2, but, in principle, this technique measures the relative availability of the indigenous trace element in soil (aged forms) compared to the availability of freshly added isotope. The availability of the indigenous (aged) trace element is usually significantly lower than the freshly added isotope when the latter is added in a soluble form.

9.1.1 Risks of Cadmium in Soil

Crop uptake of Cd and transfer to the food chain is generally considered as the critical pathway for the risk assessment of soil Cd. Acceptable increases of Cd in soil are small because Cd concentrations in agricultural crops can even exceed food standards at background soil Cd concentrations (McLaughlin et al. 1994). Major sources of Cd in agricultural soils are atmospheric deposition, fertilizers, and sewage sludge. Mass balance calculations and retrospective analyses of archived soil samples in the U.K. and Denmark have demonstrated that soil Cd concentrations have increased significantly in the 20th century (Tjell and Christensen 1985; Jones et al. 1987). This situation is likely to persist in many areas of the world, although recent mass balance calculations for Europe indicate that the current Cd accumulation rate is significantly lower than the historic rates (RAR Cd/CdO, 2001). The question whether Cd can be fixed by aging in soil is critical for the risk assessment of this element, because the Cd fixation rate could counteract the total accumulation rate of Cd in soil. In other words, the key question is whether Cd fixation in soil can reverse the slope of the predicted future trends in crop Cd at current rates of Cd input.

9.1.2 Fixation of Cadmium in Soils

Slow reactions of Cd in soil appear to be limited. Christensen (1984a) found a small, but significant decrease in solution concentration of Cd in 2 soils during 35 weeks of contact after a 20-h initial contact. However, this was not considered as an indication of fixation, because a pH increase of 0.1 unit could also explain this decrease in concentration. The high pH sensitivity of the solid–liquid equilibrium makes it difficult to observe fixation in long-term adsorption experiments. Desorption experiments were performed on the same 2 soils, previously loaded with Cd (Christensen 1984b). No irreversibility of Cd sorption was observed in the loamy sand soil, whereas in the sandy loam soil, a small hysteresis effect was observed in 10^{-3} M $CaCl_2$, but not in 10^{-2} M $CaCl_2$. However, aging the soil (stored moist) at 1 °C for 67 weeks did not cause any changes in the desorption pattern, indicating that no stronger Cd binding was developed with time.

Different techniques used to measure the labile and fixed pools are described in Chapter 1 to Chapter 3. The use of chemical extractants is convenient, but it is mostly unclear how the extracted fraction relates to the labile pool. Complexing agents (e.g., EDTA and DPTA) have been proposed to extract the labile pool. However, depending on the concentration of the extractant and metal content of the soil, the amount of Cd extracted may be either less or more than the labile pool determined by isotopic dilution (Nakhone and Young 1993; Stanhope et al. 2000). Extraction of the labile Cd pool with 1 M $CaCl_2$ was proposed, as labile Cd is brought into solution through complexation by Cl and competition for surface sites by Ca, whereas it is unlikely that fixed Cd will be extracted by $CaCl_2$ (Young et al. 2000). The Cd amount extracted by 1 M $CaCl_2$ was indeed found to agree well with the labile Cd pool determined by isotopic dilution. Repeated desorption with a neutral salt solution can also be used to assess the labile pool. However, as numerous desorptions are needed, this technique is seldom used. An alternative to this laborious technique is a desorption isotherm obtained by widening the soil/solution ratio (Filius et al. 1998). Figure 9.2 illustrates that the labile pool of Cd, derived from a desorption isotherm, corresponds well with the radiolabile Cd pool (Degryse, unpublished).

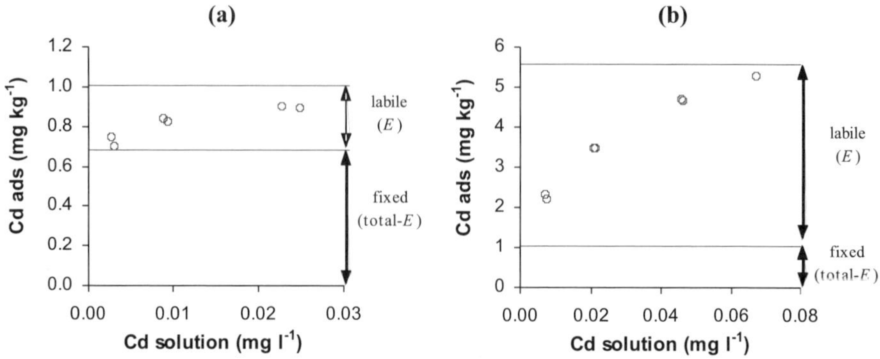

FIGURE 9.2 Desorption isotherms obtained by a widening soil/solution ratio for polluted sandy soils with (a) pH 3.9 and (b) pH 4.7. Labile and fixed Cd amounts as determined by isotopic dilution, are also shown.

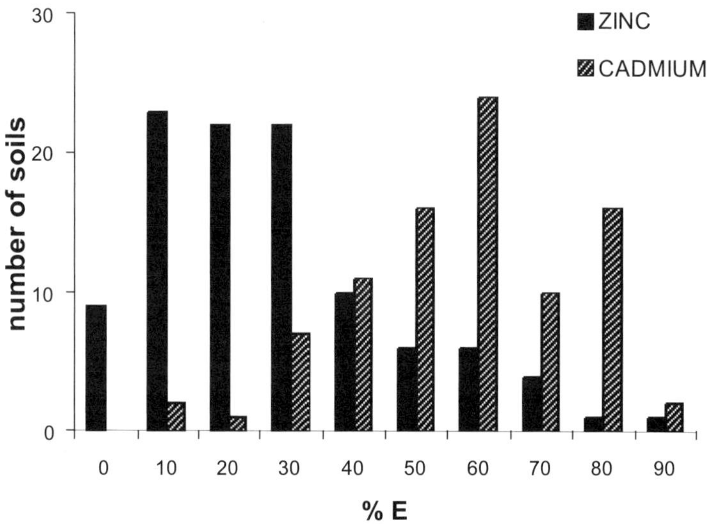

FIGURE 9.3 The percentage Cd and Zn that is radiolabile (%E) in European soils. For Zn n = 104, and for Cd n = 90. Details in Table 9.1 and Table 9.3. (Data compiled from Degryse and Smolders unpublished, and Sinaj et al. 1999.)

During the last decade, the isotopic dilution method has been used to gather data about labile Cd fractions in soils collected from the field (Figure 9.3, Table 9.1). Overall, it seems that in soils with a background content of Cd or in soils amended with Cd in a soluble form, more than half of the Cd content is in labile form (Table 9.1). When Cd is added in a partially insoluble form, radislabile fractions seem to be governed by the source of the metal (Table 9.1).

TABLE 9.1
Percentage of radiolabile Cd (%E value) in soils

Soils	n	Cd_T (mg/kg)	pH	%E[a]	Method	Reference
Sludge-amended soils	25	~0.5–52	5.8–7.1	**6–81 (34)**	$Ca(NO_3)_2$ 0.1 M, S/L 1:5; stable Cd measured and spike added after 5-d equilibration; ^{109}Cd measured 2 d after spiking	Young et al. 2000
Mine spoil soils	41	~0.5–260	4.0–7.3	**14–103 (48)**		
Sludge-amended soils	3	59–73	6.2	**29–48**		Hutchinson et al. 2000
Mine spoil soils	3	130–770	7.0–7.2	**6–15**		
Sludge-amended soils	13	0.75–343	3.4–6.5	**29–59 (46)**	KNO_3 1 M or $CaCl_2$ 0.05 M, S/L 1:3; stable Cd measured and spike added after 2-d equilibration; ^{109}Cd measured 1 d after spiking	Nakhone and Young 1993
Fertilized agricultural soils and Cd-polluted soils	8	0.4–6.5	4.5–7.1	**62–90 (74)**	$CaCl_2$ 10^{-2} M, S/L 1:10; 7-d equilibration	Smolders et al. 1999
Sludge-amended soil	1	34	6.0	**44**	^{111}Cd-spiked soil incubated for 2 weeks (θ_w 0.15); extraction with $Sr(NO_3)_2$ 0.1 M, S/L 2:15, 2 h	Ahnstrom and Parker 2001
Mine spoil soil	1	22	3.5	**13**		
Natural	1	26	7.3	**35**		
Smelter polluted	1	31	7.0	**49**		
Transect to Zn smelter	4	0.6–25	7.8–8.2	**44–70**	Water, S/L 1:2; 1-d equilibration; ^{109}Cd measured 1,4, and 10 min after spiking, extrapolated to 60 d	Gérard et al. 2000
Background	11	0.1–2.8	4.2–7.3	**51–90**	$CaCl_2$ 10^{-2} M, S/L 1:10; 3-d equilibration	Degryse and Smolders unpublished
Sludge-amended	19	0.6–55	5.6–7.5	**31–71**		
Metal salt-spiked	10	7–118	5.3–7.2	**73–90**		
Soils near metal smelter	39	0.4–104	3.3–6.9	**17–92**		
Miscellaneous	9	2–414	4.6–7.2	**9–78**		
Overall				**9–92 (60)**		

Note: n = number of soils, Cd_t = total Cd in soil, S/L = soil/ liquid ratio.

[a] Means within parentheses.

9.1.3 Biological Evidence for Cadmium Fixation

All biological evidence for Cd fixation has been obtained from Cd uptake by crops, primarily because this is the critical pathway for this element. Retrospective analyses of archived soil and crop samples are inconclusive about possible fixation of Cd in soils. Cadmium concentrations in archived grain samples of the long-term field trials in Rothamsted (U.K.) decreased between 1860 and 1986 on plots treated with farm yard manure, although soil Cd concentrations have increased on these plots (Jones and Johnston 1989). In contrast, grain Cd has increased on P-fertilizer-treated plots, whereas soil Cd concentrations increased more slowly than in farmyard manure-treated plots (Jones et al. 1987). Cadmium concentrations in grains of winter wheat have doubled between 1918 and 1980 in Sweden, but trends in soil Cd were not given (Andersson and Bingefors 1985).

Almost all Cd isotope dilution studies with plants show that a significant proportion of Cd in soil is in a nonavailable pool (%L below 100%, Table 9.2). If it is assumed that all Cd has entered the soil in a labile (soluble) form, then these data indicate that Cd fixation indeed occurs. This assumption is difficult to test, because the contamination history of the soils is generally unknown. However, there are 2 studies that applied the isotope dilution technique to compare bioavailable Cd fractions between soils with known Cd contamination histories.

The first study compared the labile Cd fractions between soils that have received continuous P fertilizer for a long time, with soils in which P fertilization was stopped 20 years before sampling (Hamon et al. 1998). The P fertilizer was the major source of Cd in these soils, as indicated by a positive correlation between the cumulative application of P fertilizer and the background-corrected Cd concentration in soil. The fraction of radiolabile Cd (%L) was significantly higher in the soils that received P fertilizer (and Cd) continuously, than in the soils in which P fertilization was stopped. A model was developed, which estimated that 1 to 1.5% of labile Cd is fixed each year, i.e., an effective half-life of labile Cd of about 46 to 69 years. The model assumed a first-order fixation without a reverse reaction; i.e., the model assumed no limit to Cd fixation during the 20-year period. This assumption may be invalid, because internal Cd binding sites in soil constituents are probably finite and all isotope dilution studies yield "aged-Cd" %L values of 50% or higher. The time course in Cd fixation in Cd-spiked soils, for example, suggests that not all Cd in soil can be fixed (e.g., see Chapter 2, Figure 2.8).

The second study applied the isotope dilution technique to soils that remained sealed for at least 200 years (under an old house) and reference soils just outside the house (Jensen and Mosbæk 1990). The Cd concentrations were lower in the "sealed soils" than in the "outside soils," which suggested that the outside soil contained more recent inputs of Cd than the sealed soils. Soluble Cd was labeled with ^{109}Cd and mixed into soil for a pot trial. The authors calculated that this soluble Cd was as available as the existing soil Cd, irrespective of soil source; i.e., both the sealed soil Cd and the outside soil Cd were as available as the soluble Cd, which indicates that no Cd fixation occurred. In their calculation, the authors assumed that a constant fraction (30%) of Cd in the laboratory-grown plants was derived from air. This assumption is questionable because uptake of Cd by plants differed over

TABLE 9.2
Relative availability to plants of indigenous soil Cd to recently added radioactively labeled Cd (%L)

Soils	Crop	Cd_T mg/kg	pH	%L	Remarks	Reference
P-fertilized plots (started 1948), various P rates						Hamon et al. 1998
Series 1: last fertilized 20 years prior to sampling	Wheat (shoots)	0.03–0.19	6.0	**60(mean)**	%L significantly different between series	
Series 2: fertilization continued till sampling		0.03–0.34	6.0	**72 (mean)**		
Sandy loam 1	Grass	0.16	7.8	**65–73**		Dalenberg and Van Driel 1990
Sandy loam 2	Spinach, carrot, wheat	0.29	6.8	**62–86**		
Soil historically enriched with sewage sludge	Indian mustard	59	6.3	**42**		Stanhope et al. 2000
Sandy loam	Canola, capeweed, clover, lettuce, chard, wheat	0.2	4.9	**20 (canola)** **36 (other species)**		Hamon et al. 1997
Belgian agricultural soils, with both background and elevated Cd content (10 soils)	Wheat	0.33–6.5	4.5–7.2	**55–109**		Smolders et al. 1999
Transect to Zn smelter	Ryegrass, Lettuce, *Thlaspi caerulescens*	0.6–25	7.8–8.2	**24–196** **44–66** **36–62**	Why %L exceeds 100% not known	Gérard et al. 2000

Note: The %L was calculated from the ratio of Cd-specific activities in soil to that in plants, Cd_T = Total Cd in soil.

tenfold between soils and it is unlikely that airborne Cd in plants would also differ over tenfold for plants grown under the same conditions. Moreover, even plants grown outside (in rural areas) rarely have more than 20% airborne Cd (Dalenberg and Van Driel 1990). Therefore, the conclusions of this study should be treated with caution (Hamon et al. 1998).

Two studies have tested the hypothesis that different plant species sample Cd from different pools. One study (Hamon et al. 1997) did not find differences in %L values between species despite 30-fold differences in leaf Cd concentration. There was one exception (canola), which could not be readily explained (Table 9.2). Another study compared the %L value between a Cd hyperaccumulator and 2 agricultural plants, but no consistent differences in %L values were found (Gérard et al. 2000).

9.1.4 Implications for Risk Assessment

Almost all Cd isotope dilution studies (E and L values) show that some part of Cd is in a nonlabile form, typically 0 to 60% of the total (Table 9.1, Table 9.2, and Figure 9.3). This observation suggests that slow reactions occur, although it cannot be excluded that these fixed fractions are related to an insoluble source of Cd. However, the studies of Young et al. (2001) and of Hamon et al. (1998) can be used as direct evidence for the slow immobilization reactions of Cd in soil. The first study (Young et al. 2001) suggested that fixation of soluble Cd is close to completion after about 3 years, after which a maximum of 50% of Cd is fixed (soils with highest pH values). The second study (Hamon et al. 1998) suggested that Cd fixation occurs at a detectable rate of 1% to 1.5% of the labile Cd per year. The results of these studies differ largely, and this may be due to soil characteristics as well as the nature of the Cd source, and the period over which the fixation rate was estimated. The first study measured E values at various times after spiking, with a maximum of 3 years. The second study used soils sampled after 50 years of known Cd input and measured the final fixed and labile fractions. Based on the difference in %L between soils that did and those that did not receive Cd during a 20-y period (Hamon et al. 1998), it is possible that Cd fixation continues beyond a few years. Whatever the exact fixation pattern, less than half of Cd added was shown to be fixed in the soils of these 2 studies, corroborating many of the %E and %L data of field samples (Tables 9.1 and Table 9.2). A fixed fraction that is 50% of its total content means that Cd availability is only twofold lower than in a soil with identical sorption properties (e.g., same pH) but where Cd is completely labile. Therefore, it seems that the extent of fixation may be too small to be detectable in long-term field data, in which, for example, annual variation in crop Cd concentrations are easily 2- to fivefold.

Fixation could be taken into account in a risk assessment model for prognosis of the average long-term effects of current Cd inputs. Mass balance modeling predicts that Cd accumulation rates in soils are generally less than 1.0% of the total or labile amount (RAR Cd/CdO, 2001) and fixation could be included to improve predictions of the sustainability of current practices. However, more information, such as the variability in fixation rate between soils and the long-term pattern of fixation (first-order or reversible first-order kinetics), needs to be gathered. In addition, we think that a risk assessment of Cd should acknowledge all difficulties in predicting future trends in crop Cd concentrations given all other factors that affect Cd bioavailability and that may change over time in an unknown way (e.g., pH).

9.2 ZINC

9.2.1 Risks of Zinc in Soil

Increasing Zn concentrations in soil can have a negative effect on soil biota such as plants, soil microbes, and soil fauna. Adverse effects of Zn on plants occur below the level at which Zn concentrations in plants are unacceptable to humans or herbivorous animals. Therefore, acceptable Zn concentrations in soil are defined by the possibility of adverse effects of Zn on the soil ecosystem.

As with other elements, fixation of Zn in soil gradually reduces Zn availability to the soil biota. This reaction may reduce Zn toxicity with increasing aging time and could be a process for natural attenuation of Zn in contaminated soils. Few data that directly show the effect of aging on Zn bioavailability in the toxic concentration range are available. However, the consequence of Zn fixation in terms of decreasing Zn availability to plants is reported in the deficient concentration range (Brennan 1990; see also Chapter 10). Zinc fixation is also important in setting criteria for acceptable Zn concentrations in soil. A gradual buildup of Zn in soil may have less toxic effects than an immediate enrichment leading to the same final soil Zn concentration. The latter situation is common for laboratory toxicity tests using soils freshly spiked with Zn^{2+} salts. Such tests are used to derive soil toxicity criteria in many risk assessment procedures, because field studies are considered too complex for interpretation. Some of these laboratory toxicity tests yield Zn toxic thresholds in the Zn background concentration range (e.g., Chang and Broadbent 1981). Soil Zn toxicity criteria based on freshly added Zn may therefore be very restrictive and, if fixation occurs, may not reflect Zn exposure and risk in the real world, where Zn usually accumulates slowly in soil.

9.2.2 Fixation of Zinc in Soils

Slow reactions appear to be more pronounced for Zn than for Cd when tested on the same soil or soil constituents. The slow reactions of Zn in soil may be related to diffusion of Zn into (hydrous) oxides; slow reactions of Zn with laboratory-prepared iron-, aluminum-, or manganese-oxides are more pronounced for Zn than for Cd (Bruemmer et al. 1988; Trivedi and Axe 2000; see also Chapter 4).

Evidence for slow reactions has been provided by prolonged isotope exchange studies. The adsorption of $^{65}Zn^{2+}$ increased 1.2- to 2-fold between 1 and 10 or 15 d after the initial reaction (Sinaj et al. 1999; Tiller et al. 1972). Adsorption of Zn (added as Zn salt) increased 1.8- to 3-fold between d3 and d30 afer addition of the Zn salt, indicative of aging reactions. The slow reactions were less pronounced at higher rates of Zn addition where adsorption isotherms plateau. Higher temperatures accelerated aging reactions (Barrow 1986).

Studies using chemical extraction methods have indicated that a fraction of Zn in soil is not labile and, more important, this fraction increases with increasing

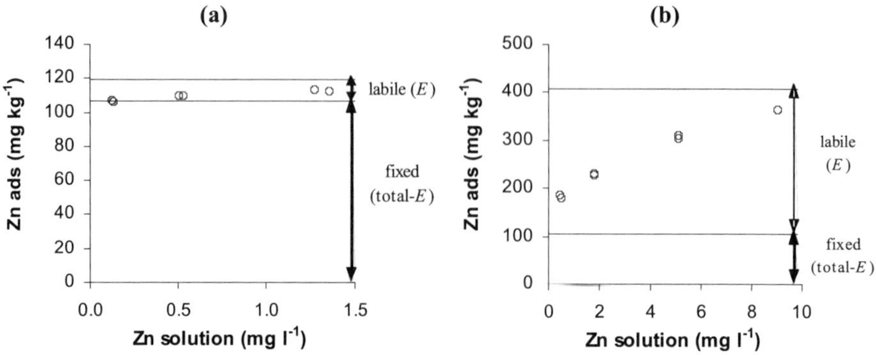

FIGURE 9.4 Desorption isotherms obtained by a widening soil/solution ratio for polluted sandy soils with (a) pH 3.9 and (b) pH 4.7. Labile and fixed Zn amounts as determined by isotopic dilution are also shown.

contact time between Zn and soil. Desorption of Zn with neutral salt solutions varying in soil/solution ratio depletes the labile pool and shows that a fraction of Zn cannot be extracted (Figure 9.4). This fraction corresponds with the fixed fraction determined by isotopic dilution on these soils. Extraction of metals using chelating agents is often used to measure Zn in soil available to plants (e.g., Lindsay and Norvell 1978). Aging reactions were identified using DTPA-extractable Zn in a calcareous soil (pH 8.5, 64% clay) amended with 3 levels of $ZnSO_4$ and aged for up to 1 year. The fixed pool of Zn increased from 0 initially to about 50% after 1 year (Ma and Uren 1997).

Aging reactions of Zn have also been identified in soils, using the isotopic dilution method. For example, in a study by Young et al. (2001), a range of soils were contaminated with Zn salt in the laboratory at 300 mg Zn/kg and the pool of labile Zn was found to change initially from 100% to values as low as 10% after 3 years of aging (See Chapter 2, Figure 2.7). Soil pH largely explained differences in slow reactions across soils; i.e., the fixation reactions were more pronounced at high pH.

Radiolabile fractions of Zn have also been measured in soils collected from the field (Figure 9.3, Table 9.3). Significant nonlabile fractions were detected in most soils but, again, this may also be due to the addition of insoluble sources of Zn. As an example, the amount of radiolabile Zn (expressed as percentage of total added) remaining in a soil (pH 6.1, 300 mg Zn/kg added) after 1 year of equilibration with free drainage was 80% if Zn was added as $ZnSO_4$ and 64% if it was added as the less soluble ZnO (Smolders and Degryse 2002). The fraction of labile Zn in the field-collected soils was almost always lower than that of Cd (Figure 9.3), indicating that aging reactions are more pronounced for Zn than for Cd or that Zn added to soils is generally in less soluble forms than Cd. The labile Zn typically varied between 10 and 40% of the total (Figure 9.3), with a median value of 29%. The lowest percentages of labile Zn were observed in uncontaminated soils with a high pH (details not shown). The labile fraction in sludge-applied soils varied between 20 and 40% (Table 9.3), a range similar to the range of %E values found by Young et al. (2000) in 25 soils contaminated by sludge addition. Higher percentages of

TABLE 9.3
Percentage of radiolabile Zn (%E) in soils collected from Belgium, U.K., France, Hungary, and Switzerland

Main source of Zn	n	pH	%C	Total Zn (mg/kg)	%E range (median)
Background	19	4.2–7.9	0.9–16.3	4–215	3.4–62 (31)
Zn^{2+} salts	5	4.9–7.3	0.9–2.8	67–294	45–100 (55)[a]
Sludge	14	5.2–7.5	0.9–10.1	29–1764	19–41 (29)
Metal smelter	48	3.3–6.9	1–23	18–34100	5.3–83 (32)
Other[b]	18	4.6–7.4	1.5–38	248–1688	14–34 (18)
All data	104				**3.4–100 (29)**

[a] %E of added Zn (not including the background), %E of total Zn is 17–95%, soils aged 1–10 years after contamination.
[b] Alluvial, industrial.

Source: Compiled from Degryse and Smolders unpublished, and Sinaj et al. 1999.

labile Zn were found in soils with background concentrations of Zn that had a low pH (<pH 5) and in some acid sandy soils affected by smelter operations.

9.2.3 Biological Evidence for Zinc Fixation

The impact of fixation on Zn bioavailability is often difficult to discern in laboratory experiments. For example, an experiment was set up to monitor trends in Zn toxicity for the springtail *Folsomia candida* in a Zn-salt-spiked soil, aged under field conditions for about 3 years (Smit et al. 1997). Posthuma et al. (1998) described similar studies with the plant *Trifolium pratense* and the enchytraeid *Enchytraeus crypticus*, using the same soil. For all 3 organisms, Zn toxicity decreased considerably after aging, but there was a parallel increase in pH of at least 1 unit, probably due to $CaCO_3$ that slowly reacted in the disturbed soil. Therefore, it remains unclear to what extent the observed decrease in Zn toxicity was due to the pH changes or to the effect of aging.

Isotope dilution can be used to verify if organisms are exposed to Zn via the labile pool only. If so, then chemical models describing Zn fixation may be used to predict future trends in Zn toxicity, assuming that changes in labile Zn over time are reflected in changes in Zn toxicity over time with all other soil properties remaining the same. However, organisms can alter the availability of Zn in soil because the chemical conditions are altered at the uptake site (e.g., the rhizosphere and the gut of invertebrates). Isotope dilution studies with ^{65}Zn (L values) have shown that plants can sample Zn from the fixed Zn pool (Smolders et al. 1999). However, the plant roots do not appear able to access the entire fixed pool of Zn in the soil. In the study by Smolders et al. (1999), the fraction of Zn in the soil that was labile according to the "plant data" (L value) was only 1.2- to 1.7-fold higher than the radiolabile fractions measured in soil suspensions (E value). The $^{65}Zn/Zn$

concentration ratio in depurated tissue of the earthworm *Eisenia fetida* exposed to a ^{65}Zn-spiked-soil was the same as that in lettuce grown on the same soil, suggesting little or no mobilization of Zn by earthworms from fixed pools (Scott-Fordsmand et al. 2004).

Biological evidence for aging may be found when comparing toxic thresholds between soils in which Zn has been aged and soils in which Zn is freshly added (see Chapter 3 for further discussion). A few studies have compared the toxicity of Zn in Zn-salt-spiked soils with the toxicity in historically contaminated field soils in the vicinity of a Zn smelter. Data interpretation is complex, mainly because soil properties affecting the biological response vary within the transects and between field- and laboratory-spiked soils. However, it is generally observed that Zn is toxic at lower concentrations in laboratory-spiked soils than in field-contaminated soil with, for example, up to tenfold differences (Spurgeon et al. 1995). Differences in labile Zn fractions may not be the only factor responsible for the differences in Zn toxicity between laboratory-spiked soils and field-contaminated soils. Spiking soils with Zn salt without appropriate leaching increases the ionic strength of the soil solution, thereby increasing the soil-solution Zn concentrations. Leaching (without lowering the total Zn concentration in soil) reduces Zn toxicity in some soils and for some organisms (Stevens et al. 2003). Recent data showed 10- to 20-fold decreases in 1 year in pore-water Zn concentrations in Zn-salt contaminated soils, aged under natural conditions with free drainage (Smolders and Degryse 2002). These decreases were not due to loss of labile Zn and could not be predicted by changes in the labile pool; i.e., the reductions in pore-water Zn in soils contaminated with Zn salts were larger than the reduction in labile pool as a function of time. A comparison of pore-water concentration of Zn between spiked artificial (Organization for Economic Co-Operation and Development, OECD) soils and field-contaminated soils revealed the former to be easily tenfold higher at the same total Zn, pH, and CEC values, especially at the highest Zn concentrations (Lock and Janssen 2001). Pore-water concentrations of Zn were a better basis than total Zn content for reconciling Zn toxicity data of laboratory-spiked soils with those of field-contaminated soils (Lock and Janssen 2001).

Aging of Zn may also affect Zn toxicity to soil microbial processes. Renella et al. (2002) showed that fresh additions of metal salts overestimate effects on respiration, but underestimate effects on biomass compared to long-term field studies (sludged soils). These data illustrate that aging of Zn in soil cannot explain the different responses to Zn in the short and long term. In response to short-term stress, the microbes may respond by, for example, changing their respiration rate, but these are responses to sudden disturbance (metal addition) only. This type of contamination almost never occurs in nature. Under field conditions, metal additions are normally slow, allowing both metal fixation and adaptation of the microbial population to take place. This means that the effects of Zn aging on soil microbial processes may be masked by the slow response of microbial populations. Results have shown that spiking soils with Zn salts greatly overestimates toxic effects of Zn in soils that have gradually become enriched with Zn from galvanized structures (electricity pylons) (Smolders et al. 2004). The great differences in toxicity could not be explained by differences in labile Zn only. Other factors such as biological adaptation (Rusk et al. 2004) and differences in concentrations of pore-water Zn between spiked and field-aged soils may explain the different response to added Zn.

9.2.4 Implications for Risk Assessment

Slow immobilization reactions of Zn reduce the fraction of Zn available to soil biota. The laboratory incubation studies of Young et al. (2001) showed that between 10 and 90% of soluble Zn can be fixed in 3 years, and that fixation increases with increasing soil pH (see Chapter 2, Figure 2.7). The fractions of Zn fixed in field-contaminated soils are sometimes even higher, but these fixed fractions may be related to the limited solubility of the Zn source in these soils.

In conclusion, despite substantial research efforts, it is still unclear if fixation of Zn can alleviate Zn toxicity. The isotope dilution studies suggest that plants and invertebrates (earthworms) have little access to fixed Zn. The only documented study on long-term effects of Zn on plants and invertebrates suggests that Zn toxicity decreases with time, but that study was confounded by increases in soil pH (Smit et al. 1997). Toxicity of recently added Zn salts to plants or soil invertebrates is found at equal or lower Zn concentrations than in field-contaminated soils, but most of these studies are also confounded by differences in soil properties between aged and freshly spiked soils. Effects of aging on the Zn toxicity affecting soil microbial processes are probably masked by the slow response of microbial populations to the Zn stress.

REFERENCES

Absalom JP, Young SD, Crout NMJ, Nisbet AF, Woodman RFM, Smolders E, Gillett AG. 1999. Predicting soil to plant transfer of radiocesium using soil characteristics. Environ Sci Technol 33:1218–1223.

Ahnstrom ZAS, Parker DR. 2001. Cadmium reactivity in metal-contaminated soils using a coupled stable isotope dilution-sequential extraction procedure. Environ Sci Technol 35:121–126.

Andersson A, Bingefors S. 1985. Trends and annual variations in Cd concentrations in grain of winter wheat. Acta Agric Scand 35:339–344.

Barrow NJ. 1986. Testing a mechanistic model. 2. The effects of time and temperature on the reaction of zinc with a soil. J Soil Sci 37:277–286.

Brennan RF. 1990. Reaction of zinc with soil affecting its availability to subterranean clover. 2. Effect of soil properties on the relative effectiveness of applied zinc. Aust J Soil Res 28:303–310.

Bruemmer GW, Gerth J, Tiller KG. 1988. Reaction-kinetics of the adsorption and desorption of nickel, zinc and cadmium by goethite. 1. Adsorption and diffusion of metals. J Soil Sci 39:37–52.

Chang FH, Broadbent FE. 1981. Influence of trace metals on carbon dioxide evolution from a Yolo soil. Soil Sci 132:416–421.

Christensen TH. 1984a. Cadmium soil sorption at low concentrations. 1. Effect of time, cadmium load, pH, and calcium. Water Air Soil Pollut 21:105–114.

Christensen TH. 1984b. Cadmium soil sorption at low concentrations. 2. Reversibility, effect of changes in solute composition, and effect of soil aging. Water Air Soil Pollut 21:115–125.

Dalenberg JW, Van Driel W. 1990. Contribution of atmospheric deposition to heavy-metal concentrations in field crops. Neth J Agric Sci 38:369–379.

Filius A, Streck T, Richter J. 1998. Cadmium sorption and desorption in limed topsoils as influenced by pH : Isotherms and simulated leaching. J Environ Qual 27:12–18.

Gérard E, Echevarria G, Sterckeman T, Morel JL. 2000. Cadmium availability to three plant species varying in cadmium accumulation pattern. J Environ Qual 29:1117–1123.

Hamon R, Wundke J, McLaughlin M, Naidu R. 1997. Availability of zinc and cadmium to different plant species. Aust J Soil Res 35:1267–1277.

Hamon RE, McLaughlin MJ, Naidu R, Correll R. 1998. Long-term changes in cadmium bioavailability in soil. Environ Sci Technol 32:3699–3703.

Hutchinson JJ, Young SD, McGrath SP, West HM, Black CR, Baker AJM. 2000. Determining uptake of 'non-labile' soil cadmium by *Thlaspi caerulescens* using isotopic dilution techniques. New Phytol 146:453–460.

Jensen H, Mosbæk H. 1990. Relative availability of 200 years old cadmium from soil to lettuce. Chemosphere 20:693–702.

Jones KC, Johnston AE. 1989. Cadmium in cereal grain and herbage from long-term experimental plots at Rothamsted, UK. Environ Pollut 57:199–216.

Jones KC, Symon CJ, Johnston AE. 1987. Retrospective analysis of an archived soil collection. II. Cadmium. Sci Tot Environ 67:75–89.

Lindsay WL, Norvell WA. 1978. Development of a DTPA soil test for zinc, iron, manganese and copper. Soil Sci Soc Am J 42:421–428.

Lock K, Janssen CR. 2001. Ecotoxicity of zinc in spiked artificial soils versus contaminated field soils. Environ Sci Technol 35:4295–4300.

Ma YB, Uren NC. 1997. The effects of temperature, time and cycles of drying and rewetting on the extractability of zinc added to a calcareous soil. Geoderma 75:89–97.

McLaughlin MJ, Palmer LT, Tiller KG, Beech TA, Smart MK. 1994. Increased soil salinity causes elevated cadmium concentrations in field-grown potato tubers. J Environ Qual 23:1013–1018.

Mück K. 1995 Longterm reduction of caesium concentration in milk after nuclear fallout. Sci Tot Environ 162:63–73.

Nakhone LN, Young SD. 1993. The significance of (radio-) labile cadmium pools in soil. Environ Pollut 82:73–77.

Posthuma L, Van Gestel CAM, Smit CE, Bakker DJ, Vonk JW. 1998. Validation of toxicity data and risk limits for soils: final report. National Institute of Public Health and the Environment (RIVM), Bilthoven, The Netherlands, RIVM report 607505004.

RAR Cd/CdO. 2001. Risk assessment. Cadmium metal and cadmium oxide. Brussels, Belgium: Ministry of Social Affairs, Public Health and the Environment, Belgian Federal Department of the Environment.

Renella G, Chaudri AM, Brookes PC. 2002. Fresh additions of heavy metals do not model long-term effects on microbial biomass and acitivity. Soil Biol Biochem 34:121–124.

Rusk JA, Hamon RE, Stevens DP, McLaughlin MJ. 2004. Adaptation of soil biological nitrification to heavy metals. Environ Sci Technol 38:3092–3097.

Scott-Fordsmand JJ, Stevens D, McLaughlin M. 2004. Do earthworms mobilize fixed zinc from ingested soil? Environ Sci Technol 38:3036–3039.

Sinaj S, Machler F, Frossard E. 1999. Assessment of isotopically exchangeable zinc in polluted and non-polluted soils. Soil Sci Soc Am J 63:1618–1625.

Smit CE, van Beelen P, Van Gestel CAM. 1997. Development of zinc bioavailability and toxicity for the springtail Folsomia candida in an experimentally contaminated field plot. Environ Pollut 98:73–80.

Smolders E, Brans K, Foldi A, Merckx R. 1999. Cadmium fixation in soils measured by isotopic dilution. Soil Sci Soc Am J 63:78–85.

Smolders E, Degryse F. 2002. Fate and effect of zinc from tire debris in soil. Environ Sci Technol 36:3706–3710.

Smolders E, Buekers J, Oliver I, McLaughlin MJ. 2004. Soil properties affecting toxicity of zinc to soil microbial properties in laboratory-spiked and field-contaminated soils. Environ Toxicol Chem 23:2633–2640.

Spurgeon DJ, Hopkin SP. 1995. Extrapolation of the laboratory-based OECD earthworm toxicity test to metal-contaminated field sites. Ecotoxicol 4:190–205.

Stanhope KG, Young SD, Hutchinson JJ, Kamath R. 2000. Use of isotopic dilution techniques to assess the mobilization of nonlabile Cd by chelating agents in phytoremediation. Environ Sci Technol 34:4123–4127.

Stevens DP, McLaughlin MJ, Heinrich T. 2003. Determining toxicity of runoff lead and zinc in soils-salinity effects on metal partitioning and phytotoxicity. Environ Toxicol Chem 22:3017–3024.

Tiller KG, Honeysett JL, De Vries MPC. 1972. Soil zinc and its uptake by plants. I. Isotopic exchange equilibria and the application of tracer techniques. Aust J Soil Res 10:151–164.

Tjell JC, Christensen TH. 1985. Evidence of increasing cadmium contents of agricultural soils. In: Lekkas TD, editor. Heavy metals in the environment. CEP consultants, Albany, UK p 391–393.

Trivedi P, Axe L. 2000. Modeling Cd and Zn sorption to hydrous metal oxides. Environ Sci Technol 34:2215–2223.

Wauters J, Sweeck E, Valcke A, Cremers A. 1994. Availability of radiocaesium in soils: a new methodology. Sci Tot Environ 157:239–248.

Young SD, Tye A, Carstensen A, Resende L, Crout N. 2000. Methods for determining labile cadmium and zinc in soil. Eur J Soil Sci 51:129–136.

Young SD, Tye A, Crout NMJ. 2001. Rates of metal ion fixation in soils determined by isotopic dilution. In: Evans L, editor. Proceedings of the 6th International Conference on the Biogeochemistry of Trace Elements (6th ICOBTE); July 2001; University of Guelph, Ontario, Canada. p 105.

10 Natural Attenuation: Implications for Trace Metal/Metalloid Nutrition

Rebecca Hamon, Samuel Stacey, Enzo Lombi, and Mike McLaughlin

10.1 INTRODUCTION

As discussed in detail in previous chapters, natural attenuation of metals by soils decreases metal bioavailability. It can therefore be highly desirable to facilitate this process in soils in which metals are present at concentrations of ecotoxicological concern. However, many metals are also essential micronutrients. In fact, micronutrient metals are probably more frequently present at low concentrations in soils than at toxic concentrations, resulting in constraints to crop growth and deficiencies for animal and human health. For example, millions of hectares of arable land are thought to be micronutrient deficient, limiting crop production (Fageria et al. 2002), with crop recovery rates for applied micronutrient fertilizers as low as 5 to 10% because of adsorption and fixation reactions in soils (Mortvedt 1994). Moreover, it has been estimated that more than 40% of the world's population suffers from some form of micronutrient malnutrition (Welch and Graham 2002) due to insufficient micronutrient uptake by crops.

This chapter reviews the current state of knowledge on attenuation of micronutrient availability in soils and the implications of this process for micronutrient nutrition and discusses potential strategies to circumvent micronutrient attenuation.

10.2 ESSENTIAL MICRONUTRIENTS

Despite the estimated importance of micronutrient deficiencies, reliable quantitative data on the global extent of specific micronutrient deficiencies in both soils and humans is actually very limited at present (Graham et al. 2001). Iron is the most widely studied micronutrient, and Fe deficiency in human populations has been found to be prevalent (e.g., Graham et al. 2001). However, Fe is not a trace metal in the context of soils; hence, discussion of this element is beyond the scope of this chapter. Of the trace metal/metalloid micronutrients, widespread deficiencies in Cu, Mo, and Zn have been recorded in crops (Fageria et al. 2002), and Co, Cu, Se, and Zn deficiencies in livestock and humans have been frequently reported (Goldhaber 2003; Lee et al. 1999). These elements form the primary focus of this chapter. Useful reviews of the essentiality of these and other trace elements to plants, livestock, and humans, and the consequences of deficiency, can be found in Fageria et al. (2002), Lee et al. (1999), and Goldhaber (2003), respectively.

10.3 IMPORTANCE OF UNDERSTANDING MICRONUTRIENT ATTENUATION

In theory, it is possible to overcome micronutrient deficiencies simply by providing micronutrient supplements to humans and livestock or supplying micronutrient fertilizers to crops. However, these may be neither efficient nor practical options in many cases. Welch and Graham (1999) have argued that a combination of economic, political, and logistical reasons prevent micronutrient supplementation from being a sustainable option for improving the health of populations in developing countries, where problems associated with micronutrient deficiency are most serious. They maintain that the only long-term solution for ensuring micronutrient sufficiency is to enhance the micronutrient density of staple crops. Schachtman and Barker (1999) noted that low availability of micronutrients in the soil solution is commonly the most limiting factor for an increase of micronutrient density in crops. Availability of micronutrients can be increased by application of fertilizers; however, these are expensive, prohibitively so in many poorer parts of the world or in soils in which rates of attenuation are high. Development of strategies for minimizing or reversing micronutrient attenuation will improve the efficiency of micronutrient fertilizer applications and should therefore be used in combination with plant breeding approaches to enhance micronutrient density in edible portions of staple crops.

Micronutrient deficiency in humans and animals impairs brain function as well as immune and reproductive systems (Graham et al. 2001) and, in crops, leads to increased disease susceptibility and decreased yield potential (Fageria et al. 2002). It is often difficult to isolate the effects of micronutrient deficiency from other factors as the causative or contributory agent of disease in humans (e.g., Hambidge et al. 1998). Hence, other indicators are often necessary to recognize micronutrient insufficiency as a concern in a population. One such indicator may be low concentrations of micronutrients in soils from which a population's food is sourced (White and Zasoski 1999). In fact, the rapidly growing discipline of medical geology is built upon the appreciation that human health and geological environs are inextricably linked. Identification of soil types with a high capacity for micronutrient fixation could provide epidemiologists with information on areas to be prioritized for assessment of micronutrient status of populations. Furthermore, the lack of good predictive soil testing protocols, and the expense associated with micronutrient testing of crops can mean that micronutrient deficiencies in crops go unrecognized. A better understanding of soil factors controlling rates of attenuation would aid in the identification of soils likely to be vulnerable to micronutrient deficiencies, and thus reveal soils which may warrant more frequent or intensive micronutrient testing protocols.

10.4 STUDIES OF MICRONUTRIENT ATTENUATION

Different methods for measuring attenuation of trace element availability have been described in Chapter 1 to Chapter 3. However, we emphasize that although a variety of methods can be applied to investigate the lability of trace elements, only chemical extraction/fractionation and spectrometric and soil isotopic exchange (E value) methods provide a surrogate assessment of actual bioavailability. This is because plants modify the chemistry of soil in the rhizosphere. In response to nutrient deficiency, plants can induce large changes in the rhizosphere pH and redox potential, and may also release metal-complexing agents to assist in nutrient acquisition (see Section 10.6). Actual bioavailability, as assessed by plant isotopic dilution (L value) or relative yield methods, may therefore be underestimated by chemical or spectrometric methods, particularly in nutrient-limited soils.

Most studies to date that have examined trace element attenuation have focused on trace metals as contaminants, not nutrients. Hence, corresponding element concentrations in soils that have been investigated are high compared to concentrations relevant to the nutritional context. It is probable that attenuation reactions are influenced by the absolute concentration of the element. For example, it is clear that at high concentrations, elements can form mineral precipitates that limit further increases in their bioavailability with increasing element concentration. However, it is not yet known whether element precipitation, or other mechanisms in operation at higher element concentrations in soils, are relevant for elements at concentrations consistent with micronutrient-limited conditions. The following section therefore summarizes the information available on attenuation of Co, Cu, Mo, Se, and Zn present only at lower (yield-limiting) concentrations in, or applied at, agronomic rates to soils.

10.4.1 ZINC

Changes in Zn availability to plants in relation to time has been studied by several authors (Boawn 1974; Armour et al. 1989; Brennan and Gartrell 1990). Boawn (1974) added $ZnSO_4$ to 2 alkaline soils (pH 7.6 + low $CaCO_3$ content, pH 7.7 + high $CaCO_3$) at rates ranging from 0- to 22.4-kg Zn ha^1 and studied Zn uptake by sweet corn under field conditions for 5 years following Zn application. Concentrations of Zn in DTPA extracts of the soils decreased over time. Concentrations of Zn in plant tops and crop uptake of Zn did not consistently decrease over time — in 1 soil there was no effect, but in the calcareous soil, Zn uptake by the crop decreased by 26% in 5 years.

Armour et al. (1989) added Zn at low rates (0–1 $mg \cdot kg^{-1}$) to a Zn-deficient soil (pH 6.5, clay content 15%, organic carbon content 3%) and grew navy beans (*Phaseolus vulgaris* L.) after incubating the Zn with the soil for 15 d (moist) at 40 °C. The dry weight of plants was stimulated up to 75% by the addition of Zn,

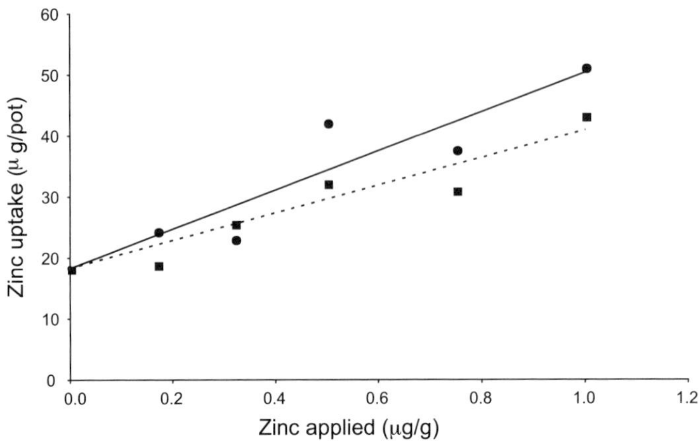

FIGURE 10.1 Effectiveness of Zn added as a fertilizer to a Zn-deficient soil either fresh (circle) or after 15-d incubation at 40 °C (square). (Drawn from Armour JD, Ritchie GSP, Robson AD. 1989. *Aust J Soil Res* 27:699–710. With permission.)

and the incubation treatment decreased the effectiveness of the supplied Zn by 29% (Figure 10.1).

The work of Brennan (1990) is perhaps the most comprehensive biological study to date of attenuation reactions of Zn in agricultural soils. Fifty-four soils varying widely in chemical properties (pH 4.8–8.6, clay 1.5–59.0%, organic Carbon 0.3–3.5%) were incubated for 30 d at 30 °C with 3 rates of Zn as $ZnSO_4$ (0, 133, and 266 $mg \cdot kg^{-1}$). Growth and uptake of Zn by subterranean clover were determined using freshly added Zn as a control. The effectiveness of incubated vs. fresh Zn (relative effectiveness) was defined as the ratio of the slopes of the linear regression relationships between Zn uptake by plants and Zn added. The relative effectiveness (RE) varied from 0.47 to 0.80, with a strong negative relationship with soil pH, i.e., greater attenuation (lower RE) at high pH values (Figure 10.2).

Multivariate analysis of the data set indicated that soil pH, clay content, calcium carbonate, and organic matter content were the major factors controlling the RE of Zn, explaining almost 90% of the variation.

10.4.2 Copper

A number of greenhouse studies have shown that the efficiency of Cu fertilizer significantly decreases with increasing contact time between soil and Cu (Brennan et al. 1980, 1983, 1984, 1986). Similar effects have not been detected in the field, but the attenuation reaction may be obscured by changes in the position of fertilizer Cu in the root zone (see the discussion that follows).

All available evidence of Cu fixation in Cu-deficient soils is based on a series of pot trials with soils from Western Australia (Brennan et al. 1980, 1983, 1984, 1986). Attenuation of Cu was indicated from differences in plant response to added Cu, between Cu freshly added and that incubated in soil after addition (Figure 10.3).

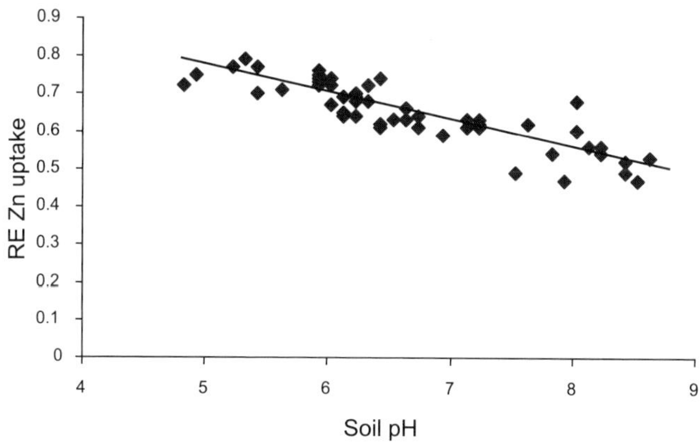

FIGURE 10.2 Relative effectiveness (RE) of 30-d-incubated (30 °C) Zn vs. fresh Zn added as $ZnSO_4$ to 54 soils as a function of soil pH. The RE is determined based on Zn uptake by clover. (Drawn from Brennan RF. 1990. Part 2. *Aust J Soil Res* 28:303–310. With permission.)

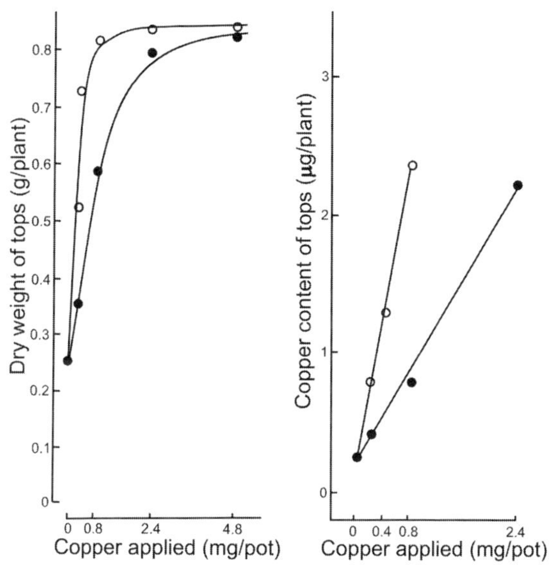

FIGURE 10.3 Effect of copper application rate and incubation time following Cu application on dry weight of wheat tops (left) and Cu content of tops (right). Fresh Cu (o) and Cu incubated for 60 d at 35 °C (•). (From Brennan RF, Gartrell JW, Robson AD. 1980. *Aust J Soil Res* 18:447–459. With permission.)

RE of Cu after 28 to 60 d of incubation was observed to range between 0.3 and 1.0 for 8 different soils (Brennan et al. 1980). The RE exhibited a steady decrease with increasing incubation time (Figure 10.4).

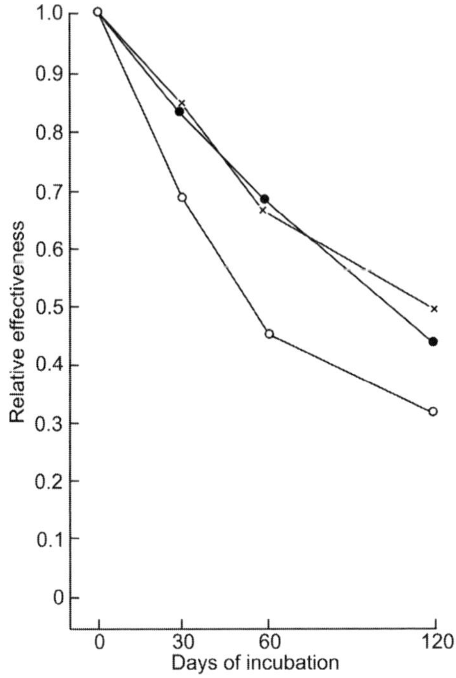

FIGURE 10.4 Effect of time of incubation (at 30 °C) following Cu application on the relative effectiveness of Cu in 3 Cu-deficient soils. (From Brennan RF, Gartrell JW, Robson AD. 1980. *Aust J Soil Res* 18:447–459. With permission.)

The first series of experiments in 8 different soils did not show strong effects of soil properties on the degree of Cu fixation. In a subsequent study (Brennan et al. 1983), effects of soil pH, soil organic matter content and soil sterilization on Cu fixation were studied. RE of Cu after 35 d of incubation was slightly lower from increased soil pH in 2 soils and was unaffected (in 1 soil), suggesting a marginal increase in Cu attenuation with increasing soil pH (tested between pH 6.7 and 7.9). The addition of straw (up to 10% by weight) almost halved RE of incubated Cu compared with freshly applied Cu. This striking response may not have field relevance given the nonrealistic dose of milled straw that increased immobilization reactions. However, it does demonstrate that biological reactions in soil related to C cycling may be important in decreasing the bioavailability of added Cu (see Chapter 7). Soil sterilization marginally increased the RE in 2 soils and had no effect in 1 soil, suggesting that the living biomass may render freshly added Cu somewhat less available upon prolonged incubation. Increasing temperature of incubation of soil with Cu (10 to 30 °C) increased the rate of Cu attenuation (Brennan et al. 1984; Figure 10.5).

Pot trials with wheat were also performed with samples of soil from a long-term Cu fertilizer field experiment (Brennan et al. 1986). The treatments in these field trials were varying rates of Cu as hydrated $CuSO_4$ and various periods since application

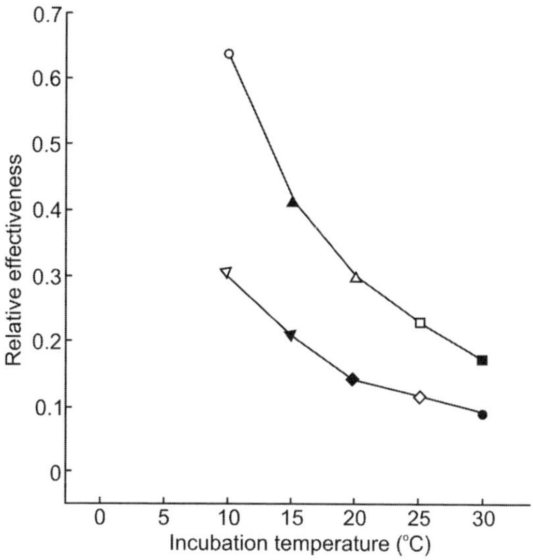

FIGURE 10.5 Effect of time of incubation and incubation temperature on the relative effectiveness of Cu in a calcareous, Cu-deficient soil. (From Brennan RF, Gartrell JW, Robson AD. 1984. Reactions of copper with soil affecting its availability to plants. III Effect of incubation temperature. *Aust J Soil Res* 22:165–172. With permission.) Upper line: 30-d incubation prior to plant growth; bottom line: 60-d incubation prior to plant growth.

(1 to 13 years). The Cu content in wheat shoots increased with increasing Cu rate and decreased with increasing time since Cu application in the field (Figure 10.6). The attenuation of Cu is reflected in the EDTA-extractable Cu in soils from these plots (McLaren and Ritchie 1993; Figure 10.7). At the lower Cu rates (up to 2-kg Cu ha^1, equivalent to 8.25-kg $CuSO_4$ ha^1), the EDTA extractability of applied Cu decreased from 90% for freshly applied Cu to only 20 to 30% for Cu aged for 20 years on these plots. At the highest rate of Cu, no such trends were found. However, this figure is somewhat misleading because the highest rate of Cu was a split application of about 3-kg Cu ha^1 20 years ago, and 2-kg Cu ha^1 at various years since that first application. This picture nevertheless suggests that attenuation of Cu may decrease at higher Cu concentrations.

Results of the pot trials with Cu-deficient soils hence invariably show attenuation of Cu bioavailability during prolonged incubation. In contrast to field data, the residual effect of fertilizer Cu applied to soils has not always been shown to decline over time. The yield response of wheat to a current Cu application drilled with the seed into a Cu-deficient soil was inferior or equal to the residual effect of fertilizer Cu applied the same way 4 to 12 years earlier (Gartrell 1980). One of the field sites on which this was demonstrated was later sampled by Brennan et al. (1986), when Cu fixation was shown in pot trials. It is suggested by Brennan et al. (1986) that initially applied Cu is positionally unavailable to the roots in field-grown plants. Fertilizer Cu moves slowly in soil, and mixing and leaching increases contact between roots

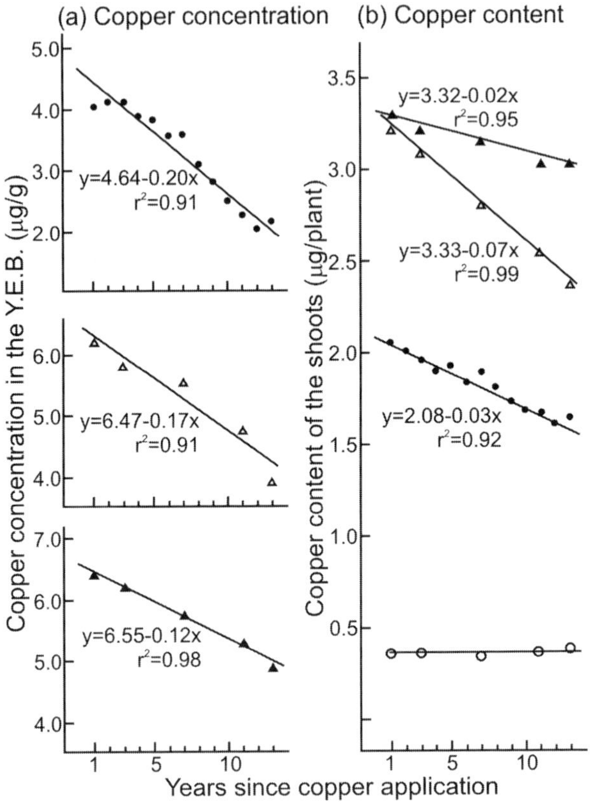

FIGURE 10.6 The relationship between time and (a) Cu concentrations in indicator leaves (youngest emerged blade, Y.E.B.) and (b) Cu content in whole wheat tops grown in a pot trial with brown lateritic sandy soil. Soil was sampled from field plots with varying Cu rate and varying years since application. No added Cu (o), 0.7-kg Cu ha^{1} (•), 1.4-kg Cu ha^{1} (Í), and 2-kg Cu ha^{1} (5). (From Brennan RF, Gartrell JW, Robson AD. 1986. *Aust J Soil Res* 37:107–113. With permission.)

and fertilizer Cu over a period of years after application, thereby obscuring effects of Cu fixation in the field. Another possibility is that attenuation of Cu may be rapid (weeks to months), so that plant growth measured over a whole growing season may not be a good indicator of Cu attenuation, because much of the attenuation process would be completed by the time the assay is concluded.

In summary, it is clear that after the addition of Cu to soils, availability to plants generally decreases markedly. This is especially so in soils of low Cu status and at low rates of Cu addition. As Cu is highly regulated in terms of plant-root uptake, and translocation from roots to shoots, it is often difficult to detect changes in Cu availability in soil through determination of plant uptake over time. Hence, attenuation reactions are more likely to be detected under conditions of deficiency, when regulation of uptake is limited.

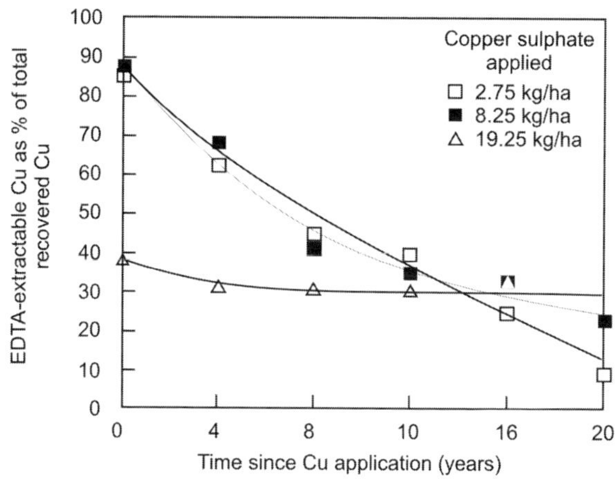

FIGURE 10.7 Effect of time since application on the EDTA extractability of applied Cu. (From McLaren RG, Ritchie GSP. 1993. *Aust J Soil Res* 31:39–50. With permission.)

10.4.3 Cobalt, Molybdenum, and Selenium

Cobalt is not a plant nutrient, but is essential for microorganisms that fix molecular N_2 (O'Hara 2001; Fageria et al. 2002). As such, the requirement of Co for crop production is low. However, Co is important in animal nutrition, being a component of vitamin B12, its deficiency commonly manifested as white liver disease (Lee et al. 1999). Hence, fertilization may be required to enhance Co content in crops used to feed animals. Information regarding natural attenuation of Co in soil is very limited. Because of the similarity of Co to Ni, it can be predicted that the availability of this element will undergo a significant attenuation over time. In particular, Mn oxides are thought to control Co availability in soil through adsorption and oxidation of Co^{2+} to Co^{3+} (McKenzie 1970). Barrow (1998) investigated the effect of time on the sorption of Co and other metals by a loamy sand soil. He reported that Ni and Co concentrations in solution continue to decrease over time and that this decrease was faster than that of Zn, suggesting that the effectiveness of Co fertilizers will decrease rapidly over time.

Molybdenum is an essential micronutrient for plant growth and, similar to Co, is also required by enzymes involved in symbiotic N_2 fixation and NO_3 reduction (O'Hara 2001; Fageria et al. 2002). Molybdenum is present as an anion in soil, so attenuation is dependent on the reactions of MoO_4^{2} with soil constituents (Martens and Westermann 1991). Riley (1987) reported that the effectiveness of Mo fertilizers varied from 1 to 15 years depending on soil characteristics. Barrow et al. (1985) showed a decrease in effectiveness of approximately 50% per year in a selection of soils. Brennan (2002) investigated the residual value of Mo (applied as Mo trioxide) for clover in an acidic sandy soil. Plant yields in field plots fertilized 10 years previously with 80-g Mo ha^{1} decreased by about 15% in comparison to freshly applied Mo. In contrast, there were no differences in terms of plant yield when Mo was applied at a rate of 320 g ha^{1}. However, the availability of Mo, when assessed

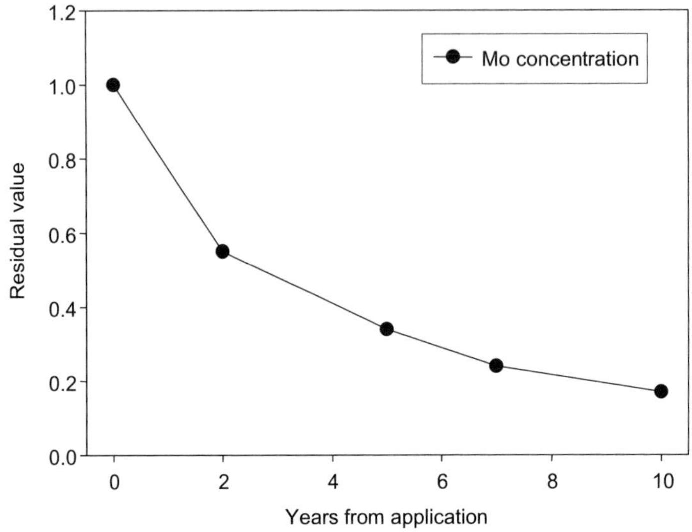

FIGURE 10.8 Effect of time since application on residual value of Mo as measured by Mo concentration (mg·kg^{-1}) and content (g ha^{1}) in plants. (Drawn from Brennan RF. 2002. *Aust J Exp Agric* 42:565–570. With permission.)

on the basis of Mo concentration in plants instead of plant yield, was shown to exhibit a continuous decline over time (Figure 10.8).

Selenium is not essential for plants but may be essential for soil organisms involved in symbiotic N_2 fixation (O'Hara 2001). However, this element does have an important role in animal and human nutrition (Lee et al. 1999; Goldhaber 2003). Selenium is present in the environment and in soils in 4 stable oxidation states (II, 0, IV, and VI) (Wright et al. 2003). Selenate ($SeO_4{}^{2}$) represents the most soluble form, and it is readily taken up by plants through mechanisms similar to those responsible for S uptake (see Chapter 8). For this reason, Se is usually applied as selenate to soils. The other forms of Se are generally strongly adsorbed on soil or show poor solubility (Wright et al. 2003). Gupta and Gupta (2002) reported that an application of Se of 10 g ha^{-1} could provide a residual effect in most crops for up to 2 years. Whelan and Barrow (1994) compared the residual effect of quick-release (Na_2SeO_4) and slow-release ($BaSeO_4$) forms of Se fertilizers in terms of Se concentrations in the blood of grazing sheep. The 2 fertilizers differed significantly in their solubility. They reported that a single application of the slow-release fertilizer could maintain an adequate Se concentration in the blood of grazing sheep for 4 years. In contrast, the highly soluble Na_2SeO_4 was effective for only 15 months. Because of the specific chemistry of Se, the attenuation of Se bioavailability in soil may be due to mechanisms different from those responsible for the natural attenuation of strongly retained cations such as Co, Cu, and Zn. In the case of Se applied as selenate, oxidation or reduction processes can decrease bioavailability (Brown et al. 1999). Furthermore, as a result of the high solubility of selenate, significant leaching of Se out of the soil profile may occur in humid environments.

10.5 ENVIRONMENTAL CONSEQUENCES

In affluent economies, where micronutrient supplementation is affordable, an unexpected consequence of micronutrient fixing of soils may be contamination of the local environment with excess micronutrient metals. For example, in order to optimize animal productivity, livestock are frequently dosed with micronutrient supplements. However, because marginal deficiencies are very difficult to recognize, the decision to provide micronutrient supplementation is often largely based on knowledge of areas susceptible to deficiency, with high-risk areas identified according to their geochemistry (Lee et al. 1999). Hence, decisions to supply micronutrients are presumably correlated, even if inadvertently, to the micronutrient attenuation capacity of soils. Because the gut adsorption efficiency of micronutrient supplements is difficult to predict and often fairly low (Lee et al. 1999), micronutrients are typically supplied in excess, a high proportion of these passing directly through the animal and ending up in the manure (Nicholson et al. 1999; McBride and Spiers 2001). Complexation of micronutrient metals, especially Cu, with organic matter in manure can increase the mobility of these metals, which could be detrimental if they are subsequently transported to more sensitive receiving environments. Similarly, to optimize crop productivity, high rates of micronutrient fertilizers may be applied to soils with a high fixing capacity for micronutrients to compensate for low plant availability. This will lead to an accumulation of the micronutrient metal in the soil. Although micronutrient bioavailability may remain low in the paddock, there can be physical displacement of the micronutrient-loaded soil, for example, through colloid-facilitated transport (Lombi et al. 2003), or erosion, to an environment with different chemical characteristics. This could result in release of the micronutrient in a bioavailable form, giving rise to contamination problems in sensitive environments. For example, Hamon et al. (2002) have demonstrated that some forms of attenuated trace elements can be released into labile pools following soil acidification. Strategies that result in increased micronutrient bioavailability could therefore minimize total micronutrient loading of highly fixing soils, decreasing the risk of off-site contamination.

10.6 STRATEGIES TO ACCESS FIXED FORMS OF MICRONUTRIENTS

Plants and microorganisms can change the soil environments they inhabit through a variety of processes. Some of these processes have a significant influence on the availability and natural attenuation of micronutrients. In particular, the biogeochemistry of trace elements may be substantially different in the rhizosphere, the few millimeters of soil surrounded and influenced by roots, in comparison to the rest of the soil (the "bulk" soil). The rhizosphere is also a focal point for soil microbial activity. Strategies that plants and microorganism use to enhance micronutrient availability include modification of soil pH and redox potential and exudation of organic and inorganic compounds in the soil (for review see Hinsinger 2001; Lombi et al. 2001).

Soil pH and redox potential are among the most important factors regulating heavy metal solubility and bioavailability. Metallic cations are typically more soluble in acidic soils, whereas metal and metalloid oxyanions are more soluble in alkaline environments (Adriano 1986). This is due to the combined effect of pH-dependent surface charge on the surface of soil minerals and organic matter, and ion competition. The rhizosphere soil pH can be significantly changed by root activities. This is generally caused by an imbalance in plant uptake of cations and anions, and the consequent efflux or uptake of H^+ or HCO_3 from the roots in order to maintain electroneutrality (Breteler 1973; Hedley et al. 1982). Changes in pH may also be due to the release of CO_2 or exudation of organic acids by plant roots and microorganisms.

In addition to their effect on pH, root exudates are also involved in micronutrient mobilization in the rhizosphere. The most studied system in this process concerns mobilization of Fe. Two different types of root response to Fe deficiency (strategy I and strategy II) have been identified in plant species. The pH in the rhizosphere of strategy-I plants is decreased by excretion of protons and a plasma membrane-bound inducible reductase is expressed in the roots (Marschner and Römheld 1994). Strategy-II plants are characterized by enhanced production and release of phytosiderophores into the rhizosphere and subsequent transport of chelated Fe by a specific uptake system at the root surface (Römheld and Marschner 1986; Kawai et al. 1988). In addition to these strategies, siderophores released by microbes have been proposed to serve as Fe sources for Fe-efficient plants from both groups (Bienfait 1988). Clearly, all these strategies can have an influence not only on Fe uptake, but also on micronutrient availability. For instance, phytosiderophores form chelates not only with Fe, but also with Zn, Cu, and Mn and therefore trigger the solubilization of these metals in calcareous soils (Takagi et al. 1988; Treeby et al. 1989; Römheld 1991). It has been shown that grasses also release phytosiderophores upon Zn deficiency (Zhang et al. 1989; Cakmak et al. 1994; Walter et al. 1994). Cakmak et al. (1996) demonstrated that enhanced release of phytosiderophores upon Zn deficiency is associated with Zn efficiency in wheat genotypes. Wiren et al. (1996) proposed 2 pathways for the uptake of Zn from Zn–phytosiderophores in grasses: 1) via the transport of the free Zn cation and, 2) the uptake of nondissociated Zn–phytosiderophore complexes.

Also, Chaignon et al. (2002) reported that under Zn starvation, a Zn-efficient genotype of wheat acquired more soil Cn. Zhao et al. (2001) showed that root exudates isolated from either Fe-deficient or Zn-deficient wheat were able to mobilize greater quantities of various trace elements in a soil compared to exudates isolated from nondeficient plants. The amount of Cu, Fe, or Zn that was mobilized per milligram of organic carbon produced as root exudates by either nutrient-sufficient, Zn-deficient, or Fe-deficient wheat plants in this study is shown in Figure 10.9.

Plants could also enhance micronutrient availability by developing symbiosis with fungi or by influencing the microbial population in the rhizosphere. Rengel (2001) suggested that Zn-efficient wheat genotypes may increase the populations of rhizosphere bacteria that enhance Zn availability. Similarly, it has been reported that the proportion of Mn-reducing bacteria in the rhizosphere of wheat increases under Mn-deficient conditions (Rengel 1997). This change in the rhizosphere microbial community may be due to a qualitative or quantitative shift in the composition of

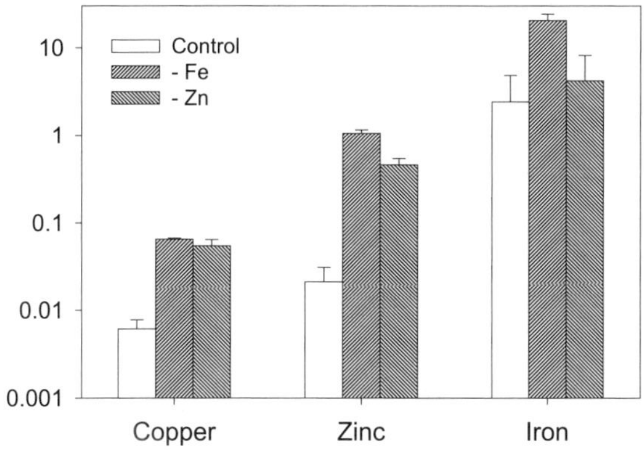

FIGURE 10.9 Amount of Cu, Zn, or Fe mobilized by exudates (measured as total organic carbon (TOC)), collected from wheat plants grown in complete nutrient solution (control) or nutrient solution deficient in either Fe (Fe), or Zn (Zn). (Data from Zhao FJ, Hamon RE, McLaughlin MJ. 2001. *New Phytol* 151:613–620. With permission.)

root exudates, which results in a change in the balance of Mn reducers and Mn oxidizers in the rhizosphere (Rengel 1997).

Mycorrhizal fungi can affect the availability of micronutrients either directly, by promoting sorption, precipitation, or dissolution reactions, or indirectly, by modifying root exudates. However, mycorrhizal effects on micronutrient mobility and uptake by plants are known to vary depending on both biological (type of plant and mycorrhiza involved, and their nutritional status) and chemical (soil and micronutrient characteristics, and metal concentrations) factors. For a review of studies concerning bioavailability of micronutrients in the mycorrhizosphere, see Leyval and Joner (2001).

10.7 STRATEGIES TO MINIMIZE FIXATION OF TRACE ELEMENTS APPLIED AS FERTILIZERS

10.7.1 Foliar Application

Foliar application of micronutrients provides direct access to the plant, bypassing the soil and hence precluding the possibility of attenuation of the applied micronutrients by the soil. Foliar applications of Cu, Fe, Mn, and Zn have been successful on a broad range of crops and pastures in southern Australia (Duncan 1967; Reuter et al. 1988; Brennan 1991). Foliar sprays are usually more efficient than soil application, in that adequate plant nutrition may be achieved with less than 4% of the recommended soil application rate (Mascagni and Cox 1985). Therefore, recommended application rates are relatively low: 100-g Cu ha^{1}, 1-kg Fe ha^{1}, 900-g Mn ha^{1}, and 240-g Zn ha^{1} (Reuter et al. 1988).

Foliar sprays have some limitations. Sharma and Katyal (1986) and Brennan (1991) found that foliar-applied Zn was less effective than banded Zn at solving moderate-to-severe deficiencies. Efficiency can be improved by targeting young plants and by applying 2 to 3 foliar sprays per growing season (Mascagni and Cox 1985; Brennan 1991). Foliar sprays cannot be applied to very young crops because of insufficient leaf cover, which may reduce their effectiveness considerably (Mascagni and Cox 1985). Other inefficiencies may result from poor penetration in leaves with thick cuticles, rapid drying of solutions and runoff from hydrophobic leaf surfaces during rainfall or heavy dew (Marschner 1995). Furthermore, foliar sprays have no residual value between crops and require annual application, whereas soil-applied micronutrients may hold residual value for up to 5 years (Duncan 1967; Boawn 1974; Reuter et al. 1988). Foliar sprays remain popular among farmers because of their low cost and ease of application. However, care needs to be taken to ensure their effectiveness in adverse conditions and in regions prone to severe micronutrient deficiencies.

10.7.2 Banding

Fertilizer banding is commonly practiced in Australian broadacre farming systems. Typically, micronutrients are coated onto macronutrient fertilizer granules (e.g., ZnO-coated DAP, MAP, and urea) to improve micronutrient distribution within the fertilizer band and increase the probability of crop-root interception. Banding is generally considered to be more effective than broadcast applications because it concentrates fertilizer near the root zone of the emerging seedling (Sleight et al. 1984). However, Wild (1993), Bailey and Grant (1990), Norvell (1988), Raun et al. (1987), and Mortvedt and Giordano (1975) argue that banding is more effective than broadcast application because it reduces fertilizer exposure to adsorption and fixation sites in the soil. However, in highly attenuating soils, insufficient nutrient diffusion (0 to 0.75 cm) may prevent the formation of a continuous nutrient band (Lombi et al. 2004). Therefore in these soils, banding will not decrease fertilizer exposure to adsorption sites in the soil. Nevertheless, banding may indirectly reduce total attenuation by increasing plant uptake of micronutrients in the early stages of crop growth.

10.7.3 Acidifying Fertilizers

A number of studies have shown that micronutrient fertilizers are more efficient when they are applied in combination with acidifying macronutrients (Hossner and Richards 1968; Mortvedt and Giordano 1975; Steckel et al. 1948). For example, Mortvedt and Giordano (1975) reported increases in Mn uptake and forage yields when Mn was applied in combination with MAP and superphosphate. Similar results were found by Hossner and Richards (1968) using APP and MAP and by Steckel et al. (1948) with superphosphate. Hossner and Richards (1968) showed that the solubility and persistence of the micronutrients were directly related to the pH of the soil surrounding the fertilizer granules, which, for APP and MAP, were pH 4.0 and pH 4.2, respectively. Soil surrounding DAP granules remained above pH 7.0, and no soluble Mn was detected in that zone (Hossner and Richards 1968). Despite this, micronutrient-coated DAP is a commonly applied fertilizer, and continues to be used in areas of highly fixing soils, including the calcareous cereal cropping soils of southern Australia (P. Flavel, Hi-Fert, Pers Comm. 8/4/03). More recently, Sanders et al. (2002) reported significant increases in Fe, Cu, and Zn uptake by corn and wheat when they were applied with acidifying sulfur compounds. The mixture of ammonium sulfate and elemental sulfur decreased soil pH in the microenvironment surrounding fertilizer granules for more than 180 d, thereby substantially increasing the solubility of micronutrients in that zone (Sanders et al. 2002).

10.7.4 Synthetic Chelates

Synthetic chelates have been used for approximately 50 years to increase the bioavailability of micronutrients in soil. They act by binding with the micronutrient metal to form a soluble complex, thus decreasing the cationic characteristics of the metal and its electrostatic attraction to adsorption sites in the soil solid phase (Wallace 1962a; Murphy and Walsh 1972; Norvell 1972; MacNaeidhe and Flemming 1988; Takkar and Walker 1993; McLaughlin et al. 1998). Chelating agents can increase micronutrient diffusion (Wallace 1983; Treeby et al. 1989) and may be able to decrease attenuation of micronutrients by soils, thereby enhancing micronutrient supply in the rhizosphere. For example, Elgawhary et al. (1970) and Prasad et al. (1976) found that addition of EDTA and DTPA to alkaline soils increased Zn diffusion by facilitating the desorption of native soil Zn (Table 10.1).

Unlike natural root exudates, synthetic chelating agents are not recognized by membrane transport proteins, nor are they readily transported into the root symplast (Parker and Pedler 1997). Most plants appear to favor the absorption of the free

TABLE 10.1
Diffusion of native soil Zn in the presence of different chelates in Zn-deficient soils

	$^{65}Zn/^{65}Zn + Zn\ SO_4$ diffused	
Treatment	**Noncalcareous sandy loam (pH 8.3)**	**Calcareous clay loam (pH 8.3)**
	With no $ZnSo_4$ fertilizer	
Control	0.012	0.009
EDTA	0.085	0.025
Fulvic acid (FA)	0.031	0.021
DTPA	0.137	0.059
	With added Zn ($ZnSo_4$)	
$ZnSO_4$	0.017	0.019
EDTA	0.206	0.083
FA	0.045	0.025
DTPA	0.267	0.259

Source: From Prasad B, Sinha MK, Randhawa NS. 1976. *Soil Sci* 122:260–266. With permission.

metal ion, which probably follows ligand exchange at the root surface, rather than the adsorption of intact metal–chelates (Parker and Pedler 1997). In fact, experiments with nutrient solutions have shown that high chelate-to-metal ratios can induce micronutrient deficiencies in plants when the chelating agents compete with roots for the free metal ions (Wallace and Wallace 1983; Wallace et al. 1983). Therefore, chelating agents, normally, only increase plant uptake of micronutrients when diffusion to the rhizosphere is the rate-limiting step for trace metal absorption, i.e., in soils that rapidly attenuate micronutrient cations (McLaughlin et al. 1998; McLaughlin 2002).

Following their application, chelated micronutrients equilibrate with the soil solution and solid phases, during which time some dissociation occurs (Norvell and Lindsay 1969). The rate of chelate dissociation is highly dependent on soil pH (Figure 10.10), the relative stability of the metal–chelate complex, and the presence of competing cations or ligands in the soil. Chelated micronutrients may be displaced (exchanged) from the chelate by cations of a higher valency or lower hydrated radius, following the Irving–Williams order:

Pb > Cu > Ni > Co > Zn > Cd > Fe > Mn > Mg (Irving and Williams 1948)

This order suggests that chelated Mn is highly unstable in soils, and results published by Norvell and Lindsay (1969) and Mortvedt (1980) have confirmed that Ca and Fe rapidly displaced Mn from MnEDTA in calcareous and acidic soils,

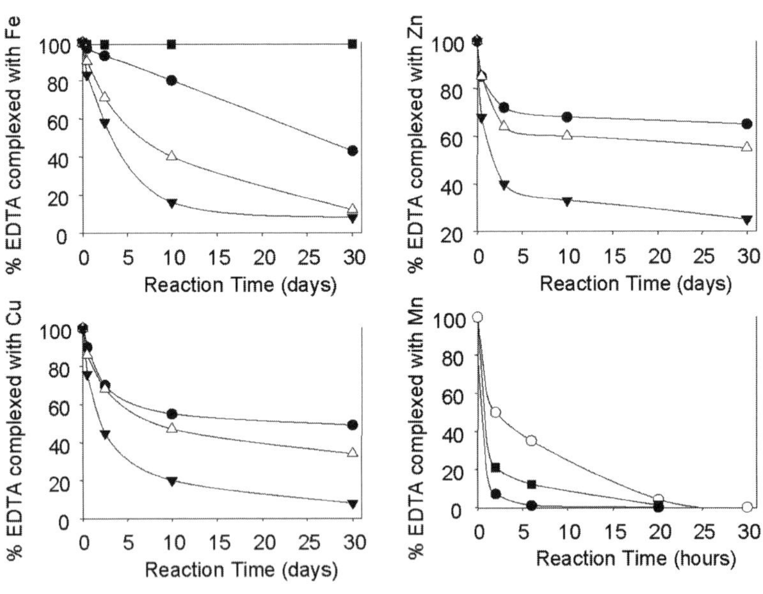

FIGURE 10.10 The dissociation of FeEDTA, ZnEDTA, CuEDTA, and MnEDTA with time, in soils of pH 5.7 (○), pH 6.1 (■), pH 6.75 (●), pH 7.3 (△), and pH 7.85 (▲). (Redrawn from Norvell WA, Lindsay WL. 1969. *Soil Sci Soc Am Proc* 33:86–91. With permission.)

TABLE 10.2
Changes in concentration of H_2O-soluble Mn in soils with time after applying Mn

	Water-soluble Mn(100) mg·kg^{-1}							
	MnEDTA alone Incubation time (d)				MnEDTA in 10-15-0[a] Incubation time (d)			
Soil	**1**	**7**	**28**	**56**	**1**	**7**	**28**	**56**
Mountview sil (pH 5.4)	40	21	18	16	71	57	52	32
Maumee ls (pH 6.8)	50	<1	<1	<1	43	23	11	<1
Hartsells fsl (pH 7.5)	73	66	4	<1	76	74	79	68
Mountview sil (pH 7.7)	70	56	16	2	73	70	71	31

[a] Fluid polyphosphate fertilizer.

Source: From Mortvedt JJ. 1980. *Soil Sci Soc Am J* 44:621–626. With permission.

respectively. However, Mortvedt (1980) also showed that MnEDTA was effective when applied in combination with fluid ammonium polyphosphate (Table 10.2). Apparently, APP reduced cation substitution for Mn in the chelate molecule, thereby reducing the rate of Mn attenuation in each of the 4 soils (Mortvedt 1980). Other studies have also shown that $MnSO_4$, $ZnSO_4$, and ZnO can be considerably more effective when applied in combination with fluid P and N fertilizers (Mortvedt and Giordano 1967; Holloway et al. 2001, 2002). However, micronutrient attenuation was not directly measured in any of these studies.

Synthetic chelates have not received widespread adoption in broadacre farming systems because their high cost has restricted use to high-value crops (Mortvedt et al. 1992). A simple cost comparison suggests that ZnEDTA would need to be 12.5 times more effective than Zn sulfate to be cost effective in broadacre farming systems. However, it may only be 2 to 5 times more effective (Mortvedt and Gilkes 1993). As a consequence, the use of synthetic chelating agents to minimize micronutrient attenuation in soil is currently greatly limited by economic factors. Moreover, they are far too expensive for widespread use in developing countries. Future research leading to the development of low-cost chelating agents would therefore be of considerable value to agricultural production globally.

10.7.5 Natural Chelating Agents

As discussed earlier, strategy-II graminaceous plants release chelating acids in response to Fe deficiency (Marschner 1995). These root exudates are termed *phytosiderophores*, although some authors prefer the more general term *phytometallophore* because the organic compounds may also chelate Mn and Zn, and enhance their transport to the crop rhizosphere (McLaughlin 2002). Nongraminaceous plants do not produce phytosiderophores, though some can utilize the metal–siderophore complexes (Hopkins et al. 1992b). In a field study, Zuo and Zhang (2003) found that peanuts absorbed twice as much Fe when they were intercropped with wheat, barley, oats, sorghum, or maize. Interestingly, the low phytosiderophore releasers such as sorghum and maize were as effective as barley, oats, and wheat at improving Zn nutrition of peanuts ($P < 5\%$). Final peanut yields were also improved by intercropping. Phytosiderophores may have improved the Fe nutrition of peanuts, though their concentration in soil was not measured (Zuo and Zhang 2003). Other studies have shown that oat and barley phytosiderophores can be utilized by tomato, muskmelon, cucumber, and some Fe-inefficient cultivars of maize and sorghum (Römheld and Marschner 1986; Hopkins et al. 1992a,b), and may also increase available Cu, Mn, and Zn in soil solutions (Hopkins et al. 1992c). However, phytosiderophores have induced Fe deficiency in soybean and in 1 cultivar of maize (Hopkins et al. 1992b,c). The availability of Fe–siderophores to strategy I plants is thought to be linked to their capacity to reduce Fe^{3+} to Fe^{2+} in the rhizosphere, which induces chelate splitting (Römheld and Marschner 1986; Bar-ness et al. 1991).

10.8 CONCLUSIONS

Micronutrient deficiency in soils is a pressing global issue. A significant contributing factor to micronutrient deficiency is the high capacity for micronutrient attenuation exhibited by many soil types and the rapidity with which this attenuation occurs following micronutrient fertilizer additions to soils. Growers can employ a number of practices to decrease natural attenuation of micronutrient fertilizers in soils. Methods proven to be effective in different farming systems include the use of synthetic chelating agents, acidifying fertilizers (for use on alkaline soils), foliar sprays, nutrient banding, and selective intercropping of complementary plant species. However, the efficiency of the above practices is dependent on production system, soil type, the nutrient being supplied, and the economic value of the crop. Further research on natural attenuation of micronutrient elements by soils is required to enable better identification of soil types vulnerable to micronutrient deficiency, as well as to optimize agronomic management strategies to improve micronutrient nutrition of crops and populations.

REFERENCES

Adriano DC. 1986. Trace elements in the terrestrial environment. New York: Springer-Verlag.

Armour JD, Ritchie GSP, Robson AD. 1989. Changes with time in the availability of soil applied zinc to navy beans and in the chemical-extraction of zinc from soils. Aust J Soil Res 27:699–710.

Bailey LD, Grant CA. 1990. Fertilizer placement studies on calcareous and non-calcareous chernozemic soils: growth, P-uptake, oil content and yield of Canadian rape. Comm Soil Sci Plant Anal 21:2089–2104.

Bar-ness E, Chen Y, Hadar Y, Marschner H and Romheld V. 1991. Siderophores of Pseudomonas putida as an iron source for dicot and monocot plants. Plant Soil 130:231–241.

Barrow NJ. 1998. Effects of time and temperature on the sorption of cadmiun, zinc, cobalt and nickel in soil. Aust J Soil Res 36:941–950.

Barrow NJ, Leaky PJ, Soutkey IN, Purser DB. 1985. Initial and residual effectiveness of molybdate fertilizer in two areas of southwestern Australia. Aust J Agric Res 36:579–587.

Bienfait F. 1988. Prevention of stress in iron metabolism of plants. Acta Bot Neerl 38:105–129.

Boawn LC. 1974. Residual availability of fertilizer zinc. Soil Sci Soc Am Proc 38:800–803.

Brennan RF. 1990. Reaction of zinc with soil affecting its availability to subterranean clover. 2. Effect of soil properties on the relative effectiveness of applied zinc. Aust J Soil Res 28:303–310.

Brennan RF. 1991. Effectiveness of zinc sulphate and zinc chelate as foliar sprays in alleviating zinc deficiency of wheat grown on zinc deficient soils in Western Australia. Aust J Exp Agric 31:831–834.

Brennan RF. 2002. Residual value of molybdenum trioxide for clover production on an acidic sandy podzol. Aust J Exp Agric 42:565–570.

Brennan RF, Gartrell JW. 1990. Reaction of zinc with soil affecting its availability to subterranean clover. 1. The relationship between critical concentrations of extractable zinc and properties of Australian soils responsive to applied zinc. Aust J Soil Res 28:293–302.

Brennan RF, Gartrell JW, Robson AD. 1980. Reactions of copper with soil affecting its availability to plants. I. Effect of soil type and time. Aust J Soil Res 18:447–459.

Brennan RF, Robson AD, Gartrell JW. 1983. Reactions of copper with soil affecting its availability to plants. II. Effect of soil pH, soil sterilization and organic matter on the availability of applied copper. Aust J Soil Res 21:155–163.

Brennan RF, Gartrell JW, Robson AD. 1984. Reactions of copper with soil affecting its availability to plants. III Effect of incubation temperature. Aust J Soil Res 22:165–172.

Brennan RF, Gartrell JW, Robson AD. 1986. The decline in the availability to plants of applied copper fertilizer. Aust J Soil Res 37:107–113.

Breteler H. 1973. A comparison between ammonium and nitrate nutrition of young sugarbeet plants grown in nutrient solutions at constant acidity. I. Production of dry matter, ionic balance and chemical composition. Neth J Agric Sci 21:227–244.

Brown GE, Foster AL, Ostergren JD. 1999. Mineral surfaces and bioavailability of heavy metals: a molecular-scale perspective. Proc Nat Acad Sci U S A 96:3388–3395.

Cakmak I, Gülüt KY, Marschner H, Graham RD. 1994. Effect of zinc and iron deficiency on phytosiderophores release in wheat genotypes different in zinc efficiency. J Plant Nutr 17:1–17.

Cakmak I, Sari N, Marschner H, Ekiz H, Kalayei M, Yilmaz A, Braun HJ. 1996. Phytosiderophore release in bread and durum wheat genotypes differing in zinc efficiency. Plant Soil 180:183–189.

Chaignon V, Di Malta D, Hinsinger P. 2002. Fe-deficiency increases Cu acquisition by wheat cropped in a Cu-contaminated vineyard soil. New Phytol 154:121–130.

Duncan O W. 1967. Correction of zinc deficiency in wheat on the Darling Downs, Queensland. Qld J Agric Animal Sci 24:287–294.

Elgawhary SM, Lindsay WL, Kemper WD. 1970. Effect of EDTA on the self-diffusion of zinc in aqueous solution and in soil. Soil Sci Soc Am Proc 34:66–70.

Fageria NK, Baligar VC, Clark RB. 2002. Micronutrients in crop production. Advan Agron 77:185–268.

Gartrell JW. 1980 Residual effectiveness of copper fertilizer for wheat in Western Australia. Aust J Expt Agric Animal Husb 20:370–376.

Goldhaber SB. 2003. Trace element risk assessment: essentiality vs toxicity. Regulat Tox Pharmacol 38:232–242.

Graham RD, Welch RM, Bouis HE. 2001. Addressing micronutrient malnutrition through enhancing the nutritional quality of staple foods: principles, perspectives and knowledge gaps. Advan Agron 70:77–142.

Gupta UC, Gupta SC. 2002. Quality of animal and human life as affected by selenium management of soils and crops. Comm Soil Sci Plant Anal 33:2537–2555.

Hambidge KM, Krebs NF, Miller L. 1998. Evaluation of zinc metabolism with the use of stable-isotope techniques: implications for the assessment of zinc status. Am J Clin Nutr 68(Suppl):410S–413S.

Hamon RE, McLaughlin MJ, Cozens G. 2002. Use of isotopic exchange techniques to determine mechanisms of attenuation of metal availability in in situ remediation studies. Environ Sci Technol 36:3991–3996.

Hedley MJ, Nye PH, White RH. 1982. Plant induced changes in the rhizosphere of rape *Brassica napus*. (var. Emerald) seedlings. I. pH changes and the increase P concentration in the soil solution. New Phytol 91:19–29.

Hinsinger P. 2001. Bioavailability of trace elements as related to root-induced chemical changes in the rhizosphere. In: Gobran GR, Lombi E, Wenzel WW, editors. Trace elements in the rhizosphere. Boca Raton, FL: CRC Pr. p 25–42.

Holloway RE, Bertrand I, Frischke AJ, Brace DM, McLaughlin MJ, Shepperd W, Graham RD. 2001. Fluid fertilisers—an efficient source of P for calcareous soils. In: Plant nutrition—food security and sustainability of agro-ecosystems. Dordrect (Netherlands): Kluwer Academic Publishers. p 832–833.

Holloway B, Frischke AJ, Brace D. 2002. How much fluid P fertiliser is enough? In: Eyre Peninsula farming systems 2002 summary. Port Lincoln, Australia: Grains Research and Development Corporation. pp 84–89.

Hopkins BG, Jolley VD, Brown JC. 1992a. Differential response of Fe-inefficient muskmellon, tomato, and soybean to phytosiderophore released by coker 227 oat. J Plant Nutr 15:35–48.

Hopkins BG, Jolley VD, Brown JC. 1992b. Plant utilization of iron solubilized by oat phytosiderophore. J Plant Nutr 15:1599–1612.

Hopkins BG, Jolley VD, Brown JC. 1992c. Variable inhibition of iron uptake by oat phytosiderophore in five soybean cultivars. J Plant Nutr 15:125–135.

Hossner LR, Richards GE. 1968. The effect of phosphorus source on the movement and uptake of band applied manganese. Soil Sci Soc Am Proc 32:83–85.

Irving H, Williams RJP. 1948. Order of stability of metal complexes. Nature 162:746–747.

Kawai S, Takagi S, Sato Y. 1988. Mugineic acid-family phytosiderophores in root-secretions of barley, corn and sorghum varieties. J Plant Nutr 11:633–642.

Lee J, Masters DG, White CL, Grace ND, Judson GJ. 1999. Current issues in trace element nutrition of grazing livestock in Australia and New Zealand. Aust J Agric Res 50:1341–1364.

Leyval C, Joner EJ. 2001. Bioavailability of heavy metals in the mycorrhizosphere. In: Gobran GR, Lombi E, Wenzel WW, editors. Trace elements in the rhizosphere. Boca Raton, FL: CRC Pr. p 165–188.

Lombi E, Wenzel WW, Gobran GR, Adriano DC. 2001. Dependency of phytoavailability of metals on indigenous and induced rhizosphere processes: a review. In: Gobran GR, Lombi E, Wenzel WW, editors. Trace elements in the rhizosphere. Boca Raton, FL: CRC Pr. p 3–24.

Lombi E, Hamon RE, McGrath SP, McLaughlin MJ. 2003. Lability of Cd, Cu and Zn in polluted soils treated with lime, beringite and red mud and identification of non-labile colloidal fraction of metals. Environ Sci Technol 37:979–984.

Lombi E, McLaughlin M, Johnston C, Armstrong RD, Holloway RE. 2004. Mobility and lability of phosphorus from granular and fluid monoammonium phosphate differs in a calcareous soil. Soil Sci Soc Am J 68:682–689.

MacNaeidhe FS, Flemming GA. 1988. A response in spring cereals to foliar sprays of zinc in Ireland. Irish J Agric Res 27:91–97.

Marschner H, Römheld V. 1994. Strategies of plants for acquisition of iron. Plant Soil 165:261–274.

Marschner H. 1995 Mineral nutrition of higher plants. 2nd ed. London, UK: Academic Pr.

Martens DC, Westermann DT. 1991. Fertiliser applications to correct micronutrient deficiencies. In: Mortvedt JJ, Cox FR, Shuman LM, Welch RM editors. Micronutrients in agriculture. Madison, WI: Soil Science Society of America.

Mascagni HJ, Cox FR. 1985. Effective rates of fertilization for correcting manganese deficiency in soybeans. Agron J 77:363–366.

McBride MB, Spiers G. 2001. Trace element content of selected fertilizers and dairy manures as determined by ICP-MS. Comm Soil Sci Plant Anal 32:139–156.

McKenzie RM. 1970. The reaction of cobalt with manganese dioxide minerals. Aust J Soil Res 5:235–256.

McLaren RG, Ritchie GSP. 1993. The long-term fate of copper fertilizer applied to a lateritic sandy soil in Western Australia. Aust J Soil Res 31:39–50.

McLaughlin M, Smolders E, Merckx R. 1998. Soil-root interface: physiochemical processes. In: Huang PM, editor. Soil chemistry and ecosystem health. Madison, WI: Soil Science Society of America, Inc p 233–277.

McLaughlin M. 2002. Bioavailability of metals to terrestrial plants. In: Allen HE editor. Bioavailability of metals in terrestrial ecosystems. Pensacola, FL: Society of Environmental Toxicology and Chemistry. p 39–68.

Mortvedt JJ. 1980. Reactions of MnEDTA applied alone or with ammonium polyphosphate fertilizer to soils. Soil Sci Soc Am J 44:621–626.

Mordvedt JJ. 1994. Needs for controlled-availability micronutrient fertilizers. Fert Res 38:213–221.

Mortvedt JJ, Gilkes RJ. 1993. Zinc fertilizers. In: Zinc in soil and plants. Robson AD, editor. Perth, Australia: Kluwer Academic Publishers. p 33–44.

Mortvedt JJ, Giordano PM. 1975. Crop response to manganese sources applied with ortho- and polyphosphate fertilisers. Soil Sci Soc Am J 39:782–787.

Mortvedt JJ, Giordano PM. 1967. Crop response to zinc oxide applied in liquid and granular fertilizers. J Agric Food Chem 15:118–122.

Mortvedt JJ, Mikkelsen RL, Kelsoe JJ. 1992. Crop response to ferrous sulphate in banded gels of hydrophilic polymers. Soil Sci Soc Am J 56:1319–1324.

Murphy LS, Walsh LM. 1972. Correction of micronutrient deficiencies with fertilizers. In: Micronutrients in agriculture. Mortvedt JJ, Giordano PM, Lindsay WL, editors. Madison, WI: Soil Science Society of America. p 347–387.

Nicholson FA, Chambers BJ, Williams JR, Unwin RJ. 1999. Heavy metal contents of livestock feeds and animal manures in England and Wales. Biores Technol 70:23–31.

Norvell WA. 1972. Equilibria of metal chelates in soil solution. In: Micronutrients in agriculture. Mortvedt JJ, Giordano PM, Lindsay WL, editors. Madison, WI: Soil Science Society of America, Inc. p 115–138.

Norvell WA. 1988. Inorganic reactions of manganese in soils. In: Graham RD, Hannan RJ, Uren NC, editors. Manganese in soils and plants. Dordrecht: Kluwer Academic Publishers, Dordrecht. p 37–58.

Norvell WA, Lindsay WL. 1969. Reactions of EDTA complexes of Fe, Zn, Mn, and Cu with soils. Soil Sci Soc Am Proc 33:86–91.

O'Hara GW. 2001. Nutritional constraints on root nodule bacteria affecting symbiotic nitrogen fixation: a review. Aust J Expt Agric 41:417–433.

Parker DR, Pedler JF. 1997. Reevaluating the free-ion activity model of trace metal availability to higher plants. Plant Soil 196:223–228.

Prasad B, Sinha MK, Randhawa NS. 1976. Effect of mobile chelating agents on diffusion of zinc in soils. Soil Sci 122:260–266.

Raun WR, Sander DH, Olson RA. 1987. Phosphorus fertilizer carriers and their placement for minimum till corn under sprinkler irrigation. Soil Sci Soc Am J 51:1055–1062.

Rengel Z. 1997. Root exudation and microflora populations in the rhizosphere of crop genotypes differing in tolerance to micronutrient deficiency. Plant Soil 196:255–260.

Rengel Z. 2001. Genotypic differences in micronutrient use efficiency in crops. Comm Soil Sci Plant Anal 32:1163–1186.

Reuter DJ, Cartwright B, Judson GJ, McFarlane JD, Maschmedt DJ, Robinson JB. 1988. Trace elements in South Australian agriculture, Technical Report No. 139. Department of Agriculture, South Australia.

Riley MM. 1987. Molybdenum deficiency in wheat in western Australia. J Plant Nutr 10:2117–2123.

Römheld V. 1991. The role of phytosiderophores in acquisition of iron and other micronutrients in graminaceous species: an ecological approach. Plant Soil 130:127–134.

Römheld V, Marschner H. 1986. Evidence for a specific uptake system for iron phytosiderophores in roots of grasses. Plant Physiol 80:175–180.

Sanders L, Kimmerly M, Murphy LS. 2002. New methods for influencing nutrient availability. In: Schlegel AJ, editor. Proceedings of the Great Plains soil fertility conference, Denver, CO, March 5–6, 2002. p 293–304.

Schachtman DP, Barker SJ. 1999. Molecular approaches for increasing the micronutrient density in edible portions of food crops. Field Crop Res 60:81–92.

Sharma BD, Katyal JC. 1986. Evaluation of amounts, methods and sources of zinc application to wheat in flood plain soils. J Agric Sci 106:41–44.

Sleight DM, Sander DH, Peterson GA. 1984. Effect of fertilizer phosphorus placement on the availability of phosphorus. Soil Sci Soc Am J 48:336–340.

Steckel JE, Bertramson BR, Ohlrogge AJ. 1948. Manganese nutrition of plants as related to applied superphosphate. Soil Sci Soc Am Proc 13:108–111.

Takagi S, Nomoto K, Takemoto T. 1988. Physiological aspect of mugineic acid, a possible phytosiderophore of graminaceous plants. J Plant Nutr 7: 469–477.

Takkar PN, Walker CD. 1993. The distribution and correction of zinc deficiency. In: Robson AB, editor. Zinc in soils and plants. Perth, Australia: Kluwer Academic Publishers. pp 151–165.

Treeby M, Marschner H, Romheld V. 1989. Mobilization of iron and other micronutrient cations from a calcareous soil by plant-borne, microbial, and synthetic metal chelators. Plant Soil 114:217–226.

Wallace A. 1962a. Metal chelates in plant nutrition—a revision on major questions and answers on their use. In: Wallace A, editor. A decade of synthetic chelating agents in inorganic plant nutrition. Los Angeles, CA: Edwards Brothers, Inc. p 105–112.

Wallace A. 1962b. Metal chelation and mechanisms of metal function in biological systems. In: Wallace A, editor. A decade of synthetic chelating agents in inorganic plant nutrition. Los Angeles, CA: Edwards Brothers, Inc.

Wallace A, Wallace GA. 1983. Use of synthetic chelating agents in experimental and commercial nutrient solutions. J Plant Nutr 6:513–525.

Wallace A, Wallace GA, Alexander GV. 1983. Effect of excess chelating agent in nutrient solution at low levels of iron, zinc, copper and manganese. J Plant Nutr 6:507–511.

Walter A, Römheld V, Marschner H, Mori S. 1994. Is the release of phytosiderophores in zinc-deficient plants a response to impaired iron utilization? Physiol Plant 92:493–500.

Welch RM, Graham RD. 1999. A new paradigm for world agriculture: meeting human needs. Productive, sustainable, nutritious. Field Crop Res 60:1–10.

Welch RM, Graham RD. 2002. Breeding crops for enhanced micronutrient content. Plant Soil 245:205–214.

Whelan BR, Barrow NJ. 1994. Slow-release selenium fertilizers to correct selenium deficiency in grazing sheep in Western Australia. Fert Res 38:183–188.

White JG, Zasoski RJ. 1999. Mapping soil micronutrients. Field Crop Res 60:11–26.

Wild A. 1993. Soils and the environment: an introduction. Cambridge: Cambridge University Pr.

Wiren N, Marschner H, Römheld V. 1996. Roots of iron-efficient maize also absorb phytosiderophore-chelated Zn. Plant Physiol 111:1119–1125.

Wright MT, Parker DR, Amrhein C. 2003. Critical evaluation of the ability of sequential extraction procedures to quantify discrete forms of selenium in sediments and soils. Environ Sci Technol 37:4709–4716.

Zhang F, Römheld V, Marschner H. 1989. Effect of zinc deficiency on the release of zinc and iron mobilizing root exudates. Z Pflanzenernaehr Bodenk 152:205–210.

Zhao FJ, Hamon RE, McLaughlin MJ. 2001. Root exudates of Thlaspi caerulescens do not enhance mobility of metals in soils. New Phytol 151:613–620.

Zuo Y, Zhang F. 2003. Effects of peanut intercropping with different gramineous species and their intercropping model on iron nutrition of peanut. Scientia Agric Sinica 36:300–306.

11 Use of Soil Amendments to Attenuate Trace Element Exposure: Sustainability, Side Effects, and Failures

Michel Mench, Jaco Vangronsveld, Nick Lepp, Ann Ruttens, Petra Bleeker, and Wouter Geebelen

11.1 INTRODUCTION

The extent of the hazard posed by trace elements in the soil to organisms depends on their concentrations and chemical speciation in the solid, liquid, and gaseous phases. Limiting the exposure pathways will help to decrease acute and chronic risks. Potentially the most cost-effective strategies to achieve this involve the so-called "mild" remediation techniques (Vangronsveld and Cunningham 1998). Mild remediation options for reducing trace element exposure include deep ploughing, phytoremediation, and chemical immobilization (Osté 2001). Deep ploughing can work if only the upper soil layer is contaminated and the underlying soil has a sufficient fertility and buffer capacity to function as a topsoil. Phytoextraction uses plants to remove soil contaminants by translocating them into plant tissue. The treatment duration and lack of commercially available plant strains with high biomass currently limit its implementation. Furthermore, in soils contaminated with a number of metals, 1 or more elements may limit the phytoextraction potential, e.g., high soil Cu content can reduce the growth of *Thlaspi caerulescens*, and thus Zn and Cd phytoextraction. Chemical immobilization, also called "in situ immobilization," "inactivation," or "attenuation," comprises several methods that aim at reducing potential exposure via the soil or wastes. Mench et al. (1994), Vangronsveld and Cunningham (1998), and Singh and Osté (2001) provide definitions. In this chapter, *chemical immobilization* is defined as an amendment added to the contaminated soil or waste that renders trace elements less bioavailable by altering chemical forms so that toxicity is reduced via a range of exposure pathways, for example, the soil solution, the gaseous phase, or the ingested particles. These technologies need to result in suitable conditions for living organisms. At least 2 options can be adopted. The first option is to promote naturally occurring processes that can alter both speciation and concentration in solid phases and the soil solution. This may be time consuming or even impossible. An alternative option is to introduce 1 or several compounds via an amendment, leading, for example, to new solid reactive phases or the presence of an essential element for the transformation process. Different mechanisms can be involved, including sorption, acid–base reaction, precipitation, oxidation–reduction, and demethylation.

If the bioavailability of contaminants is decreased through attenuation, it may be possible to restore a vegetative cover (i.e., phytostabilization) to the soil with beneficial effects on wind erosion, water transport, and leaching. Alternatively, an increase in biodiversity may be the aim. One challenge is to enhance processes which minimize exposure to nonessential trace elements, without inducing nutrient deficiencies or introducing any further unwanted contaminants (Geebelen et al. 2003b). Many short-term experiments have demonstrated that it is possible to use chemical immobilization methods to attenuate the exposure of plant species and other organisms to metals and metalloids (Brown et al. 2000; Lepp et al. 2000; McBride and Martinez 2000; Mench et al. 2000; Vangronsveld et al. 2000a). However, long-lasting demonstrations are necessary to gain public acceptance and also to establish the endurance of the remediation techniques. Accordingly, several field experiments and mesocosms have been established to assess the feasibility to use natural and synthetic soil amendments for attenuating trace element exposures. This review will pay special attention to the successes, failures, and side effects indicated by these experiments in relation to the primary production of plant species, metal concentrations in plant parts (especially those consumed by animals and humans), the biodiversity, the presence of genotoxic effects, and the risk of off-site contaminant transport. There is a clear need for long-term attenuation and phytostabilization experiments, to stimulate private initiatives and cofinancing, and to convince the general public and legislators. New soil remediation policy based on risk assessment, bioavailability and risk reduction, and site-specific circumstances, needs to be promoted.

11.2 TYPES OF SOIL AMENDMENTS

A key question for remediation by chemical immobilization is how to select an appropriate amendment. An ideal amendment will rapidly decrease the mobility and bioavailability of the contaminant, preventing leaching, plant uptake, etc. A long lasting, if not permanent, effect is required. Other characteristics to be considered are price, commercial availability, ease of application and safety to workers, lack of disruptive or adverse effects, especially on the soil structure and fertility, compatibility with plants used for revegetation, suitability for several contaminants, and compliance with regulations. Amendment addition should also not result in additional environmental concerns (see the following discussion). The soil amendment should be suitable for combination with other techniques, if possible adaptable to agricultural management, and suitable for different soil types. Osté (2001) stated that the material must have a high metal-binding capacity at common soil pH (about 4 to 8) and needs to be durable under the environmental conditions of the soil. These prerequisites may be overly constraining, as materials such as zerovalent Fe and steelshots have no initial binding capacity, but their alteration in soil can lead to sorption processes or chemical transformations. Aging and crystallization of materials such as Fe oxides, allophanes, and silica gels may increase the attenuation of trace elements.

Numerous materials have been selected for both their lack of negative influences on key aspects of soil health and their efficacy in decreasing the solubility and bioavailability of trace elements in contaminated soils (Knox et al. 2000; McBride and Martinez 2000; Mench et al. 2000; Osté 2001) (Table 11.1). Physical and chemical properties of these amendments have been reviewed (Mench et al. 1998; Knox et al. 2000). These materials decrease exposure by 1 or more processes, e.g., sorption, redox reaction, precipitation, ion exchange, complexation, excesses of competing elements, and humification (Mench et al. 1998).

Materials such as lime, phosphates, and organic matter have been used for a long time in agriculture. Some of the solid phases and processes used for the remediation of metal-contaminated soils are naturally occurring. Amorphous Al, Fe, and Mn oxide minerals are ubiquitous in soils as both discrete particles and surface coatings. Properties of these phases, which contribute significantly to metal attenuation, include high porosity, micropores, high surface area, and a large number of adsorption sites. Intraparticle surface diffusion is a rate-limiting mechanism in the sorption process when metal ions diffuse in hydrated micropores (see Chapter 4). In this chapter, attention is focused on the following materials, which have been

TABLE 11.1
Natural and anthropogenic soil amendments that have been used to attenuate inorganic contaminant exposure

Alkaline materials	**Fe, Mn, Al oxides/hydroxides**
Calcium, magnesium carbonates	Ferrihydrite
Calcium, magnesium oxides	Hematite
Limed sludges, alkaline biosolids	Lepidocrocite
Fluidised bed coal fly ashes	Magnetite
	Maghemite
Phosphate minerals	Mud from water treatment
Apatites	Fines from ferrous smelter
Hydroxyapatite	Waste from TiO_2 treatment (Fe-rich)
Basic slags	Red mud (bauxite, Al)
Phosphoric acid	Hydrous manganese oxides
Phosphates salts (K, Na, Ca, NH_4, etc.)	Birnessite
Natural phosphates	Chalcophanite
	Iron grit (steelshot)
Alumino-silicates	Zerovalent Fe
Clays (smectites, bentonite, etc.)	
Gravel sludge	**Zeolites**
Coal fly ashes, e.g., beringite	Natural zeolites
	Synthetic zeolites
Organic matter (compost, farmyard manure, etc.)	**Salts** (Cu, FE sulfate, Zn, etc.)

tested in long-term field experiments and outdoor mesocosms: lime, zeolites, phosphate compounds, metal oxides, biosolids, and industrial by-products.

Abbreviations used within the text are as follows: Unt (Untreated contaminated material), C (compost), B (beringite), SS (steelshot), Z4A (zeolite 4A), CB (compost and beringite), CSS (compost and steelshot), CBSS (compost combined with beringite and steelshot), and DOM (dissolved organic matter).

11.3 ENDPOINTS FOR TESTING EFFICACY OF ATTENUATION

There are 3 main routes of exposure to soilborne trace metals. Soluble contaminants migrate with soil water, are taken up from this medium by plants and aquatic organisms, and, in some cases, can be lost from solution through volatilization. The other pathways are direct ingestion and dermal contact with contaminated particulates. Consequently, any thorough evaluation of the overall effectiveness of amendments for chemical immobilization should combine physicochemical and biological methods. For ecotoxicological evaluation, it is important to survey several endpoints, e.g., biodiversity, bioaccumulation in living organisms, changes in metabolite and protein patterns, as well as genotoxicity, at different trophic levels, with a progression from a well-defined battery of tests to more complex conditions representative of real ecosystems. A combined use of chemical extraction, microbial biosensors, phytotoxicity and zootoxicity tests was reported by Vangronsveld, Mench et al. (2000). However, there is currently no consensus on the battery of tests that should be used to assess the extent of hazard reduction.

11.4 BACKGROUND TO EXPERIMENTAL SITES

Selected long-term field trials, as well as small-scale semifield experiments (i.e., large outdoor mesocosms, which mimic a field trial) in Europe, the U.S., and Australia that feature in this chapter are shown in Table 11.2. Soil characteristics from selected sites in Europe are shown in Table 11.3. One series of long-term field trials not described in this chapter, but important to note, consists of those established in the mid-19th century at Rothamsted Experimental Station in the U.K. Data from these field trials have proven particularly useful in understanding the relationship between land use, acidification, lime use, and metal mobilization (Goulding and Blake 1998).

Lommel-Maatheide, the site of 1 of the oldest field trials on soils contaminated by industrial fallout, is located in Kempen, Limburg Province, Belgium, a region where the Zn industry operated several smelters for more than 100 years. The sandy soil of this region is characterized by an acid pH (4.5 to 6.0), a low organic matter content (<2%), and a low cation exchange capacity (<2 cmol kg^{-1}). In 1990, 3 ha of a highly metal-contaminated soil were treated with a combination of beringite (a type of cyclonic ash, B, 5%) and compost (C, 5%) (Vangronsveld, Van Assche, Clijsters, 1995; Vangronsveld, Colpaert, Van Tichelen 1996; Vangronsveld et al. 2000a; Vangronsveld et al. 2000b). Field remediation trials and small-scale semifield

TABLE 11.2
Long-term field trials and outdoor mesocosms which have used soil amendments to attenuate trace element bioavailability and mobility

Site	Jales	Reppel	Overpelt	Louis Fargue	Maatheide	Northampton	Staffordshire	South Australia
Country	Portugal	Belgium	Belgium	France	Belgium	UK	UK	Aus
Project	PHY	PHY	PHY	PHY	LUC			
Source	Gold mine	As smelter	Zn/Cd smelter	Sewage sludge	Zn/Pb smelter	Cd oxides	Sewage farm	Fertilizers Cd impurities
Metals	Zn, Cu, Cd, Pb		Zn, Cu, Cd, Pb	Cd, Ni	Zn, Cd, Pb, Cu	Cd	Cd	Cd
Metalloids	As	As						
Set up	1998	1997	1998	1995	1990	1998	1998	
Soil type	Sandy soil	Sandy soil	Sandy soil	Sandy soil	Sandy soil	Loamy soil		Alfisols
Field trial	On site		On site	On site	On site	On site	On site	On site
Outdoor mesocosms	Bordeaux	Bordeaux	Hasselt					
Treatments	C, CB, CBSS, CSS	C, B, SS, BSS	C, B, SS, CB, CSS, CBSS	B, SS	B	L, $FeSO_4$, SS, Z4A	L, Fe-oxide SS, Z4A	Cu, Zn, L, Al-WTR $MgCO_3$, FRF, Clay
References[a]	1,2,3	4,5	6	7	8	9	9	10

Note: C: compost (5%), B: cyclonic ashes beringite (5%), SS: iron grit (steelshots), L: lime, Z4A: zeolite 4A, FRF: Fe-rich fines from a ferrous smelter, Al-WTR: alum-based water treatment residual. PHY: Phytorehab, LUC: Limburgs Universitair Centrum.

[a] 1) Mench et al. 2000, 2) Mench et al. 2003, 3) Bleeker et al. 2002, 4) Boisson, Mench, and Chartier 1998, 5) Boisson 1999, 6) Ruttens et al. 2006, 7) Mench et al. 2000, 8) Vangronsveld 1998, 9) Lepp et al. 2000, 10) McLaughlin et al. 2000.

TABLE 11.3
Soil characteristics of selected long-term experiments in Europe

Site	pH (water)	OM g kg^{-1}	Org.C g kg^{-1}	CEC cmol kg^{-1}	As	Cd	Cu	Pb	Zn	Ni
					mg·kg^{-1} dry weight					
Jales	4.1		0.4	0.9	1325	3.8	15.2	170	165	<2
Louis Fargue	6.6	18		3.7		29				55
Lommel-Maatheide	4.5–6	19.5		3.3		6–54	206–1393	375–2158	1078–11425	
Northampton	5.3		165		72	47	489	630	1301	87
Overpelt	4.1		1.4	1.3		60	319	756	1620	
Reppel			17.6	3.8	113	1.1		61.8	122	
Staffordshire	6.6				8	16	63	213	225	22

experiments have also been established at 5 additional sites, i.e., Louis Fargue (France), Jales (Portugal), Overpelt (Belgium), Pyhäsalmi (Finland), and Reppel (Belgium) (Verkleij et al. 1999). Steelshot (SS), an iron grit containing mainly zerovalent Fe, and beringite (B) were the primary amendments tested, both separately and in combination. Compost (C) was also added to increase soil fertility. The Zn smelter at Overpelt, also located in Limburg province, Belgium, produced 250 tons of Cd waste in 1950, half of which was released as atmospheric emissions (340 kg of Cd d^{1}). The smelter converted to an electrolytic process between 1969 and 1974, but even though emissions were significantly reduced, there were still elevated concentrations of Cd, Pb, and Zn in the local environment. The Jales mine, north-eastern Portugal, operated from 1933 to 1993. Metal-enriched mine spoils cover 14.4 ha. The vein contained, besides quartz, FeS_2FeAs, FeS_2, PbS, CuFeS, and sulphur salts of silver and gold (Santos Oliveira and Freira Avila 1995). Four tons of Jales soil were collected (0 to 0.3 m top soil layer) and transported in January 1998, to the INRA Bordeaux-Aquitaine Centre (Verkleij et al. 1999). A sandy soil (0 to 0.3 m depth layer) sampled at the INRA Pierroton Centre, Gironde, France, was used as an uncontaminated control soil (Control). Reppel, a village in Limburg province, Belgium, is the site of a former As(III) refinery, which operated from 1910 to 1965. The on-site storage of As and Zn products resulted in high-level contamination of approximately 8 ha of the site. Lysimeters were filled with a soil sampled in the adjacent agricultural area contaminated by fallout from this refinery and placed outside as a small-scale semifield experiment (Boisson 1999).

For each experiment described in the following text, time zero is the day on which the amendment was applied. The amount of amendment applied is always expressed as a percentage by soil air-dried weight. Concentrations are all expressed on a dry weight basis.

11.5 CHEMICAL TESTS AND SPECIATION

Soils were sampled at Lommel-Maatheide experiment 5 years following amendment with a mixture of beringite (5%) and compost (5%). Analysis of the treated soil showed a change in pH from acid to slightly alkaline (7.3 to 7.9), together with increases in organic matter content and cation exchange capacity. The ratio of water-extractable vs. total Zn was highest in untreated soils (Table 11.4). This ratio decreased (up to 70 times) in the CB-treated soils, and its value was below that in the control. Similar results were found for Cd. The mechanisms responsible for decreasing metal solubility are considered to be an increase in soil pH, precipitation, and sorption on Ca-phosphates, ferrihydrite, allophanes, and other minerals added in the CB amendment. However, EXAFS and x-ray diffraction indicated that silicates, such as hemimorphite ($Zn_4Si_2O_7(OH)_2.H_2O$) and willemite (Zn_2SiO_4), were also present in the CB-treated soil and could be involved in the attenuation process (Hargé 1997).

Soluble and exchangeable Cd and Zn fractions in the small-scale Overpelt experiment were assessed 3 years after initial treatment, using a 0.01 M $Ca(NO_3)_2$ extraction (Semane 2001) (Figure 11.1). Soil pH values were as follows: 6.5 (control),

TABLE 11.4
Ratio of water-extractable Zn vs. total Zn concentration in soil measured 5 years after soil treatment at the Lommel-Maatheide experiment

Treatments	Total Zn (aqua regia, mg·kg^{-1})	Water-extracted Zn (mg·kg^{-1})	Ratio (%)
Uncontaminated garden soil	106 (7)	0.7 (0.06)	0.660
Untreated	11425 (506)	14 (11)	1.234
CB	7639 (455)	2.2 (0.22)	0.029

Note: Standard deviation in parentheses; CB: compost (5%) + beringite (5%).

Source: From Vangronsveld J, Colpaert JV, Van Tichelen KK. 1996. *Environ Pollut* 94:131–140. With permission.

6.4 (C), 6.9 (CB), 6.9 (BSS), 7 (CBSS), and 6.1 (Unt). The lowest Cd concentration among treated soils was found for CBSS (up to 4 times lower than in Unt soil). Despite a similar pH, CBSS more efficiently decreased extractable Cd than CB, demonstrating an additional mechanism, probably due to steelshots oxidation. The CBSS and BSS treatments showed the greatest decrease in extractable Zn (4 times compared to the Unt treatment).

As discussed in Chapter 1, attenuation can be assessed by using sequential fractionation techniques to examine metal redistribution between different solid phases. An example of this is shown for data from the small-scale Reppel experiment, 4 years following soil amendment. Zinc redistribution among organic and mineral phases was associated with a decrease in Zn exposure. Comparison of B-treated and Unt soils showed that the percentage of Zn associated with the organic compounds decreased, whereas Zn associated with amorphous minerals and other compounds increased (Figure 11.2) (Pannetier 2000).

The effect of soil amendments on attenuation of metalloids in soils has also been studied, but considerably fewer data are available compared to metals. In soil solutions sampled from the small-scale Reppel experiment, As(V) was the primary As species present in the (1%) SS-treated soil, both before and after the Fe in this material was oxidized (Boisson 1999). A range of iron oxide-producing systems, including Fe grit, Fe II sulfate + lime, and Fe III sulfate + lime, and goethite + lime, demonstrated a high sorption affinity for As in solution in 3 soils originating from northwestern England (Hartley, Edwards, Lepp 2001). These selected data provide evidence that the use of amendments can decrease metal or metalloid exposure via the soil solution (soil–plant system–animal pathway) in several different types of contaminated soils. But does a decrease in trace element concentration in the soil solution necessarily indicate a decrease in trace element mobility (leaching) to groundwater?

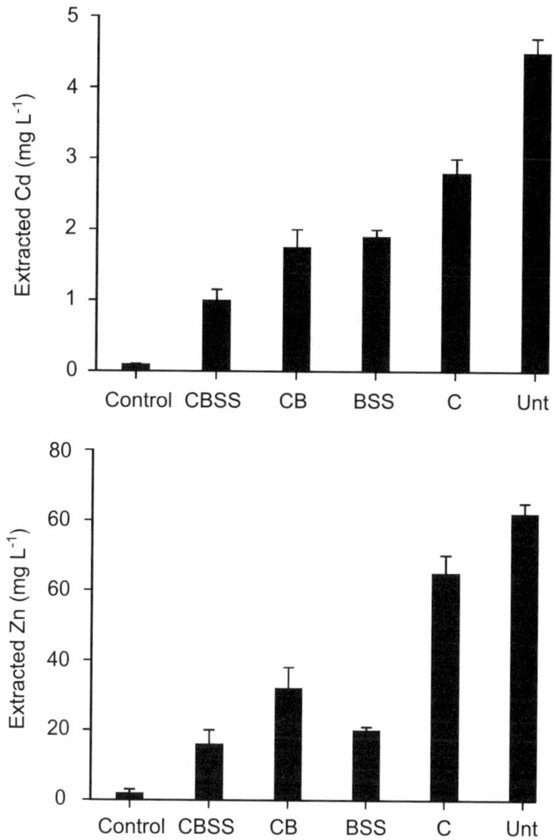

FIGURE 11.1 Cadmium and zinc extracted by 0.1-M calcium nitrate in soils sampled at the small-scale Overpelt experiment 3 years after soil treatment (Unt: untreated; Control: control soil; C: compost (5%); CB : C + beringite (5%); BSS: beringite (5%) + steelshot (1%); CBSS: C + beringite (5%) + steelshot (1%). (From Semane B. 2001. With permission.)

11.6 LEACHING

Time-dependant changes in metal and As concentrations in water percolating through amended soil layers are reported in some long-term experiments. The small-scale Overpelt experiment showed that all the amendments decreased the leachability of Cd and Zn in comparison to the control, with the amount of these elements leaching over a 12-month period decreasing in the following order: Unt > C > CB > CBSS (Figure 11.3). However, in contrast, compost addition resulted in an increase in Cu and Pb leachability in comparison to the control (Figure 11.3). As discussed in Chapter 6, increases in soil pH and organic matter may change the solid–solution partitioning of the organic matter, increase DOM concentration in the soil solution, and enhance the presence of colloids able to transport metals. An additional

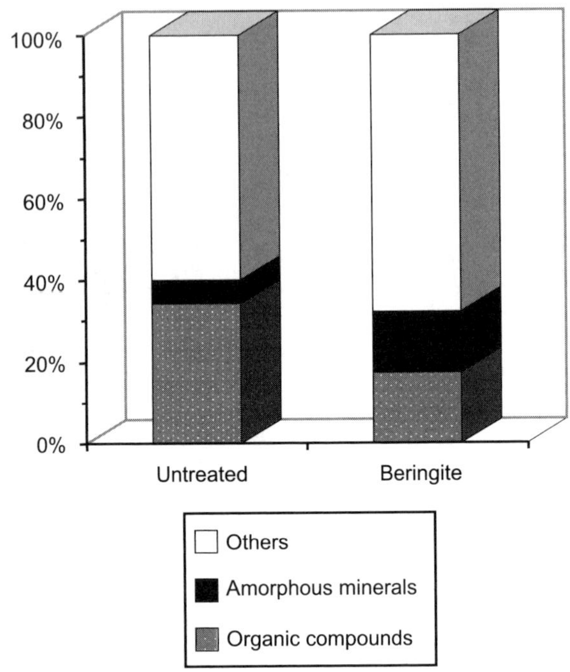

FIGURE 11.2 Distribution of Zn in soil fractions of the beringite-treated and untreated soils at the small-scale Reppel experiment. (From Pannetier S. 2000. Etude de l'incidence des amendements de béringite sur l'abondance des minéraux amorphes dans les sols: conséquences sur l'immobilisation des cations métalliques. Report, IUP EGID, Bordeaux III. Talence, France. With permission.)

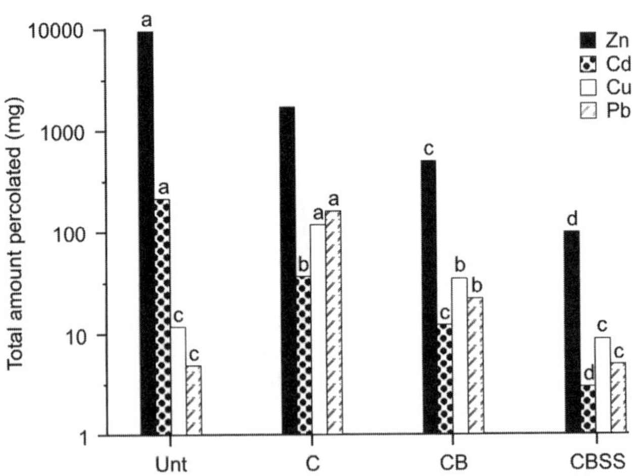

FIGURE 11.3 Total percolated metal content (mg) at the small-scale Overpelt experiment, during the 12-month period after soil treatment. For each metal, mean values with the same letter are not statistically different at the 5% level.

approach, e.g., phosphates added in the amendment combination, may solve this side effect when these metals are of primary concern at a contaminated site. The effect of amendment with beringite on metal leachability in a kitchen garden soil from the Lommel-Maatheide site was investigated (Vangronsveld 1998; Vangronsveld et al. 2000b). The simulated annual amounts of percolated Cd and Zn were found to sharply decrease (i.e., by 90% Cd and 80% Zn) with the beringite treatment (Vangronsveld 1998). Data for leaching of As, Cd, Cu, Pb, and Zn are available from the small-scale Jales Experiment (Mench et al. 2003). The pH of percolates from the unamended soil was acidic and decreased further over 3 years. An increase in SO_4^{2-} concentration in percolated water supported the occurrence of sulfide oxidation. All amendments increased the percolate pH and significantly decreased the leachability of the metals compared to the untreated soil. In contrast, leached As increased more than 50-fold in C- and CB-amended soils compared with the unamended treatment and also increased, but to a lesser extent, in the CBSS and CSS soils (De Koe, Bleeker, Assunção 1998). Possible mechanisms following compost addition that could trigger the increase in As mobility include the increase in soil pH, As binding to DOM, and competition for solid phase binding sites with DOM or with inorganic anions such as phosphates. Boisson, Mench, and Chartier (1998) monitored As leaching over a 9-month period at the small-scale Reppel Experiment. Arsenic concentration in leachates decreased 8.6 times and 12 times in the SS and BSS treatments, respectively, compared with the unamended control. We conclude from the monitoring of small-scale Overpelt, Jales, and Reppel experiments that amendments such as CB, CBSS, and BSS have beneficial effects in decreasing Cd and Zn leaching. Leaching of other cationic metals such as Ni, which has a pH-dependent solubility, would also probably be attenuated. However, the addition of organic carbon may increase leachability of metals such as Cu and Pb, which have a high affinity for organic materials. Attention must be paid to As leaching, and to other trace elements that form anionic species in solution such as Mo, Cr(VI), and Se, as their mobility can be enhanced by the same amendments that attenuate mobility of cationic metals.

11.7 EFFECTS OF DIFFERENT AMENDMENTS ON PLANT GROWTH AND CONTAMINANT UPTAKE

11.7.1 Biosolids Combined with Liming

Use of biosolids and similar organic wastes (e.g., papermill sludge) alone and in combination with other materials, such as limestone and cyclonic ashes that display a high calcium carbonate equivalent, is a long-standing practice by which many contaminated sites have been restored (Sopper 1993). Applications of this type provide the organic matter necessary to improve soil physical properties, water infiltration, and water holding capacity. They deliver micro- and macronutrients necessary for plant growth, and decrease bulk density. Several experiments have investigated the efficacy of biosolids and calcium carbonate mixtures to restore a vegetative cover.

11.7.1.1 Pronto Mine Experiment, Canada

Rio Algom's Pronto Waste Management Area, near Elliot Lake, Ontario, was a former uranium mine, in operation until 1960, where Cu ore was processed until 1970 (Tisch et al. 2000). Factors contributing to the lack of vegetation establishment on the fine-grained Cu tailings (23 ha) included low existing pH and continued acid generation, high levels of soluble metals, compaction, erosion, poor nutritional status, lack of organic matter, and a high water table. Test plots were constructed on these tailings using papermill sludge in combination with lime as a cover material, with and without a capillary barrier (gravel and blast furnace slag), or through direct incorporation. Over a 3-year period, only cover materials incorporated directly led to acceptable establishment of vegetation. Elemental concentrations in the grass shoots were similar to values listed for vegetation growing on Ontario background soils.

11.7.1.2 Leadville Experiment, Colorado

Pyrite-rich wastes from historic Pb and Zn mine tailing piles entered the Arkansas River and contaminated areas along an 18-km stretch. Oxidation of the pyrite has resulted in acidic soil pH (1.5 to 4.5) (Compton et al. 2001). Several contaminated areas were amended with a mixture of municipal biosolids and agricultural limestone, and then seeded with a mixture of native grasses or annual rye grass. In year 1, the amendment increased pH from 3.9 up to 6.4. Calcium nitrate-extractable Cd, Pb, and Zn decreased significantly in the surface horizon. Metal concentrations in annual rye grass also decreased. Initial results indicate that the amendment was effective in restoring a plant cover, but metal concentrations in plant tissues remained too high for animal exposure. Chemical and biological parameters measured over time suggest that ecosystem function has been restored to the amended tailings, but that these systems are not yet in equilibrium (Brown et al. 2005).

11.7.1.3 Bunker Hill Experiment, Idaho

Mining and smelting of Pb and Zn ores resulted in extensive metal contamination of the surrounding hillsides and waterways, leaving more than 500 ha barren of vegetation. Surface application of biosolids (112 Mt ha^{1}) in 1997, in combination with wood ash (220 Mt ha^{1}) and log yard debris (20% by volume), was able to restore a vegetative cover on the metal-contaminated materials for 3 years (Brown et al. 2000; Brown, Henry, Chaney 2001). Metal concentrations in plant tissues were within "normal" concentration ranges for all treatments during the 3-year period. In a wetland located at the same site, surface applications of a biosolid compost (60% DW) and wood ash (40% DW) to a depth of 15 cm were sufficient to enable a volunteer plant community to reestablish in the treated area. In this case, metal concentrations of reeds (*Typha latifoli*) were within the normal range.

11.7.1.4 Palmerton Experiment, Pennsylvania

The Palmerton Zinc Superfund site surrounds a former Zn smelting facility located in Palmerton, PA, which operated from 1898 until 1980. The site has been treated on a large scale with biosolids and coal fly ash. Amendments were initially tested from 1988 to 1990. At Blue Mountain, application of high-Fe and high-lime sewage sludge compost into the soil decreased Zn toxicity to grasses. The metal-attenuating effect of the biosolids-compost soil treatment remained stable over 6 years (Li et al. 2000).

11.7.2 Cyclonic Ashes (Beringite): Lommel-Maatheide and Overpelt Experiments, Belgium

Data for the following experiments as well as those described in the following subsections, summarizing the degree of soil contamination and application rates used for the soil amendments, are shown in Table 11.2 and Table 11.3.

Cyclonic ashes (beringite) were found to decrease plant exposure to metals and to restore vegetative cover. Twelve years after the CB treatment (5% C + 5% B) the vegetation (mainly *Agrostis capillaris* and *Festuca rubra*) was found to be healthy and regenerating by both vegetative means and seeds at the Lommel-Maatheide experiment (Vangronsveld, Colpaert, Van Tichelen 1996; Vangronsveld 1998). On untreated soils, growth of test plants (e.g., *Phaseolus vulgaris*) was strongly inhibited (Vangronsveld 1998). Contaminated soil in the playground area of the Lommel school was also treated with CB. The vegetation immediately recovered and was still developing after 3 years (Vangronsveld et al. 2000a, 2000b). Following evidence of metal attenuation in the small-scale Overpelt experiment at Limburgs Universitair Centrum (LUC), a field experiment was established at Overpelt. In year 3, vegetation was well established in the B-, SS-, CSS-, and CBSS-treated plots. Plant biomass production was highest in the CBSS plots.

The efficacy of beringite amendments in decreasing Cd and Zn uptake in vegetables was investigated. Test plots in 10 kitchen gardens, which consisted of sandy soils contaminated by aerial deposition from the former Lommel-Maatheide zinc smelter, were treated with beringite (100 ton ha^{1}). Comparison of Cd contents in edible parts of plants grown on both untreated and B-treated plots showed marked (2 to 4 times) reductions in Cd content in plants from the beringite-treated plots.

11.7.3 Metal Oxides

Three field trials in the U.K. were established to investigate the efficacy of Fe oxide as an immobilizing agent for As (Alloway et al. 2001). These were an agricultural field adjacent to a derelict As smelter in Cornwall, long-term sludge-spreading plots at a sewage works in Northampton, and a domestic garden in St Helens, Merseyside. Ferrous sulfate (commercial grade), which would be oxidized to Fe oxide in the soil, was applied as a single treatment before the first crops were planted with no further amendment for the duration of the trial. Lime was also added to compensate for the acidifying effect of the sulfate. Treatments with 0.2% Fe caused a significant decrease in As transfer to calabrese leaf, cauliflower, and radish at the Cornwall site. However, the amendment had an inconsistent effect with potato. There were no significant treatment responses found at either the Northampton or Merseyside sites. Results from these trials illustrate the problems of using prescriptive amendments to reduce As mobility across widely differing sites.

11.7.4 Zerovalent Fe-Related Compounds Combined with Cyclonic Ashes

11.7.4.1 Louis Fargue Experiment, Domaine INRA de Couhins, France

Sewage sludges were applied to a coarse sandy soil from 1976 to 1980. In 1995, 1 block was amended with beringite (5%) and another 1 with steelshot (1%) (Mench et al. 2000). Corn ears were better filled in the SS plot, whereas those from the B-treated plots showed poor filling, and a corresponding yield decrease, due to Mn deficiency, which resulted from Mn attenuation by beringite in the sandy soil. The SS addition resulted in a sustained decrease in Cd phytoavailability to maize. In 2000, Cd concentration in corn grain from the SS plot was approximately 40% lower than concentrations found from 1980 to 1994 before the soil was treated. The data indicated increased amelioration over time, suggesting that Cd was becoming occluded in the newly formed Fe oxides. Both SS and B amendments resulted in a sustained decrease in Ni concentration in corn grain, probably from an increase in soil pH. In 2001 (year 6 after soil treatment), metal toxicity to and metal uptake by lettuce was investigated. Dry matter yield showed a beneficial effect of SS amendment (+32% compared with Unt), whereas a detrimental effect of B (27%) was again observed (Table 11.5). Lettuce shoots revealed elevated Cd and Ni and low Mn when the untreated plots were compared with the uncontaminated plots. Lettuce grown on the SS plots showed decreases in Cd (40%), Ni (57%), Zn (21%), and P (30%) concentrations, and elevated Mn concentration (Table 11.5). Copper and Pb concentrations were not affected by SS amendment.

11.7.4.2 Jales Experiments

Previous attempts to establish vegetative cover on the fine-grained spoil at the former Jales gold mine, northeastern Portugal, were unsuccessful. Colonization by plants was limited to a few isolated spots. The grasses *Holcus lanatus*, *Agrostis castellana*,

TABLE 11.5
Yield DM and shoot metal concentrations in lettuce cultivated at the Louis Fargue experiment, 6 years after soil treatment

		Untreated	Beringite	Steelshots	Uncontaminated
Yield	Mg ha^{1}	37.2 ± 7.9	27.2	49.4	63.9
Cd	$mg \cdot kg^{-1}$	56.9 ± 10.4	68.8	34.7	1.6
Cu	$mg \cdot kg^{-1}$	8.0 ± 0.4	5.4	8.3	9.3
Mn	$mg \cdot kg^{-1}$	8.7 ± 1.6	9.5	26.2	40.1
Ni	$mg \cdot kg^{-1}$	15.5 ± 3.4	7.6	6.7	1.6
Pb	$mg \cdot kg^{-1}$	0.33 ± 0.11	0.37	0.43	0.88
Zn	$mg \cdot kg^{-1}$	70.5 ± 2.4	71.8	55.8	82.1
P	g kg^{-1}	8.0 ± 0.2	5.6	5.6	8.6

Note: Beringite (5%), steelshot (1%).

TABLE 11.6
Arsenate (μmol g^{-1} DW) in above-ground biomass of *Holcus lanatus, Agrostis castellana, Cytisus striatus,* and *Betula alba* grown on untreated soil (Unt) and soil treated with steelshot (SS, 1%), organic matter (C, 5%), and/or beringite (B, 5%) in presence and absence of P

	H. lanatus			*A. castellana*		*C. striatus*	*B. alba*
	Green	*Green* fertilized	*Senescent* fertilized	*Green*	*Green* fertilized	*Green*	*Green*
Unt	n.g.	1.39 ± 0.12	2.11 ± 0.23	n.g.	0.63 ± 0.10	n.g.	0.90 ± 0.15
SS	n.g.	0.89 ± 0.15	1.73 ± 0.12	n.g.	0.36 ± 0.04	0.38 ± 0.04	0.91 ± 0.28
SSB	0.47 ± 0.05	0.34 ± 0.09	0.81 ± 0.08	0.27 ± 0.05	0.43 ± 0.05	0.27 ± 0.02	1.34 ± 0.16
CSS	0.39 ± 0.03	0.23 ± 0.02	0.76 ± 0.06	0.25 ± 0.01	0.28 ± 0.01	0.32 ± 0.03	0.94 ± 0.15
CSSB	0.34 ± 0.03	0.25 ± 0.01	0.72 ± 0.06	0.17 ± 0.04	0.28 ± 0.05	0.19 ± 0.02	1.18 ± 0.10
CB	0.38 ± 0.05	0.33 ± 0.01	0.89 ± 0.07	0.29 ± 0.03	0.30 ± 0.05	0.21 ± 0.01	1.35 ± 0.43
C	1.06 ± 0.04	0.54 ± 0.03	1.27 ± 0.09	0.50 ± 0.12	0.50 ± 0.16	0.32 ± 0.01	1.44 ± 0.14
B	0.61 ± 0.04	0.44 ± 0.06	1.20 ± 0.09	0.59 ± 0.03	0.40 ± 0.07	0.34 ± 0.03	0.86 ± 0.06

Note: $n = 8$ (+SE); n.g. indicates no plant growth.

Source: From Bleeker PM, Assunção GL, Teiga PM, de Koe T, Verkleij JAC. 2002. *Sci Total Environ* 300:1–13. With permission.

and *A. delicatula* were the sole colonizers, growing in small isolated tufts (De Koe 1994). The effectiveness of the following amendments in promoting plant growth was assessed in 8 plots established on 1 of the Jales spoil terraces: Unt, B (5%), SS (1%), C (commercial garden compost, 5%), BSS, CSS, CB, and CBSS, with and without P fertilization (Bleeker et al. 2002). Material was incorporated into the top 30 cm of the soil. Despite a small pH increase in SS and CSS treatments, water-soluble As was similar to Unt. In contrast, the CBSS treatment resulted in a threefold higher water-soluble As concentrations compared to the control. The use of tolerant grasses in combination with soil treatments resulted in a rapid and effective revegetation of the contaminated soils. Colonization and reproduction of both *H. lanatus* and *Agrostis castellana* were most successful when additives were combined and soil supplemented with P fertilizer (Bleeker et al. 2002). Lowest As concentrations in *H. lanatus* and *A. castellana* were found in CBSS, CSS, CB, and BSS (Table 11.6).

11.7.4.3 Small-Scale Reppel Experiment

In this contaminated agricultural soil, incorporation of B and BSS soil amendments had sustained beneficial effects on yield of corn, lettuce, and radish (Table 11.7). A decrease in tissue concentrations of Cd and Zn was found for the B and BSS treatment (Table 11.8). Only the BSS treatment resulted in a decrease in plant concentrations of As and Pb (Table 11.8). The SS treatment either had no effect or increased the plant concentrations of metals compared to Unt. The concentration of Cu was increased in all amended soils compared to Unt; however, the concentrations

TABLE 11.7
Plant biomass and soil pH at the small-scale Reppel experiment, Bordeaux, France

Year	Corn ear[a] 4	Soil pH 5	Lettuce shoot[b] 5	Radish whole plant[b] 5
Control soil	204[c]	4.9	113[c]	6.1[cd]
Unt	157[c]	4.9	168[c]	5.7[c]
SS	230[cd]	5.15	246[cd]	8.5b[e]
B	264[d]	6.35	283[e]	7.9[d]
BSS	249[d]	6.9	293[de]	10.5[e]

Note: In each column, mean values followed by the same letter are not statistically different.

[a] g DW ear^{-1}.
[b] g FW $plant^{-1}$.

TABLE 11.8
Trace element concentrations ($mg \cdot kg^{-1}$) in corn third-leaf at the small-scale Reppel experiment 4 years after soil treatment

	As	Zn	Cd	Cu	Pb
Control soil	<0.5[a]	69[b]	0.065[a]	<3[a]	0.41[c]
Unt	2.16[c]	69[bc]	0.668[d]	4.6[b]	0.28[a]
SS	2.16[c]	86[c]	0.705[d]	8.2[d]	0.51[d]
B	1.97[c]	43[a]	0.315[c]	5.2[c]	0.38[b]
BSS	1.09[b]	48[a]	0.233[b]	5.5[c]	0.28[a]

Note: In each column, mean values followed by the same letter are not statistically different.

of Cu were not at the phytotoxic level for corn. Soil CEC increased from 3.8 to 5.3 cmol kg^{-1} in B-treated soils, which was higher than expected, based on the 22 cmol kg^{-1} CEC of beringite. This may be due to the formation of minerals. Chemical extractions inferred that B addition induced *de novo* formation of ferrihydrite and allophane in Reppel soil and that the proportion of Cu, Mn, and Zn associated with amorphous minerals increased (Pannetier 2000).

11.7.5 Zeolites

Field trials were established at Northampton and Staffordshire to evaluate the efficacy of soil amendments, including zeolite, in decreasing Cd transfer from soil to vegetable crops (Lepp et al. 2000). The application of sewage sludge resulted in elevated soil Cd at the Northampton site (mean total Cd content 47 $mg \cdot kg^{-1}$). At the Staffordshire site (mean total Cd content 16 $mg \cdot kg^{-1}$), deposition of Cd oxide particles from an adjacent pigment manufacturer had contaminated the soil of a domestic garden. The Staffordshire site was only Cd contaminated, whereas the Northampton soil contained elevated contents of other metals and As. Plots were treated in 1998 with the following amendments: $FeSO_4$ 1% (equal to 1% Fe_2O_3 plus 5.7% lime), iron grit 1%, lime 3%, and zeolite 4A 1%. The results demonstrated that although transfer of both As and Cd from soils to crops (lettuce, spinach, radish, and red beet) was generally low, incorporation of a range of amendments, including zeolite, had little further influence on reducing risk associated with this transfer.

11.7.6 Red Muds

Red mud is a by-product of the alumina industry that is alkaline and rich in Al/Fe-oxides. Several pot experiments have demonstrated its efficacy for remediation of metal-contaminated soils over relatively long time periods (Mench et al. 2000). Red mud (1%) performed well in a 15-month pot study carried out by Friesl et al. (2001). Müller and Pluquet (1998), who used red mud to treat a harbor dredging site contaminated with Cd and Zn, also showed that red mud could reduce metal mobility and availability.

11.7.7 Phosphate Compounds

Phosphates react with many metals, metalloids, and radionuclides. Precipitates formed can be stable over a wide range of geochemical conditions. In particular, conversion of soil Pb to pyromorphite, an insoluble lead phosphate [$Pb_5(PO_4)_3$(OH, Cl, F, ...)], could immobilize soil Pb, decrease its bioavailability, especially in the gastrointestinal tract (Yang et al. 2002), and decrease Pb leaching in soil. A range of compounds has been evaluated including mineral apatite, synthetic hydroxyapatite, phosphoric acid and diammonium phosphate materials.

A field trial was installed in a vacant city lot at Joplin, MO, in 1997 to evaluate different techniques for Pb inactivation in contaminated soils. Lead was mainly present as Pb carbonates. The efficacy of TSP, rock phosphate, phosphoric acid, compost, and an Fe-rich by-product from titanium processing was compared (Berti, Cunningham, Cooper 1998; Brown, Chaney, Berti 1999). Lime was also added to each plot to bring the pH to 7. Seed (K31 tall fescue) was hand-scattered over the plot surface. The relative plant uptake of Pb in relation to total soil Pb was calculated. This ratio ranged from 0.0013 (3.2% P) to 0.0085 (2.5% Fe + 1% P). Lead availability measured by both relative plant Pb uptake and *in vitro* accessible Pb indicated 3.2% P as the most effective amendment. Reductions in Pb availability were also evident with 10% compost + 0.32% P and 2.5% Fe + 0.32% P.

11.7.8 Clays

Addition of clay to soil can produce physical and chemical changes that could affect contaminant fate and transport. These include increased cation exchange capacity, increased mineral surface area, and sorption within the clay interlayer. Illitic materials may be effective stabilizing agents for Cs^+ because of their ability to fix this cation under a range of moisture conditions (Seaman, Meehan, Bertsch 2001).

Gravel sludge, a waste product of the gravel industry, contains illite (29%), calcite (30%), and quartz (18%). Its efficacy as an in situ immobilizing additive was investigated at 2 application rates in 3 field trials with sandy loam soils at Dottikon, Rafz, and Giornico, Switzerland, contaminated by Cd, Cu, and Zn (Krebs et al. 1999). Gravel sludge application increased pH in all 3 topsoils by up to 0.6 units. In the Dottikon soil, Cu and Zn concentrations in ryegrass were reduced by more than 35%. Lettuce Cd and Zn tissue concentrations decreased by 22 to 48% at Giornico and Dottikon, whereas no effect was found at Rafz. The efficacy of gravel sludge in attenuating metal bioavailability was highest in soils with high $NaNO_3$-extractable metals, and higher for ryegrass than for lettuce.

11.7.9 Competitive Uptake at the Root Surface and Competitive Transfer into Plant Parts

One mechanism for attenuation of trace element exposure may be a decrease in uptake at the biological membrane due to competition with other ions or competition during internal transfer to organs (McLaughlin et al. 2000). There are several examples of this type of interaction, e.g., Cd/Zn, Cu/Fe, and As/P (Oliver et al. 1994; Boisson 1999), and these may be of use in treating contaminated agricultural soils. Several amendments, i.e., natural clay, alum-based water treatment residuals (Alum WTR, 15 t ha^{1}), lime (15 t ha^{1}), magnesite ($MgCO_{3,}$ 15 t ha^{1}), clay, Fe-rich fines from a ferrous smelter (FRF), zinc sulphate (25 kg ha^{1}), and copper sulphate (10 to 50 kg ha^{1}), were evaluated under field conditions at 3 sites in Australia established on Alfisols (clay mineralogy dominated by kaolinite) with pH values between 5 and 7 (McLaughlin et al. 2000). Cadmium concentrations in potato tubers grown on soils amended with a mixture of Cu/Zn salts and with natural clay were lower (18%) than those amended with lime. All other treatments at all sites were ineffective in decreasing Cd concentrations in tubers. However, results were not sufficiently large to warrant recommendation of the use of Cu/Zn salt application to decrease crop Cd concentrations.

11.8 IMPACTS ON AND UPTAKE BY OTHER ORGANISMS

11.8.1 Soil Microorganisms

The presence and behavior of microbial communities are 1 bottleneck in the process of restoring soil functions that enhance phytostabilization. Metal exposure to microorganisms was assessed at the Louis Fargue experiment using a biosensor kit (Biomet™) (Van der Lelie et al. 2000). In this assay, bacterial luminescence intensity is related to metal exposure in the soil. The Cd/Zn specific indicator strain of *Ralstonia metallidurans* CH34 showed a decrease in Cd exposure in B- and SS-treated soils in both the farmyard manure and 300 Mg ha^{1} sewage sludge plots. For the 50 Mg ha^{1} plots, Cd exposure to microorganisms was slightly decreased in the B-treated soil. The Ni strain showed a general decrease in Ni exposure for both

B- and SS-amended soils compared with the Unt soil. Lowest Ni exposures occurred in SS-treated soils in the farmyard manure and 50-Mg sewage sludge ha[1] plots. Consequently, a beneficial effect on microbial communities was likely to be expected in either B-treated or SS-treated soils. Accordingly, soils were sampled at 1 long-term experiment, Louis Fargue (year 7), and 2 small-scale experiments, Reppel (year 6), and Jales (year 5). At Louis Fargue, both SS and B treatments restored enzyme activities inhibited in the 50 Mg ha[1] sludged soil, SS being the most efficient soil treatment (Renella et al. 2005). Soils were potted, inoculated with a bean rhizospheric solution, and then cultivated with dwarf bean. *Rhizobium* nodules on roots were enumerated (Mench, Solda, Recalde 2003). The number of *Rhizobium* nodules was enhanced from 15% to 50% (100% based on control soil) by SS amendment in the 50 Mg ha[1] Louis Fargue sludged plot. Unexpectedly, beringite addition did not result in restoration of symbiosis. *Rhizobium* symbiosis was inhibited in all Jales and Reppel soils, except for a 33% restoration in the BSS-Reppel soil.

An As Biomet strain was used to quantify As exposure in treated and untreated soils from the small-scale Jales experiment, but no luminescence was found for all soils (Corbisier, personal communication). The As exposure may not be toxic for this strain, despite the As contamination in this soil and increase in water-soluble As after CB and C addition into the soil.

11.8.2 Earthworms and Mites

Earthworms are thought to be highly exposed to soil metals as their diet consists of organic material in soil. Toxicity testing using earthworms is a well-developed method for studying bioavailability and toxicity of soil contaminants. Amendments (B, SS, and BSS) were found to decrease the toxicity of the Reppel soil to earthworms (*Lumbricus terrestris*), but the As, Cd, and Zn concentrations in depurated earthworms were not decreased (Mench, Solda, Recalde 2003). Consequently, this may be a route of metal exposure for biota that consume earthworms.

Soil ecosystem development on 4 vegetated tailings sites at Copper Cliff, Ontario, 0, 8, 20, and 40 years after rehabilitation, was assessed in terms of mite (Acari) populations and compared with those from 4 control sites (John et al. 2002). Mite density on older and more botanically diverse tailings sites was similar to that on control sites, but species richness of oribatids and mesostigmatics was lower. Species richness and diversity on tailings were lower at less botanically diverse sites regardless of age. The similarity of tailings mite communities to control-site communities increased with age, but it was always less than 60%. A few colonizing species dominated mite assemblages on tailings, whereas control sites had a diverse assemblage of species.

11.8.3 Mammals and Birds

Potential pathways of animal exposure to metals from contaminated soils include "soil to flora to animal" (herbivore exposure), "soil to fauna to animal" (carnivore, insectivore exposure), and "soil to animal" (direct ingestion of soil and dermal contact). The effect of soil amendments on each of these pathways needs to be considered in terms of assessing the overall effectiveness of any chemical immobilization treatments for soil remediation. For example, if the remediation treatment results in greater survival of earthworms and other soil fauna in the contaminated soils, then insectivorous mammals can inadvertently be exposed to a greater risk than if the fauna was unable to survive in the soil (Brown et al. 2002). To date, there has been very little full-scale ecosystem monitoring at remediated sites.

Direct soil ingestion by grazing animals or by children, resulting from hand-to-mouth actions, can be a significant soil contaminant exposure pathway (Basta et al. 2001; Basta, Armstrong, Hanke 2001). Human dosing to establish the efficacy of amendments in decreasing exposure through this route is not feasible, but batch tests can deliver some information. One such test used to predict metal availability is the so-called *in vitro* physiologically based extraction test (PBET) (Ruby et al. 1996; Brown and Chaney 1999).

A number of studies have been conducted to assess PBET-extractable As and Pb in treated soils in pot trials (Hettiarachchi, Pierzynski, Ransom 2001; Geebelen 2002; Geebelen et al. 2003a). The results of these studies demonstrate that metal and As exposure via soil ingestion, can be decreased by application of different types of amendments to contaminated soils.

11.9 BIODIVERSITY AND GENETIC ADAPTATION OF ORGANISMS

There is little information available on the effect of amendment applications to contaminated soils on biodiversity. This aspect was however investigated at Lommel-Maatheide experiment. Diversity of higher plant species and saprophytic fungi was extremely low in the untreated soil because of the high soil toxicity and the absence of metal-tolerant ecotypes of plants and fungi (Table 11.9) (Vangronsveld 1998). Only 1 plant species was prevalent (*Agrostis capillaris*), the frequency of the 2 other species present (*Betula pendula, Stelleria media*) being less than 1%. In contrast, 2 plant species (*A. capillaris, Festuca rubra*) were mainly present in the CB-treated plots, and several nonmetal-tolerant perennial forbs colonized the remediated area (e.g., *Cerastium fontanum*, *Centaurium erythraea*, *Plantago media*) (Table 11.9). Most of these species belong to mycotrophic families; so the presence of a mycorrhizal network in the soil promotes their establishment. In Unt plots, the arbuscular mycorrhiza (AM) infection percentage of grass roots ranged from 0 to 42% (Table 11.10). This percentage was greatly increased in treated soils, ranging from 37 to 81%, even though total metal concentrations in soils were similar or even higher than the Unt plots. Also, the functional diversity of soil bacterial populations, measured as the capacity to metabolize a number of different substrates, almost doubled after soil treatment (Bouwman et al. 2001).

In treated soils of the small-scale Jales experiment, volunteer plant species established themselves from year 2 onward. These included trees (*Salix caprea* L.), vascular plants (*Erigeron canadensis* L.), and bryophytes (*Funaria hygrometrica* Hedw). *Salix caprea* colonized the CBSS, CSS, and C, but not the Unt, uncontaminated control, or CB soil. Birch (*Betula* sp.) was present on CBSS, and maple (*Acer*

TABLE 11.9
Increase in number of plant species at the Lommel–Maatheide experiment and the small-scale Jales experiment

Experiments	Soil treatments					Untreated soil	Control
	CB	CBSS	CSS	CB	C		
Lommel–Maatheide							
Frequency in the vegetated quadrats (%)							
90	2						
1–10	3					1	
<1	8						
Presence	13					2	
Jales							
Number of plant species		8	5	5	5	1	4
Soil covered by *H. lanatus*[a]		249	399	499	432	32	100

[a] Expressed in percentage compared with the control soil.

Source: From Vangronsveld J, Colpaert JV, Van Tichelen KK. 1996. *Environ Pollut* 94:131–140. With permission. Mench unpublished data.

negundon L.) on CB, C, and CBSS. *Gnaphalium sp.* was identified on B soil and *E. canadensis* grew well in the CBSS, CSS, and C soils. Fungi (i.e., *Coprinus*) developed in year 2 on CBSS- and CB-treated soils. From year 3 onward, carpophores of ectomycorrhizal fungi, e.g., *Rhizopogon roseolus* (Corda) Th. M. Fries associated with *P. pinaster*, were observed in the CSS and CBSS soils (Mench, Guinberteau, Recalde 2003). *Hebeloma leucosarx* Orton and *Hebeloma mesophaeum* (Fries) Quélet developed in association with *S. caprea* in the CBSS soils.

11.10 FAILURES, SIDE EFFECTS, AND LIMITATIONS OF CHEMICAL IMMOBILIZATION METHODS FOR SOIL REMEDIATION

11.10.1 Failures

Data from the 5- to 12-year-old experiments described earlier demonstrate that attenuation can be sustained, but that a given amendment can give different responses in different soils. The methods used to apply and incorporate the amendment can be very important for the attenuation process. For example, in the Northampton experiment described earlier, Fe grit was applied to the soil surface, but not immediately mixed with the soil. A lack of clear effects of this material on plant Cd uptake in this experiment may have been due to the fact that the surface of Fe grit oxidizes very rapidly, and newly formed oxides may not have been appropriately located to immediately react with Cd in the soil solution.

TABLE 11.10
AM infection percentages in the roots of grasses along the transect lines in the Beringite-treated and untreated areas at the Lommel-Maatheide experiment

Distance (m)	AM infection (%)	pH	Zn	Cd	Cu
			(mg·kg^{-1})		
		CB-treated			
50	81	7.3	12750	18	475
100	69	7.9	3600	12	895
150	76	7.4	1168	8	206
200	38	7.6	9875	53	495
250	37	7.9	13250	55	765
300	65	7.6	4750	15	1393
		Untreated			
25	0	4.5	2400	16	160
108	0	5.1	2080	8	80
150	3	5.5	2720	14	400
200	42	5.7	4960	61	395
250	24	5.9	4160	56	405
300	14	5.9	800	12	80

Note: pH and total concentrations of Zn, Cd, and Cu in the corresponding soil cores.

We have learned that quantifying attenuation in laboratory simulations is necessary, but field demonstrations are also required. For example, a decrease with time in the efficacy of both B and SS amendments for attenuating As exposure was observed in the small-scale Reppel experiment. Similarly, amendments to reduce Cd bioavailability (i.e., lime, clay, Fe-rich fines, Cu and Zn sulfate, and alum-water treatment residuals) performed poorly under field conditions in 3 Cd-contaminated agricultural soils in South Australia, despite good results obtained in the laboratory (McLaughlin et al. 2000). The application rate of the amendment is also very important. A continuing decrease in 0.01 M $Ca(NO_3)_2$-extractable Cd, Ni, and Zn was observed with increasing application rates of SS from 0.1 up to 10% (Sappin-Didier 1995). However, the 10% application rate was expensive and produced undesirable side effects; e.g., sorption of P, increased Mn exposure.

There is a lack of information on the potential of attenuated sites to revert to their preattenuated soil chemistry, particularly under field conditions. Long-term degradation, weathering processes, acidification, and reducing conditions can influence attenuation, potentially resulting in enhanced transport and exposure (Hamon et al. 2002). Consequences of soil acidification on amended soils have been investigated in laboratory-based batch experiments. Biosolids-induced metal immobilization became destabilized as acidic soil conditions developed ($pH < 6$) (Basta et al. 2001; Basta, Armstrong, Hanke 2001). Immobilization products of rock phosphates (i.e., metal pyromorphites) were more stable as soils acidified ($pH < 5$), even though they have less effect on metal extractability and phytoavailability compared with biosolids. Use of carbonate-rich materials (e.g., ground limestone) has had some success in decreasing solubility and crop uptake of Cd and Zn in acid soils, but because of leaching, these responses are often not sustainable. For example, it was noted in the Pronto experiment described earlier that applied limestone has been more or less consumed as a result of oxidation of the Cu tailings and associated soil acidification. It would be interesting to follow the consequences for established vegetation and microbial communities if additional lime is not administered.

11.10.2 Side Effects

Among potential unwanted side effects of soil amendments, the following aspects should be considered (Osté 2001):

1. Toxicity of the amendment material
2. Toxicity of contaminants in the amendments
3. Imbalance in nutrients as a result of reactions induced by the amendments, either amendment-induced nutrient deficiency, or excess fertility as in the case of P or biosolids addition
4. Negative effects due to changes in soil conditions, such as release of trace elements through amendment-induced dissolution of different phases
5. Negative effects due to impact on soil structure and soil organic matter

Some examples of these aspects are as follows. In the Reppel soil, a transient adverse effect was found on bacteria and earthworms following the rapid oxidation of steelshot in soil and the release of Fe and Mn to soil-pore water before the newly formed Fe and Mn oxides. Several potentially useful amendments such as incinerator ashes, coal fly ashes, biosolids, and Fe-rich muds can contain either inorganic or organic contaminants. Many coal fly ashes contain high concentrations of boron and SO_4 that may rapidly affect plant growth. The sodicity or radioactivity of red muds can be an immediate problem. Steelshot contain Ni, which raised the Ni content of maize grain in the Louis Fargue experiment upon application.

The amendment application can decrease solubility of essential trace elements and macronutrients. Maintaining sufficient P is sometimes difficult because of P fixation by Fe oxides present in soil amendments such as red muds. Phosphate fixation can be counteracted by adjusting the application rate and by repeated application, but competition with As and induction of As leaching must also be considered in soils co-contaminated with As. Beringite was effective in decreasing plant exposure to Ni in the Louis Fargue experiment. Over time, however, Mn deficiency and low P availability developed, both affecting plant yield (Mench et al. 2000). Attenuation of Mn availability induced by beringite addition to the soil was also observed at both the small-scale Jales and Reppel experiments. This could be a frequently occurring side effect using this material, especially in sandy soils.

Increasing the soil fertility is often necessary to enable establishment of a vegetation cover in contaminated soils with low nutrient and water-retentive capacity. This can be achieved using compost. The increase in dissolved organic matter from this source may result in reaction with solid phases or an elevated labile pool of contaminants. In the small-scale Jales experiment, compost increased As leaching and plant exposure, and addition of other materials such as beringite was unable to counterbalance this side effect. Higher leaching and organism exposure to Cu and Pb can also occur after organic matter application. Arsenic mobility can increase after hydroxyapatite addition, as was observed in the Overpelt soil (Boisson et al. 1999). Increased soil pH, resulting from alkaline amendments and manganese oxide amendments (K-birnessite), can increase leaching of nonessential metals because of increases in dissolved organic matter (McBride and Martinez 2000; Chapter 6). Some long-term experiments therefore recommend caution in the use of combinations of fly ash and biosolids (Bhumbla, Sekhon, Sajwan 2001). DTPA-extractable Cu, Fe, Mn, Pb, and Zn were higher in acidic mine soils receiving fly ash and biosolid mixtures. Alkaline additives poor in Ca, such as some zeolites, can strongly increase the dissolved organic matter concentration, resulting in increased metal leaching (Osté 2001).

Reducing conditions may occur with high-volume applications of biologically active amendments such as sewage sludge (biosolids). Several trace elements, e.g., As, Cr, Hg, Mo, Se, and V may change their oxidation state and hence toxicity within the range of redox conditions commonly encountered in soils (Chapter 8). In addition, redox reactions may induce changes in the mineral composition, structure, and stability of soil solid phases. Labile organic acids produced during the anaerobic decomposition of organic matter may form metal complexes, increasing metal concentration in the soil solution. Newly formed solid phases such as Fe(III) and Mn(IV) oxy-hydroxides, resulting from the oxidation of steelshot or related zerovalent Fe compounds, may undergo reductive dissolution and release adsorbed and coprecipitated trace elements in the soil solution under reducing conditions. For example, the reductive dissolution of birnessite by oxidizable organic ligands such as catechol was rapid, independent of pH, and essentially complete within seconds under conditions of excess of catechol at pH 4 to 6 (Matocha et al. 2001).

High application rates of some additives can affect soil properties. Synthetic zeolite in Na form can damage the soil structure (Osté 2001). High application rates of steelshot (over 10% soil weight) reduced the soil porosity (Sappin-Didier 1995). When papermill waste was incorporated into tailings, a good ground cover did not establish. This was probably due to salt accumulation at the surface, which created an osmotic environment unfavorable for the establishment of glycophyte seedlings, although seedlings of volunteer halophytes did start to colonize the area (Tisch et al. 2000).

11.10.3 Limitations

At polluted sites, soils are usually contaminated with several trace elements. For effective remediation, the amendments used need to be able to attenuate the range of elements present, but this is often not possible. For example, MnO_2 can bind metals such as As, Cd, Pb, and Zn but appears less effective in Cu-contaminated soils (McBride and Martinez 2000). Calplus (clay-aluminum hydroxides) was found effective for Cu and Zn but not for Cd (Osté 2001). Synthetic zeolites can have a high affinity for Ca; hence, addition to calcareous soils may decrease their metal binding capacity and increase leaching of organic matter (Osté 2001). Beringite combined with compost decreases Cd and Zn exposure, whereas it can increase As exposure and leaching (Mench et al. 2003).

In the majority of sites where chemical immobilization has been used, treatments have been confined to the uppermost regions of the soil; however, contamination, and plant roots, may extend to some depth especially when trees are used for revegetation. Experience from the Lommel-Maatheide experiment showed that trees died as their roots came into contact with soil solution from untreated contaminated lower soil layers.

11.11 CONCLUSIONS

Soil amendments that can enhance natural attenuation of trace elements and lead to a decrease in trace element exposure are available.

Several current small-scale semifield trials and field experiments show that successful attenuation and phytostabilization can be sustained for over a decade for a range of contaminated soils (Table 11.11). Successful phytostabilization, equally protective of human health and the environment, is frequently based on a combination of several additives. Some materials such as steelshot combined with cyclonic ashes (beringite) have proven effective over time. The oldest field experiments confirm that enhanced attenuation can be a cost-effective technique to initiate a healthy and diverse ecosystem in contaminated soils.

Preselection of additives in laboratory experiments is useful but may not forecast some long-term effects. Materials found to be promising under laboratory or glasshouse conditions can be less effective in the field. A decrease in effectiveness can occur over time. Failures and side effects such as induced deficiencies can arise and tend to be site specific. To avoid failures, a thorough evaluation of materials must be conducted, preferably in field lysimeters or in plots at each site. Year-to-year variation in the concentrations of trace elements in crops generally occurs (McGrath and Johnston 2001). This prevents clear trends from being detected with short-term experiments. A long-term monitoring program to examine changes in speciation, leaching, and ecotoxicity should be conducted prior to implementation of any large-scale site treatment

Because of the lack of biogeochemical models for amendments, the sustainability of attenuation is not currently predictable. Environmental models are needed to help identify the limiting factors, which can be biotic or abiotic. Rapid and reliable exposure tests, for example using biosensors, are needed for routine low-cost site monitoring. A range of biosensors, specific for different key contaminants need to be developed. The monitoring program should take into account attenuation in different exposure pathways, for example, soil solution, direct ingestion, and the gaseous phase. It should assure human health and environmental protection. Most studies to date have focused on the effects of chemical immobilization treatments on plant contamination via the soil solution and contaminant mobility to the groundwater. More long-term studies dealing with soil ingestion, dermal contact, and consumption of trace element enriched–products are needed. As discussed in Chapter 7, microorganisms may influence metal attenuation; hence, their activity deserves further consideration. Concerns

TABLE 11.11
Summary of beneficial and side effects induced by soil amendments demonstrated in field trials and outdoor mesocosms

Sites	Jales	Reppel	Overpelt	Louis Fargue	Maatheide	Northampton	Stafford-shire	Corn-wall
Monitoring duration	4	5	5	7	12	3	3	3
Most effective treatments	CSS CBSS	CBSS	CBSS	SS	CB	Z4A		
				Beneficial effects				
Decrease in metal exposure	+	+	+	+	+			
Decrease in As exposure	+	+		+				+
Decrease in leaching	+	+	+		+			
Increase in plant growth	+	+	+	+	+			
Increase in plant diversity	+		+		+			
Decrease in metal content in plants	+	+	+	+	+			
Promote microbial communities	+	+		+	+			
Decrease in animal exposure					+			
				Side effects				
Induced deficiency	+			+				
Enhanced metal leaching	+							
Enhanced As leaching	+							
Increase in As exposure	+							
Increase in Ni exposure				+				
Increase in metal exposure								+

Note: C: compost (5%); B: cyclonic ashes, beringite (5%); SS: steelshot (iron grit); Z4A: zeolite 4A.

about the long-term effectiveness of amendments in attenuating contaminants can be addressed by ensuring that monitoring will provide protection and that alternative remedies will be implemented as needed (Swindoll and Firth 1998).

Regulatory acceptance of enhanced natural attenuation as a strategy for risk reduction in contaminated sites is related to the evolving scientific and social consensus over how bioavailability should be measured and which organisms we specifically seek to model. Present environmental regulations for trace element-contaminated soils are frequently based on total contaminant concentrations. However, from ecological, toxicological, and health viewpoints, the bioavailable fraction in exposure pathways should be considered. Although attenuation may be attractive in some locations, it should not be viewed as an exclusive remedial strategy. It may be combined with other options, in particular, with the use of tolerant plant species to help decrease the risk of off-site contaminant transport through erosion or leaching processes (phytostabilization). Most phytostabilization experiments currently involve crops that could directly enter food chains. However to minimize consumer exposure, it would be necessary to demonstrate the effectiveness of alternative crops that represent a sustainable, low-hazard economic use for remediated sites. These may include plants used for fuel, fiber, oil, and construction materials. Woody and herbaceous plant materials that fall on the surface can be removed to accelerate the remediation. Knox, Kaplan, and Hinton (2001) proposed the use of a mineral-containing mat (geomat) deployed at the ground surface for immobilizing contaminants released from decomposing plant materials. Among several materials tested in the geomat, metallic iron and an Fe oxide waste were the most efficient for lowering the aqueous Ba, Co, Cr, Eu, Hg, Pb, and U concentrations. Unterköfler et al. (2001) have investigated the immobilization of metals leached from fallen leaves and OM (surface layer) from *Salix caprea*, *Populus tremula,* and *Betula pendula,* on-site using a 2-cm layer of vermiculite. This material was able to adsorb and immobilize more than 99% of the leached metals.

Enhancing natural attenuation of trace elements using soil amendments and phytostabilization of contaminated sites remains a matter for further experimentation. Facing the challenge to attenuate trace element exposure, to restore a vegetation cover and a microbial community, or to reestablish the foodstuff compliance mean recognizing that each site has its own unique problem and potential solutions. "Adapt, not adopt," commented Peters (1995).

REFERENCES

Alloway BG, Warren N, Lepp B, Singh F, Bochereau, Penny C. 2001. Remediation of arsenic and cadmium contaminated soils with adsorptive minerals. In: Evans L, editor. Proceedings of the 6th International Conference on the Biogeochemistry of Trace Elements (6th ICOBTE); July 2001; University of Guelph, Ontario, Canada. p 281.

Basta NT, Gradwohl R, Snethen KL, Schroder JL. 2001. Chemical immobilization of lead, zinc, and cadmium in smelter-contaminated soils using biosolids and rock phosphate. J Environ Qual 30:1222–1230.

Basta NT, Armstrong FP, Hanke EM. 2001. Effect of chemical remediation of contaminated soil on arsenic mobility and gastrointestinal availability. In: Evans L, editor. Proceedings of the 6th International Conference on the Biogeochemistry of Trace Elements (6th ICOBTE); July 2001; University of Guelph, Ontario, Canada. p 40.

Bhumbla DK, Sekhon BS, Sajwan KS. 2001. Trace elements bioavailability in mine soils treated with sewage sludge and fly ash mixtures. In: Evans L, editor. Proceedings of the 6th International Conference on the Biogeochemistry of Trace Elements (6th ICOBTE); July 2001; University of Guelph, Ontario, Canada. p 368.

Bleeker PM, Assunção GL, Teiga PM, de Koe T, Verkleij JAC. 2002. Revegetation of the acidic, as contaminated Jales mine spoil tips using a combination of spoil amendments and tolerant grasses. Sci Total Environ 300:1–13.

Berti WR, Cunningham SD, Cooper EM. 1998. Case studies in the field-In-place inactivation and phytorestoration of Pb-contaminated sites. In: Vangronsveld J, Cunningham SD, editors. Metal contaminated soils: *In situ* inactivation and phytorestoration. Berlin: Springer-Verlag. p 235–248.

Boisson J, Mench M, Chartier S. 1998. Limited soil-plant transfer of As by using immobilizing soil additives: a semi-field study. Proceedings of the International Soil Science Society, Montpellier, France.

Boisson J. 1999. Réhabilitation de sols pollués en éléments traces par des amendements minéraux. Faisabilité et durabilité d'après la mobilité des éléments et la phytotoxicité du sol [PhD thesis]. Institut National Polytechnique de Lorraine, Nancy.

Boisson J, Ruttens A, Mench M, Vangronsveld J. 1999. Immobilization of trace metals and arsenic by different soil additives. Evaluation by means of chemical extractions. Commun Soil Sci Plant Anal 30:365–387.

Bouwman L, Bloem J, Römkens PFAM, Boon GT, Vangronsveld J. 2001. Beneficial effect of the growth of metal tolerant grass on biological and chemical parameters in copper- and zinc contaminated sandy soils. Minerva Biotech 13:19–26.

Brown SL, Chaney R. 1999. A rapid in-vitro procedure to characterize the effectiveness of a variety of in-situ lead remediation technologies. In: Wenzel WW, Adriano DC, Alloway B, Doner HE, Keller C, Lepp NW, Mench M, Naidu R, Pierzynski GM, editors. Proceedings of the 5th International Conference on the Biogeochemistry of Trace Elements (5th ICOBTE), Vienna. p 419.

Brown SL, Chaney R, Berti B. 1999. Field test of amendments to reduce the in situ availability of soil lead. In: Wenzel WW, Adriano DC, Alloway B, Doner HE, Keller C, Lepp NW, Mench M, Naidu R, Pierzynski GM, editors. Proceedings of the 5th International Conference on the Biogeochemistry of Trace Elements (5th ICOBTE), Vienna. p 506.

Brown SL, Henry CL, Compton H, Chaney RL, De Volder P. 2000. Using municipal biosolids in combination with other residuals to restore zinc and lead contaminated mining areas. In: Luo YM, McGrath SP, Cao ZH, Zhao FJ, Chen YX, Xu JM, editors. Proceedings of the International Conference on Soil Remediation (SoilRem2000), October 15–19; Hangzhou, China. p 285–289.

Brown SL, Henry CL, Chaney RL. 2001. Restoration of large-scale metal contaminated sites using biosolids and other residuals. In: Evans L, editor. Proceedings of the 6th International Conference on the Biogeochemistry of Trace Elements (6th ICOBTE); July 2001; University of Guelph, Ontario, Canada. p 282.

Brown SL, Chaney RL, Sprenger M, Compton H. 2002. Soil remediation using biosolids. BioCycle 43(6):41–44.

Brown SL, Sprenger M, Maxemchuk A, Compton H. 2005. Ecosystem function in alluvial tailings after biosolids and lime addition. J Environ Qual 34:139–148.

Compton H, Brown S, Henry C, Sprenger M. 2001.Use of biosolids and lime to restore a metal affected ecosystem in Leadville, CO. In: Evans L, editor. Proceedings of the 6th International Conference on the Biogeochemistry of Trace Elements (6th ICOBTE); July 2001; University of Guelph, Ontario, Canada. p 35.

De Koe T. 1994. Arsenic resistance in submediterranean *Agrostis* species [PhD thesis]. Vrije Universiteit, Amsterdam, NL.

De Koe T, Bleeker PM, Assunção GL. 1998. Field experiments at the Jales minespoil. In: Verkleij JAC, editor. Strategies for rehabilitation of metal polluted soils: *in situ* phytoremediation, immobilization and revegetation, a comparative study (PHYTOREHAB). Progress report no. 5 ENV4-CT95-0083, EU DGXII Environment & Climate programme, Vrije Universiteit Amsterdam, The Netherlands. p. 44–67.

Friesl W, Lombi E, Horak O, Wenzel WW. 2001. Use of amendments to reduce trace elements mobility. In: Evans L, editor. Proceedings of the 6th International Conference on the Biogeochemistry of Trace Elements (6th ICOBTE); July 2001; University of Guelph, Ontario, Canada. p 217.

Geebelen W. 2002. Remediation of Pb contaminated soils by phytoextraction and amendment induced immobilization: biological aspects [PhD thesis]. Limburgs Universitair Centrum, Diepenbeek, Belgium.

Geebelen W, Adriano DC, Van der Lelie D, Mench M, Carleer R, Clijsters H, Vangronsveld J. 2003a. Selected bioavailability assays to test the efficacy of amendment-induced immobilization of lead in soils. Plant Soil 249:217–228.

Geebelen W, Adriano DC, Mench M, Clijsters H, Vangronsveld J. 2003b. Amendment induced immobilization of Pb in contaminated soils: effect on Pb, Cu, Zn, Cd, Ni, Fe and Mn phytoavailability and phytotoxicity. In: Gobran G. editor. Proceedings of the 7th International Conference on the Biogeochemistry of Trace Elements (7th ICOBTE), Uppsala, Sweden.

Goulding KWT, Blake L. 1998. Land use, liming and the mobilisation of toxic metals. Agric Ecosys Environ 67:135–144.

Hamon RE, McLaughlin MJ, Cozens G. 2002. Use of isotopic exchange techniques to determine mechanisms of attenuation of metal availability in *in situ* remediation studies. Environ Sci Technol 36:3991–3996.

Hargé JC. 1997. Spéciation comparée du zinc, du plomb et du manganèse dans des sols contaminés [PhD thesis]. Univ. J. Fourier, Grenoble, France.

Hartley W, Edwards R, Lepp NW. 2001. A study of novel methods for the *in situ* remediation of arsenic contaminated soils. In: Evans L, editor. Proceedings of the 6th International Conference on the Biogeochemistry of Trace Elements (6th ICOBTE); July 2001; University of Guelph, Ontario, Canada. p 38.

Hettiarachchi GM, Pierzynski GM, Ransom MD. 2001. In situ stabilization of soil lead using phosphorus. J Environ Qual 30:1214–1221.

John MG St, Bagatto G, Behan-Pelletier V, Lindquist EE, Shorthouse JD, Smith IM. 2002. Mite (Acari) colonization of vegetated mine tailings near Sudbury, Ontario, Canada. Plant Soil 245:295–305.

Knox AS, Seaman JC, Mench MJ, Vangronsveld J. 2000. Remediation of metal- and radionuclides-contaminated soils by *in-situ* stabilization techniques. In: Iskandar IK, editor. Environmental restoration of metals-contaminated soils. Boca Raton, CL: CRC Pr LLC, Lewis Publishers. p 21–60.

Knox AS, Kaplan DI, Hinton TG. 2001. Remediation of metals and radionuclides by phytoextraction and sequestration. In: Evans L, editor. Proceedings of the 6th International Conference on the Biogeochemistry of Trace Elements (6th ICOBTE); July 2001; University of Guelph, Ontario, Canada. p 314.

Krebs R, Gupta SK, Furrer G, Schulin R. 1999. Gravel sludge as immobilizing agent in soils contaminated by heavy metals: a field study. Water Air Soil Pollut 11: 465–479.
Lepp NW, Alloway B, Penny C, Warren G, Bochereau F. 2000. The use of synthetic zeolites as in situ soil amendments to reduce metal transfer from soils to vegetables. In: Luo YM, McGrath SP, Cao ZH, Zhao FJ, Chen YX, Xu JM, editors. Proceedings of the International Conference on Soil Remediation (SoilRem2000), October 15–19, Hangzhou, China. p 280–284.
Li YM, Chaney RL, Siebielec G, Kerschner BA. 2000. Response of four turfgrass cultivar to limestone and biosolids-compost amendment of a zinc and cadmium contaminated soil at Palmerton, Pennsylvania. J Environ Qual 29:1440–1447.
Matocha CJ, Sparks DL, Amonette JE, Kukkadapu RK. 2001. Kinetics and mechanism of birnessite reduction by catechol. Soil Sci Soc Am J 65:58–66.
McBride MB, Martinez CE. 2000. Copper phytotoxicity in a contaminated soil: remediation tests with adsorptive materials. Environ Sci Technol 34:4386–4391.
McGrath SP, Johnston J. 2001. Long term trends in metals in agroecosystems. In: Evans L, editor. Proceedings of the 6th International Conference on the Biogeochemistry of Trace Elements (6th ICOBTE); July 2001; University of Guelph, Ontario, Canada. p 303.
McLaughlin MJ, Nardecchia D, Maier NA, Smart MK, Cozens GD. 2000. Remediation of cadmium-contaminated soils. In: Luo YM, McGrath SP, Cao ZH, Zhao FJ, Chen YX, Xu JM, editors. Proceedings of the International Conference on Soil Remediation (SoilRem2000), October 15–19, Hangzhou, China. p 275–279.
Mench MJ, Didier V, Löffler M, Gomez A, Masson P. 1994. A mimicked in situ remediation study of metal-contaminated soils with emphasis on cadmium and lead. J Environ Qual 23: 58–63.
Mench M, Vangronsveld J, Lepp NW, Edwards R. 1998. Physico-chemical aspects and efficiency of trace element immobilisation by soil amendments. In: Vangronsveld J, Cunningham S, editors. In situ inactivation and phytorestoration of metal-contaminated soils. Berlin: Springer-Verlag. p. 151–182.
Mench M, Vangronsveld J, Clijsters H, Lepp NW, Edwards R. 2000. In situ metal immobilisation and phytostabilisation of contaminated soils. In: Terry N, Banuelos G, editors. Phytoremediation of contaminated soil and water. Boca Raton, FL: Lewis Publishers. p 323–358.
Mench M, Bussière S, Boisson J, Castaing E, Vangronsveld J, Ruttens A, De Koe T, Bleeker P, Assunção A, Manceau A. 2003. Progress in remediation and revegetation of the barren Jales gold mine spoil after *in situ* treatments. Plant Soil 249:187–202.
Mench M, Guinberteau J, Recalde N. 2003. Ectomycorrhizal fungi in the contaminated Jales soil after *in situ* treatment and phytostabilisation. In: Gobran G. editor. Proceedings of the 7th International Conference on the Biogeochemistry of Trace Elements (7th ICOBTE), Uppsala, Sweden. p 180–181.
Mench M, Solda P, Recalde N. 2003. Plant, earthworm, and *rhizobium* responses to natural remediation in sludged-plots contaminated by trace elements. In: Gobran G. editor. Proceedings of the 7th International Conference on the Biogeochemistry of Trace Elements (7th ICOBTE), Uppsala, Sweden. p. 316–317.
Müller I, Pluquet E. 1998 .Immobilization of heavy metals in sediment dredged from a seaport by iron bearing materials. Water Sci Technol 37:379–386.
Oliver DP, Hannam R, Tiller KG, Wilhelm NS, Merry RH, Cozens GD. 1994. The effects of zinc fertilization on cadmium concentration in wheat grain. J Environ Qual 23:705–711.

Osté LA. 2001. *In situ* immobilization of cadmium and zinc in contaminated soils: fiction or fixation? [PhD thesis]. Wageningen Universiteit, The Netherlands. ISBN 90-5808-434-5.

Pannetier S. 2000. Etude de l'incidence des amendements de béringite sur l'abondance des minéraux amorphes dans les sols: conséquences sur l'immobilisation des cations métalliques. Report, IUP EGID, Bordeaux III. Talence, France.

Peters TH. 1995. Revegetation of the Copper Cliff tailing area. In: Gunn JM, editor. Restoration and recovery of an industrial region. New York: Springer-Verlag. p 123–134.

Renella G, Mench M, Gelsomino A, Landi L, Nannipieri P. 2005. Biochemical parameters and bacterial species richness in soils contamined by sludge-borne metals and remediated by inorganic soil amendments. In: Lombi E, editor. Proceedings of the 8th International Conference on the Biogeochemistry of Trace Elements (8th ICOBTE), Adelaide, Australia. p 188.

Ruby MV, Davis A, Schoof R, Eberle S, Sellstone CM. 1996. Estimation of lead and arsenic bioavailability using a physiologically based extraction test. Environ Sci Technol 30:422–430.

Ruttens A, Colpart JV, Mench M, Boisson J, Carleer R., Vangronsveld, J. 2006. Phytostabilitzation of a metal contaminated sandy soil: II: Influence of compost and/or inorganic metal immobiizing soil amendments on metal leaching. Envrion Pollut. Submitted for publication.

Santos Oliveira JM, Freira Avila EP. 1995. Avaliação do impacto quimico ambiental provocado por uma exploração mineira. Um caso de estudo na Mina de Jales. Estudos Nota e Trabalhos 37:25–50.

Sappin-Didier V. 1995. Contrôle des flux de métaux dans les agrosystèmes par apport d'un composé du fer [PhD thesis]. Université Bordeaux I, ENSCPB, Bordeaux, France.

Seaman JC, Meehan T, Bertsch PM. 2001. Immobilization of cesium-137 and uranium in contaminated sediments using soil amendments. J Environ Qual 30:1206–1213.

Semane B. 2001. Evolution d'un sol pollué aux métaux lourds en mésocosme. Importance des amendements pour la phytostabilisation. Report Maîtrise Biologie des Populations et des Ecosystèmes, Université Bordeaux I, Talence, France.

Singh BR, Osté L. 2001. In situ immobilization of metals in contaminated or naturally metal-rich soils. Environ Rev 9:81–97.

Sopper WE. 1993. Municipal sludge use for land reclamation. Ann Arbor, MI: Lewis Publishers.

Swindoll CM, Firth MJ. 1998. Phytorestoration—regulatory, industry, and public concerns. In Vangronsveld J, Cunningham S, editors. Metal-contaminated soils: *in situ* inactivation and phytorestoration. Berlin: Springer-Verlag. p 249–259.

Tisch B, Beckett P, Okonski A, Gordon C, Spiers G. 2000. Remediation and revegetation of barren copper tailings using paper mill sludge; an overview. CLRA annual meeting, Edmonton, Alberta Canada.

Unterköfler J, Wenzel WW, Adriano DC, Wieshammer G, Sommer P, Fitz W. 2001. Integrated phytoextraction and (physico-)chemical immobilization—a new approach to remediate contaminated soils. In: Evans L, editor. Proceedings of the 6th International Conference on the Biogeochemistry of Trace Elements (6th ICOBTE); July 2001; University of Guelph, Ontario, Canada. p 411.

Van der Lelie D, Tibarzawa C, Corbisier P, Vangronsveld J, Mench M. 2000. Bacterial biosensors to quantify bioavailable concentration of heavy metals in polluted soils and to predict their risk of transfer to the food chain. International Conference on Heavy Metals in the Environment, Session 25: bioadsorption and biomonitoring, Toronto.

Vangronsveld J, Van Assche F, Clijsters H. 1995. Reclamation of a bare industrial area, contaminated by non-ferrous metals: *in situ* metal immobilization and revegetation. Environ Pollut 87:51–59.

Vangronsveld J, Colpaert JV, Van Tichelen KK. 1996. Reclamation of a bare industrial area contaminated by non-ferrous metals: physico-chemical and biological evaluation of the durability of soil treatment and revegetation. Environ Pollut 94: 131–140.

Vangronsveld J. 1998. Case studies in the field-industrial sites. Phytostabilisation of zinc-smelter contaminated site: The Lommel-Maatheide case. In: Vangronsveld J, Cunningham SD, editors. Metal contaminated soils: *in situ* inactivation and phytorestoration. Berlin: Springer-Verlag. p 211–216.

Vangronsveld J, Cunningham SD. 1998. Metal contaminated soils: *in situ* inactivation and phytorestoration, Berlin: Springer-Verlag.

Vangronsveld J, Ruttens A, Colpaert J, Van der Lelie D. 2000a. In situ fixation and phytostabilization of metals in polluted soils. In: Luo YM, McGrath SP, Cao ZH, Zhao FJ, Chen YX, Xu JM, editors. Proceedings of the International Conference on Soil Remediation (SoilRem2000), October 15–19, Hangzhou, China. p 262–267.

Vangronsveld J, Mench M, Lepp NW, Boisson J, Ruttens A, Edwards R, Penny C, Van der Lelie D. 2000b. In situ inactivation and phytoremediation of metal- and metalloid contaminated soils: field experiments. In: Wise J, Trantolo D, Cichon E, Yang H, Stotmeister U, editors. Bioremediation of contaminated soils. New York: Marcel Dekker Inc. p 859–884.

Verkleij JAC, Karenlampi S, De Koe T, Mench M, Vangronsveld J. 1999. Strategies for rehabilitation of metal polluted soils: in situ phytoremediation, immobilization and revegetation, a comparative study (PHYTOREHAB). Final report ENV4-CT95-0083, EU DGXII Environment & Climate programme, Vrije Universiteit Amsterdam, The Netherlands.

Yang J, Mosby DE, Casteel SW, Blanchar RW. 2002. *In vitro* lead bioaccessibility and phosphate leaching as affected by surface application of phosphoric acid in lead-contaminated soil. Arch Environ Contam Tox 43:399–405.

Index

C

D

E

F

G

H

I

J

L

M

P

R

S

Other Titles from the Society of Environmental Toxicology and Chemistry (SETAC):

Working Environment in Life-Cycle Assessment
Poulsen and Jensen, editors
2005

Life-Cycle Assessment of Metals
Dubreuil, editor
2005

Life-Cycle Management
Hunkeler, Saur, Rebitzer, Finkbeiner, Schmidt, Jensen, Stranddorf, Christiansen
2004

Scenarios in Life-Cycle Assessment
Rebitzer and Ekvall, editors
2004

Life-Cycle Assessment and SETAC: 1991–1999
(15 LCA publications on CD-ROM)
2003

Code of Life-Cycle Inventory Practice
de Beaufort-Langeveld, Bretz, van Hoof, Hischier, Jean, Tanner, Huijbregts, editors
2003

Life-Cycle Assessment in Building and Construction
Kotaji, Edwards, Shuurmans, editors
2003

Community-Level Aquatic System Studies—Interpretation Criteria (CLASSIC)
Giddings, Brock, Heger, Heimbach, Maund, Norman, Ratte, Schäfers, Streloke, editors
2002

Interconnections between Human Health and Ecological Variability
Di Giulio and Benson, editors
2002

Life-Cycle Impact Assessment: Striving towards Best Practice
Udo de Haes, Finnveden, Goedkoop, Hauschild, Hertwich, Hofstetter, Jolliet, Klöpffer, Krewitt, Lindeijer, Müller-Wenk, Olsen, Pennington, Potting, Steen, editors
2002

Silver in the Environment: Transport, Fate, and Effects
Andren and Bober, editors
2002

Test Methods to Determine Hazards for Sparingly Soluble Metal Compounds in Soils
Fairbrother, Glazebrook, van Straalen, Tararzona, editors
2002

Avian Effects Assessment: A Framework for Contaminants Studies
Hart, Balluff, Barfknecht, Chapman, Hawkes, Joermann, Leopold, Luttik, editors
2001

Bioavailability of Metals in Terrestrial Ecosystems: Importance of Partitioning for Bioavailability to Invertebrates, Microbes, and Plants
Allen, editor
2001

Ecological Variability: Separating Natural from Anthropogenic Causes of Ecosystem Impairment
Baird and Burton, editors
2001

Impacts of Low-Dose, High-Potency Herbicides on Nontarget and Unintended Plant Species
Ferenc, editor
2001

Risk Management: Ecological Risk-Based Decision-Making
Stahl, Bachman, Barton, Clark, deFur, Ells, Pittinger, Slimak, Wentsel, editors
2001

Development of Methods for Effects-Driven Cumulative Effects Assessment Using Fish Populations: Moose River Project
Munkittrick, McMaster, Van Der Kraak, Portt, Gibbons, Farwell, Gray, authors
2000

Ecotoxicology of Amphibians and Reptiles
Sparling, Linder, Bishop, editors
2000

Environmental Contaminants and Terrestrial Vertebrates: Effects on Populations, Communities, and Ecosystems
Albers, Heinz, Ohlendorf, editors
2000

Evaluation of Persistence and Long-Range Transport of Organic Chemicals in the Environment
Klecka, Boethling, Franklin, Grady, Graham, Howard, Kannan, Larson, Mackay, Muir, van de Meent, editors
2000

Multiple Stressors in Ecological Risk and Impact Assessment: Approaches to Risk Estimation
Ferenc and Foran, editors
2000

Natural Remediation of Environmental Contaminants: Its Role in Ecological Risk Assessment and Risk Management
Swindoll, Stahl, Ells, editors
2000

Evaluating and Communicating Subsistence Seafood Safety in a Cross-Cultural Context: Lessons Learned from the Exxon Valdez Oil Spill
Field, Fall, Nighswander, Peacock, Varanasi, editors
1999

SETAC

A Professional Society for Environmental Scientists and Engineers and Related Disciplines Concerned with Environmental Quality

The Society of Environmental Toxicology and Chemistry (SETAC), with offices currently in North America and Europe, is a nonprofit, professional society established to provide a forum for individuals and institutions engaged in the study of environmental problems, management and regulation of natural resources, education, research and development, and manufacturing and distribution.

Specific goals of the society are:

- Promote research, education, and training in the environmental sciences.
- Promote the systematic application of all relevant scientific disciplines to the evaluation of chemical hazards.
- Participate in the scientific interpretation of issues concerned with hazard assessment and risk analysis.
- Support the development of ecologically acceptable practices and principles.
- Provide a forum (meetings and publications) for communication among professionals in government, business, academia, and other segments of society involved in the use, protection, and management of our environment.

These goals are pursued through the conduct of numerous activities, which include:

- Hold annual meetings with study and workshop sessions, platform and poster papers, and achievement and merit awards.
- Sponsor monthly and quarterly scientific journals, a newsletter, and special technical publications.
- Provide funds for education and training through the SETAC Scholarship/Fellowship Program.
- Organize and sponsor chapters to provide a forum for the presentation of scientific data and for the interchange and study of information about local concerns.
- Provide advice and counsel to technical and nontechnical persons through a number of standing and ad hoc committees.

SETAC membership currently is composed of more than 5,000 individuals from government, academia, business, and public-interest groups with technical backgrounds in chemistry, toxicology, biology, ecology, atmospheric sciences, health sciences, earth sciences, and engineering. If you have training in these or related disciplines and are engaged in the study, use, or management of environmental resources, SETAC can fulfill your professional affiliation needs.

All members receive a newsletter highlighting environmental topics and SETAC activities, and reduced fees for the Annual Meeting and SETAC special publications. All members except Students and Senior Active Members receive monthly issues of *Environmental Toxicology and Chemistry* (*ET&C*) and *Integrated Environmental Assessment and Management* (*IEAM*), peer-reviewed journals of the Society. Student and Senior Active Members may subscribe to the journal. Members may hold office and, with the Emeritus Members, constitute the voting membership.

If you desire further information, contact the appropriate SETAC Office.

1010 North 12th Avenue
Pensacola, Florida 32501-3367 USA
T 850 469 1500 F 850 469 9778
E setac@setac.org

Avenue de la Toison d'Or 67
B-1060 Brussels, Belgium
T 32 2 772 72 81 F 32 2 770 53 83
E setac@setaceu.org

www.setac.org

Environmental Quality Through Science®